Lecture Notes in Computer Science 16043

The series Lecture Notes in Computer Science (LNCS), including its subseries Lecture Notes in Artificial Intelligence (LNAI) and Lecture Notes in Bioinformatics (LNBI), has established itself as a medium for the publication of new developments in computer science and information technology research, teaching, and education.

LNCS enjoys close cooperation with the computer science R & D community, the series counts many renowned academics among its volume editors and paper authors, and collaborates with prestigious societies. Its mission is to serve this international community by providing an invaluable service, mainly focused on the publication of conference and workshop proceedings and postproceedings. LNCS commenced publication in 1973.

Panos K. Chrysanthis · Kjetil Nørvåg ·
Kostas Stefanidis · Zheying Zhang
Editors

Advances in Databases and Information Systems

29th European Conference, ADBIS 2025
Tampere, Finland, September 23–26, 2025
Proceedings

Springer

Editors
Panos K. Chrysanthis
University of Pittsburgh
Pittsburgh, PA, USA

Kjetil Nørvåg
Norwegian University of Science and Technology
Trondheim, Norway

Kostas Stefanidis
Tampere University
Tampere, Finland

Zheying Zhang
Tampere University
Tampere, Finland

ISSN 0302-9743 ISSN 1611-3349 (electronic)
Lecture Notes in Computer Science
ISBN 978-3-032-05280-3 ISBN 978-3-032-05281-0 (eBook)
https://doi.org/10.1007/978-3-032-05281-0

This Springer imprint is published by the registered company Springer Nature Switzerland AG
The registered company address is: Gewerbestrasse 11, 6330 Cham, Switzerland

Preface

This volume contains the main track papers presented at the 29th European Conference on Advances in Databases and Information Systems (ADBIS 2025 - https://adbis2025.github.io/), held September 23–26, 2025, in Tampere, Finland.

The first ADBIS Conference was held in Saint Petersburg, Russia (1997). Since then, ADBIS has been continuously organized as an annual event. Its previous editions were held in: Bayonne, France (2024); Barcelona, Spain (2023); Turin, Italy (2022); Tartu, Estonia (2021); Lyon, France (2020); Bled, Slovenia (2019); Budapest, Hungary (2018); Nicosia, Cyprus (2017); Prague, Czech Republic (2016); Poitiers, France (2015); Ohrid, North Macedonia (2014); Genoa, Italy (2013); Poznań, Poland (2012); Vienna, Austria (2011); Novi Sad, Serbia (2010); Riga, Latvia (2009); Pori, Finland (2008); Varna, Bulgaria (2007); Thessaloniki, Greece (2006); Tallinn, Estonia (2005); Budapest, Hungary (2004); Dresden, Germany (2003); Bratislava, Slovakia (2002); Vilnius, Lithuania (2001); Prague, Czech Republic (2000); Maribor, Slovenia (1999); and Poznań, Poland (1998). The official ADBIS portal at http://adbis.eu provides up-to-date information on all ADBIS Conferences, persons in charge, publications, and issues related to the ADBIS community.

The program of ADBIS 2025 included keynote talks, research papers, a panel, tutorials, demos, workshops, a doctoral consortium school (DC), as well as a diversity, equity, and inclusion (DEI) panel and keynotes. The main conference attracted 66 paper submissions (61 papers and 5 demos). All the papers went through a process of rigorous single-blind reviewing by at least three reviewers. Eventually, the Program Committee selected 14 submissions as full contributions, appearing in this Springer LNCS proceedings volume, and 14 short papers and 3 demo papers, appearing in Springer's Communications in Computer and Information Science (CCIS) series. The 14 papers accepted as full contributions amount to an acceptance rate of 23%.

The selected papers span a large spectrum of topics in the broader field of data management. We have organized the papers in six sessions: (1) Query Optimization, (2) Explainable AI, (3) Entity Resolution & Integration, (4) Data & Machine Learning, (5) Spatio-Temporal and Graph Data, and (6) Data Sharing & Synthesis.

Following a successful tradition, selected best papers of ADBIS 2025 will be invited for a special issue of Information Systems (Elsevier). The PC chairs would like to express their sincere gratitude to the Editors-in-Chief of Information Systems for their approval regarding this special issue.

For this ADBIS event we had keynote talks by experts in different fields of the broader area of data management. In particular (in alphabetical order):

- Eero Hyvönen (Aalto University, Finland) delivered a talk on "Digital Humanities on the Semantic Web: from Infrastructure to Practical Applications, AI-based Knowledge Discovery, and Web of Wisdom"
- Evaggelia Pitoura (University of Ioannina, Greece) delivered a talk on "Explainability, Fairness and Their Interplay"

- Felix Naumann (University of Potsdam, Germany) delivered a talk on "Data Quality in the Age of AI"
- Yannis Ioannidis (National and Kapodistrian University of Athens, Greece) delivered a talk on "Interactive Digital Storytelling"

In addition, four excellent tutorials were also included in the program, which were delivered by:

- Alejandra Josiowicz (State University of Rio de Janeiro, Brazil) and Genoveva Vargas-Solar (CNRS, LIRIS Laboratory, France) who presented a tutorial entitled "Graph Analytics for Bridging Human and Data Sciences",
- Henrietta Dombrovskaya (DRW Holdings, USA) who presented a tutorial entitled "Data Warehousing: The Industrial Perspective",
- Toni Taipalus (Tampere University, Finland) and Jiaheng Lu (University of Helsinki, Finland) who presented a tutorial entitled "Vector Representations of Multi-Modal Data", and
- Valter Uotila (University of Helsinki, Finland), Soror Sahri (Université Paris Cité, France), and Sven Groppe (University of Lübeck, Germany) who presented a tutorial entitled "Utilizing Quantum Computing to Improve the Quality of Data".

ADBIS 2025 hosted a panel titled "Do LLMs Solve the Data Integration Problem Once and for All?" The panel moderator was Matthias Boehm (TU Berlin & BIFOLD, Germany), and our distinguished panelists were:

- Yannis Ioannidis (University of Athens, Greece)
- Oscar Romero Moral (UPC Barcelona, Spain)
- Genoveva Vargas-Solar (CNRS, France)
- Robert Wrembel (Poznań University of Technology, Poland)

The ADBIS 2025 conference hosted seven affiliated workshops:

- 7th Workshop on Modern Approaches in Data Engineering and Information System Design (MADEISD),
- 6th Workshop on Intelligent Data - From Data to Knowledge (DOING),
- 4th Workshop on Knowledge Graphs Analysis on a Large Scale (K-GALS),
- 1st Workshop on Cooperative AI Models and Applications (CAIMA),
- 1st Workshop on Entity Resolution and Graph Alignment (ERGA),
- 1st Workshop on Fairness Exploration in Heterogeneous Data and Algorithms (FEHDA), and
- 1st Workshop on Information Technology for Tourism and Culture Industries (IT4TOCI).

All seven workshops addressed cutting-edge topics that call for extensive research from both a foundational and an application perspective to progress the state of the art. In total, 44 papers were submitted to these workshops, out of which 22 were selected for presentation and publication, yielding an overall acceptance rate of 50%. The accepted workshop papers are published in the accompanying proceedings volume published in the CCIS series by Springer.

The ADBIS 2025 conference participates in the Database Diversity, Equity and Inclusion (DEI) initiative aiming to guide researchers in our community to adopt a more inclusive mindset. Specifically, authors and participants were encouraged to follow DEI-aware communication guidelines when preparing papers and presentations to ensure inclusive communication. The DEI stream offered a keynote address by:

- Maija Hirvonen (Tampere University, Finland) who delivered a talk on "From restriction to resource: Fostering (computing) workplace with diverse abilities".

DEI also hosted a panel titled "Repositioning DEI: The Urgent Reimagining of DEI in an Era of Retrenchment". The panel moderators were Panagiota Fatourou (University of Crete, Greece) and Genoveva Vargas-Solar (CNRS, France), and our renowned panelists were:

- Maija Hirvonen (Tampere University, Finland)
- Yannis Ioannidis (University of Athens, Greece)
- Kaisa Väänänen (Tampere University, Finland)

We want to thank everyone who contributed to the ADBIS 2025 conference. The list is long, and we resort to grouping our thanks to the following groups of persons:

- Authors, without their hard work the conference would not be possible.
- Keynote speakers, Tutorial presenters, and Panelists for sharing their wisdom with the conference audience.
- Members of the Program Committee and External Reviewers, for all their hard work in diligently reviewing, commenting, and discussing the submitted papers and for their useful suggestions to the authors.
- Members of the ADBIS Steering Committee for their trust and support.
- Members of the Organizing Committee for taking care of several different tasks related to local affairs, without whom the conference would have not been such a pleasant and inspiring experience.
- The CMT staff for hosting the process of paper submission, reviewing, and collection of the final material.
- Springer staff for making it possible to publish the proceedings of the ADBIS 2025 conference in the LNCS and CCIS series.

Thanks are also due to the sponsors of the ADBIS 2025 conference, namely RINNOCO, Tampere University, Tampere City, and Springer.

September 2025

Panos K. Chrysanthis
Kjetil Nørvåg
Kostas Stefanidis
Zheying Zhang

Organization

General Chairs

Kostas Stefanidis	Tampere University, Finland
Zheying Zhang	Tampere University, Finland

Program Chairs

Panos Chrysanthis	University of Pittsburgh, USA
Kjetil Nørvåg	NTNU, Norway

Demonstration Chairs

Johann Eder	University of Klagenfurt, Austria
Dimitris Sacharidis	Université Libre de Bruxelles, Belgium

Tutorial Chairs

Georgia Koloniari	University of Macedonia, Greece
Robert Wrembel	Poznań University of Technology, Poland

Panel Chair

Matthias Boehm	Technische Universität Berlin, Germany

Workshop Chairs

Elisa Quintarelli	Università degli Studi di Verona, Italy
Ester Zumpano	University of Calabria, Italy

Doctoral Consortium Chairs

Enrico Gallinucci	University of Bologna, Italy
Genoveva Vargas-Solar	French Council of Scientific Research, France

Diversity and Inclusion Chairs

Panagiota Fatourou	University of Crete, Greece
Genoveva Vargas-Solar	French Council of Scientific Research, France

Publicity Chairs

Barbara Catania	Università degli studi di Genova, Italy
Javier A. Espinosa Oviedo	University of Lyon 1, France

Sponsorship Chairs

Tony Lindgren	Stockholm University, Sweden
George Papadakis	Agricultural University of Athens, Greece
Jaakko Peltonen	Tampere University, Finland

Special Issues Chair

Mirjana Ivanovic	University of Novi Sad, Serbia

Proceedings Chair

Vasilis Efthymiou	Harokopio University of Athens, Greece

Web Chair

Emilia Lenzi	Politecnico di Milano, Italy

Volunteers Chairs

Maria Stratigi	Tampere University, Finland
Toni Taipalus	Tampere University, Finland

Program Committee

Alberto Abella	Universitat Politècnica de Catalunya, Spain
Maribel Acosta	Université Libre de Bruxelles, Belgium
Cristina D. Aguiar	Universidade de São Paulo, Brazil
Bernd Amann	Sorbonne University, CNRS, France
Witold Andrzejewski	Poznań University of Technology, Poland
Sylvio Barbon Junior	University of Trieste, Italy
Nelly Barret	University of Novi Sad, Serbia
Andreas Behrend	University of Bonn, Germany
Ladjel Bellatreche	Universidad de Alicante, Spain
Sandro Bimonte	INRAE, France
Paweł Boiński	University of Orléans, France
Drazen R. Brdjanin	University of Banja Luka, Bosnia and Herzegovina
Richard Chbeir	Université de Pau et des Pays de l'Adour, France
Silvia Chiusano	Politecnico di Torino, Italy
Antonio Corral	University of Almería, Spain
Jerome Darmont	University of Lyon 2, France
Anton Dignös	Free University of Bozen-Bolzano, Italy
Christos Doulkeridis	University of Pireaus, Greece
Johann Eder	Université Lumière Lyon 2, France
Markus Endres	Technische Universität München, Germany
Javier A. Espinosa-Oviedo	Charles University, Czech Republic
Georgios Evangelidis	University of Macedonia, Greece
Bentayeb Fadila	Université Lumière Lyon 2, France
George Fakas	Uppsala University, Sweden
Johann Gamper	Universität Klagenfurt, Austria
Matteo Golfarelli	University of Augsburg, Germany
Vincenzo Gulisano	Chalmers University of Technology, Sweden
Herodotos Herodotou	Cyprus University of Technology, Cyprus
Tomas Horvath	Eötvös Loránd University, Hungary
Mirjana Ivanovic	Université Paris-Dauphine – PSL, France
Stefan Jablonski	University of Bayreuth, Germany
Pokorny Jaroslav	Universitat Politècnica de Catalunya, Spain
Petar Jovanovic	Poznań University of Technology, Poland

Alexandros Karakasidis	University of Macedonia, Greece
Zoubida Kedad	University of Versailles, France
Georgia Koloniari	University of Macedonia, Greece
Georgia Koutrika	Athena Research Center, Greece
Yannis Manolopoulos	University of Nicosia, Cyprus
Maude Manouvrier	University of Bologna, Italy
Patrick Marcel	University of Ioannina, Greece
Jose-Norberto Mazon	Free University of Bozen-Bolzano, Italy
Sara Migliorini	University of Verona, Italy
Alex Mircoli	Università Politecnica delle Marche, Italy
Angelo Montanari	University of Udine, Italy
Boris Novikov	Unaffiliated, Finland
Boussaid Omar	Université Lumière Lyon 2, France
George Palis	University of Cyprus, Cyprus
George Papastefanatos	Athena Research Center, Greece
Veronika Peralta	University of Tours, France
Giuseppe Polese	University of Salerno, Italy
Elisa Quintarelli	Università degli Studi di Verona, Italy
Franck Ravat	Institut de Recherche en Informatique de Toulouse, France
Gunter Saake	University of Magdeburg, Germany
Mahmoud A. Sakr	ISAE-ENSMA, France
Ingo Schmitt	Brandenburg University of Technology, Germany
Sergey A. Stupnikov	Russian Academy of Sciences, Russia
Bernhard Thalheim	Christian Albrecht University of Kiel, Germany
Goce Trajcevski	Iowa State University, USA
Panos Vassiliadis	Politecnico di Milano, Italy
Yannis Velegrakis	Utrecht University, Netherlands
Goran Velinov	Ss. Cyril and Methodius University in Skopje, North Macedonia
Szymon Wilk	Poznań University of Technology, Poland
Robert Wrembel	Poznań University of Technology, Poland
Demetrios Zeinalipour-Yazti	University of Cyprus, Cyprus
Yongluan Zhou	University of Copenhagen, Denmark
Ester Zumpano	University of Calabria, Italy

Steering Committee Chair

Yannis Manolopoulos	University of Nicosia, Cyprus

Steering Committee

Andreas Behrend	TH Köln, Germany
Ladjel Bellatreche	ENSMA Poitiers, France
Maria Bielikova	Kempelen Institute of Intelligent Technologies, Slovakia
Barbara Catania	University of Genoa, Italy
Tania Cerquitelli	Politecnico di Torino, Italy
Richard Chbeir	IUT de Bayonne et du Pays Basque, France
Silvia Chiusano	Politecnico di Torino, Italy
Jérôme Darmont	University of Lyon 2, France
Johann Eder	Alpen-Adria-Universität Klagenfurt, Austria
Johann Gamper	Free University of Bozen-Bolzano, Italy
Tomáš Horváth	Pavol Jozef Šafárik University, Slovakia
Mirjana Ivanović	University of Novi Sad, Serbia
Manuk Manukyan	Yerevan State University, Armenia
Kjetil Nørvåg	Norwegian University of Science and Technology, Norway
Boris Novikov	Unaffiliated, Finland
George Papadopoulos	University of Cyprus, Cyprus
Jaroslav Pokorny	Charles University in Prague, Czech Republic
Oscar Romero	Universitat Politècnica de Catalunya - BarcelonaTech, Spain
Sergey Stupnikov	Russian Academy of Sciences, Russia
Bernhard Thalheim	Christian Albrechts University of Kiel, Germany
Goce Trajcevski	Iowa State University, USA
Valentino Vranić	Pan-European University, Slovakia
Tatjana Welzer	University of Maribor, Slovenia
Genoveva Vargas-Solar	French Council of Scientific Research, France
Robert Wrembel	Poznań University of Technology, Poland
Ester Zumpano	University of Calabria, Italy

In Memory of Professor Tadeusz Morzy

15 Aug, 1954 - 21 Jan, 2025

It is with profound sadness that we write this obituary for our dear friend and colleague, and founding member of the ADBIS Steering Committee, Professor Tadeusz Morzy. Tadeusz left us in January 2025, after a long and heroic fight with an illness.

The academic journey of Tadeusz Morzy started in Poznań, Poland, where he completed his secondary education at a leading school. He then earned a Master's degree in Electrical Engineering from Poznań University of Technology (PUT). After briefly working in the IT industry, Professor Morzy's dedication to research led him back to PUT, where he obtained his doctoral and post-doctoral degrees. His significant research contributions in the field of databases were recognized in 2005 with the conferral of the title of Professor by the President of Poland. For at least two decades, he led a research group at the Institute of Computing Science at PUT. The research interests of Tadeusz focused broadly on data processing technologies, with a strong focus on databases, data warehousing, transaction processing, data mining, and machine learning. His research acumen led to his appointment to the National Science Center as an expert evaluating research grants and a long-term membership in the Committee on Informatics of the Polish Academy of Sciences, where he led a database research group. The group, composed of 16 staff members, works steadily under a new chair. Four of Tadeusz's former Ph.D. students become university professors and work at PUT.

Professor Tadeusz Morzy's research career began with foundational work in concurrency control for databases, published in top-tier venues such as VLDB, ICDE, and IEEE Transactions on Software Engineering, as well as in the book "Concurrency Control in Distributed Database Systems" (co-authored with Prof. W. Cellary and Prof. E. Gelenbe). Subsequently, his focus shifted to query optimization algorithms in relational databases, with key publications in EDBT, COMAD, and ICDT. His research interests then evolved to encompass data mining and machine learning, resulting in numerous contributions to leading conferences like KDD, PAKDD, and PKDD, as well as journals such as Information Systems, Information Systems Frontiers, Knowledge Information Systems, and Knowledge Engineering Review. These interests naturally intersected with data engineering and data warehousing, leading to further publications in journals like Information & Software Technology and inter alia international conferences ADVIS, CAiSE, DaWaK, DB&IS, DEXA, DOLAP, EDBT, ER, ICEIS, MEDES, SAC, and WIDM. Demonstrating his strong commitment to ADBIS, Tadeusz contributed 11 research papers and shaped 6 volumes of its proceedings.

Professor Morzy's impact extended beyond research - as a practitioner, he initiated the development of an information system for managing educational processes in universities. The system was deployed in multiple universities throughout the country.

Tadeusz's dedication to top-tier teaching was evident in his flagship courses: database technologies, data exploration, and data mining. His book in Polish entitled "Data exploration. Methods and algorithms" became an indispensable academic resource, serving as the de facto standard in most Polish universities and receiving prestigious recognition from the Polish Academy of Sciences. He also established and successfully managed two distinct post-graduate programs in Databases and Data Warehouses for several years.

Within the Poznań University of Technology, he held significant roles as a member of the Academic Senate, the Senate Committee for Research Staff Promotions, and the Director of the Doctoral Study Program.

Professor Morzy was deeply involved in international academic partnerships. His long-standing cooperation with Klagenfurt University (and Prof. Johann Eder) led to the establishment of a double Master's degree program, a likely early example in Poland. Furthermore, he initiated student and staff exchanges with Ostfalia University of Applied Sciences (Germany) and twice served as a Visiting Professor at Loyola University in New Orleans, USA.

In recognition of his significant contributions to research and higher education, he was honored with the Medal of the Commission of National Education (2002), the Gold Cross of Merit (2004), and the Knight's Cross of the Order of Polonia Restituta (2013).

Prof. Tadeusz Morzy was involved in the development of the ADBIS conference. Since the very beginning, he was a member of the ADBIS Steering Committee and hosted the conference twice in Poznań (1998, 2012). In addition, he served as Program Committee Chair several times and contributed as an advisor to several ADBIS Ph.D. workshops.

The 29th European Conference on Advances in Databases and Information Systems is dedicated to his memory.

ADBIS Steering Committee
Amdreas Behrend, Ladjel Bellatreche, Maria Bielikova,
Barbara Catania, Tania Cerquitelli, Richard Chbeir,
Silvia Chiusano, Jérôme Darmont, Johann Eder,
Johann Gamper, Tomáš Horváth, Mirjana Ivanovic,
Yannis Manolopoulos, Manuk Manukyan, Kjetil Norvag,
Boris Novikov, George Papadopoulos, Jaroslav Pokorny,
Oscar Romero, Sergey Stupnikov, Bernhard Thalheim,
Goje Trajcevski, Genoveva Vargas-Solar, Valentino Vranic,
Tatjana Welzer, Robert Wrembel, Ester Zumpano

Contents

Keynote Talk

Data Quality in the Age of AI 3
Felix Naumann, Lisa Ehrlinger, Hazar Harmouch, Sedir Mohammed, and Divesh Srivastava

Query Optimization

Independence Rules: Analysis of Nine Simple Cardinality Estimators and Their Impact on Plan Quality 11
Daniel Flachs and Guido Moerkotte

Vector Database Benchmarking: A Face Retrieval Use Case 28
Teodor Sakal Francišković, Nikola Todorović, Jelena Hrnjak, and Vladimir Dimitrieski

PUL: Pre-load in Software for Caches Wouldn't Always Play Along 44
Arthur Bernhardt, Sajjad Tamimi, Florian Stock, Andreas Koch, and Ilia Petrov

Explainable AI

Responsible AI: Training Deep Learning Model Efficiently 63
Shwetha Salimath, Francesca Bugiotti, and Sylvain Wlodarczyk

Exploring Local Feature Influences with Hierarchical Explanation Trees 79
Emmanuel Doumard, Julien Aligon, Paul Monsarrat, Nicolas Labroche, and Alexandre Chanson

Towards Trace Variant Explainability 95
Iuliana Malina Grigore, Gabriel Marques Tavares, Vincenzo Pasquadibisceglie, Thomas Seidl, and Sylvio Barbon Junior

Evaluating Knowledge Generation and Self-refinement Strategies for LLM-Based Column Type Annotation 111
Keti Korini and Christian Bizer

Entity Resolution and Integration

LSBlock: A Hybrid Blocking System Combining Lexical and Semantic Similarity Search for Record Linkage 131
Dimitrios Karapiperis, Christos Tjortjis, and Vassilios Verykios

Evaluating Quality of Disparate Data Sources: A Discord-Driven Approach 147
Yeasmin Ara Akter, Alberto Abelló, Petar Jovanovic, Tomer Sagi, and Katja Hose

Pasteur: Scaling Privacy-Aware Data Synthesis 164
Antheas Kapenekakis, Daniele Dell'Aglio, Martin Bøgsted, Minos Garofalakis, and Katja Hose

Generating Semantically Enriched Mobility Data from Travel Diaries 181
Maja Schneider, Charini Nanayakkara, Matthias Mohn, Peter Christen, and Erhard Rahm

Data and Machine Learning

Learning a Distance for the Clustering of Patients with Amyotrophic Lateral Sclerosis 201
Guillaume Tejedor, Veronika Peralta, Nicolas Labroche, Patrick Marcel, Hélène Blasco, and Hugo Alarcan

Designing Efficient and Scalable Substructure Discovery Algorithms for Multilayer Networks 217
Arshdeep Singh, Abhishek Santra, and Sharma Chakravarthy

Bitemporal Property Graphs: Dealing with Both Valid and Transaction Time 234
Luca Anselma, Marco Ballerini, Paolo Giordano, Erica Raina, and Paolo Terenziani

Tutorials

Networks of Knowledge: Graph Analytics for Bridging Human and Data Sciences: *ADBIS-2025 Tutorial - Exploring Intellectual Influence and Epistemic Violence* 253
Genoveva Vargas-Solar and Alejandra Josiowicz

Data Warehousing: The Industrial Perspective 264
Henrietta Dombrovskaya

Vector Representations of Multi-modal Data ... 272
Toni Taipalus and Jiaheng Lu

Utilizing Quantum Computing to Improve the Quality of Data ... 280
Valter Uotila, Soror Sahri, and Sven Groppe

Correction to: Independence Rules: Analysis of Nine Simple Cardinality Estimators and Their Impact on Plan Quality ... C1
Daniel Flachs and Guido Moerkotte

Correction to: Bitemporal Property Graphs: Dealing with Both Valid and Transaction Time ... C2
Luca Anselma, Marco Ballerini, Paolo Giordano, Erica Raina, and Paolo Terenziani

Author Index ... 289

Keynote Talk

Data Quality in the Age of AI

Felix Naumann[1(✉)], Lisa Ehrlinger[1], Hazar Harmouch[2], Sedir Mohammed[1], and Divesh Srivastava[3]

[1] Hasso Plattner Institute, University of Potsdam, Potsdam, Germany
{felix.naumann,lisa.ehrlinger,sedir.mohammed}@hpi.de
[2] University of Amsterdam, Amsterdam, The Netherlands
h.harmouch@uva.nl
[3] AT&T Chief Data Office, Bedminster, USA
divesh@research.att.com

Abstract. Data quality comprises a large set of dimensions, each describing a specific aspect. The assessment of these dimensions requires the collection of simple statistics, the identification of syntactic problems and factual errors, as well as organizational and business aspects to be considered. With the current trend in data-oriented sciences and the increasing reliance on machine learning methods and AI systems, the challenges of poor data quality are ever more apparent. Even recent legislation, such as the European AI Act, mentions data quality requirements for training data; with it, the notion of data quality extends to novel dimensions, such as fairness, diversity, or explainability. We present a modular approach to the assessment of data quality, based on five key facets: data, source, system, task, and human.

Keywords: Data quality · information quality · data quality dimensions · data quality assessment · data profiling

1 Data Quality

Data quality issues are as old as data collection itself. Data quality research in computer sciences has a long tradition and can be categorized into three areas: definition, assessment, and improvement. The first area aims at identifying and defining data quality in its many dimensions, which include objective measures, such as completeness, accuracy, or timeliness, but also subjective measures, such as readability, trustworthiness, or transparency, and measures concerning the systems that manage the data, such as accessibility, cost, or ease of manipulation.

The second area, data quality assessment, is concerned with measuring the quality of a given dataset, along various dimensions and in aggregate. Assessment extends the concept of measurement by evaluating the measurement results and drawing a conclusion about the object of assessment, for instance whether it is of sufficient quality to solve a downstream task. Often, such assessment is not possible on the data itself, but needs to consider also the data source, the data system, the task at hand, and the human facet, as we explain in the next section.

P. K. Chrysanthis et al. (Eds.): ADBIS 2025, LNCS 16043, pp. 3–7, 2026.
https://doi.org/10.1007/978-3-032-05281-0_1

Finally, the third area, data quality improvement, which includes data preparation and data cleaning, involves active data manipulation to correct errors or select only high-quality data.

With the rise of machine learning techniques and artificial intelligence systems, which heavily rely on data to train their models, data quality is increasingly on the minds of data scientists, researchers, and practitioners. We have examined the effects of poor data quality on traditional machine learning models, reconfirming the well-known garbage-in-garbage-out principle [8]. The same concerns are true for developers of large language models (LLMs), who filter training data, deduplicate it, check for toxicity, diversity, and age, and perform many other steps to ensure high-quality models [12].

Creating useful models from data is not only in the interest of the data providers, but is recently also facing the scrutiny of lawmakers, who want to ensure that only safe models are deployed and used. For instance, recent EU legislation, the AI Act [6], specifically mentions "Data sets for training, validation and testing, including the labels, should be relevant, sufficiently representative, and to the best extent possible free of errors and complete in view of the intended purpose of the system." The law thus mentions the quality dimensions of *relevancy*, *representativity*, *accuracy*, and *completeness* of the data, which the developers and deployers of AI models will need to ensure and, if necessary, prove. To do so requires the ability to assess data quality along its dimensions.

2 Data Quality Assessment

The key to effective data quality management is the ability to assess data quality in all its dimensions. As Lord Kelvin (supposedly) stated already in 1883, "If you can not measure it, you can not improve it." Research in data quality assessment is yet in its infancy, as the authors of the 2025 Cambridge Report on Database Research [1] also recognize: "Key research gaps in responsible data management include developing methods to assess data quality in relation to specific downstream tasks and socially meaningful metrics e.g., fairness, accuracy, and robustness."

Each of the diverse data quality dimensions is in need of a formal definition, a method to implement this definition, and a system to efficiently and effectively execute the method. Some dimensions are already well understood and described [5], such as completeness or accuracy, while others have received less attention. To address the diversity of the field, in [9] we identify five key facets or "ingredients" to measure data quality – data, source, system, task, and human – and show which facet is relevant for which of a collection of 29 data quality dimensions. Developing methods to tap into these various facets will contribute to a general data quality assessment framework, covering many dimensions simultaneously.

Apart from the facet-specific challenges to assess quality in its various dimensions, building a framework that supports data quality assessment and management along the entire pipeline gives rise to further challenges [9]:

Definition ambiguity The many attempts to define data quality dimensions, such as in [3,10], highlight the challenge of formally grasping this broad concept. Along with the definitions, also the meanings, measures, and assessment methods of these dimensions are inherently ambiguous [2].

Score composition Having determined a single data quality score per dimension and data item, the appropriate scores need to be selected, normalized, weighted, and aggregated to an overall data quality score, also taking into account the uncertainty that comes with this aggregation.

Task adequacy Data quality measurement is only a first step: the question remains whether the data is of *sufficient* quality or *adequate* for the downstream task at hand.

Regulatory compliance The impetus to ensure that data are of sufficient quality often stems from an overarching objective, like fulfilling organizational data governance goals or legal compliance, such as GDPR or the AI Act. To document compliance, the assessment procedure should be transparent and auditable.

Explainability Due to their ambiguity, data quality assessment results must be explainable to consumers. The results should be traceable to their root cause, allowing steps to improve quality.

Efficiency Assessment effort and time should be low from the user perspective. Data consumers might be unable or unwilling to wait for assessment results, and experts might not have much time to complete questionnaires or help in data quality assessment. The ability to trade off efficiency and assessment accuracy will support the adoption of data quality management systems.

Metadata management Deploying the DQ assessment procedure requires effective mechanisms to store and query vast and diverse metadata in standardized formats. An example solution and its challenges are discussed in [11].

3 Research Challenges

While the previous section lays an outline for the rather specific use case of assessing the quality of static, structured data, the current AI landscape demands a data quality methodology for many other use cases. The following challenges paint a broader picture, acknowledging that various individual research contributions have already addressed some of these challenges.

Foundation model quality Foundation models and multimodal AI systems operate on enormous scales of data diversity, requiring novel quality assessment methods that can handle text, images, audio, and structured data simultaneously.

Synthetic data The rise of synthetic data generation for training AI models introduces questions about how to assess the quality of artificially generated data and its relationship to real-world distributions. New metrics are needed to evaluate fidelity, diversity, and utility of synthetic datasets.

Changing data As data evolve, quality assessment must be performed incrementally to avoid expensive re-assessments of the entire datasets. Simultaneously, knowledge of data changes can help assess (and improve) data quality, as we explore in the JANUS project [4].

Data streams Data can be generated at high velocity, necessitating distributed and online quality assessment and monitoring techniques. Real-time quality assessment must balance accuracy with computational efficiency.

Non-structured data Data other than structured relations, such as nested data, text documents, and images lack predefined schemas, making traditional data quality metrics like completeness and consistency difficult to apply directly. Quality assessment for such data requires domain-specific approaches that do not rely on metadata that is typical for tables.

Human-in-the-loop Many data quality dimensions heavily rely on human judgments, necessitating human-computer collaboration and possibly crowd-sourced quality assessment.

AI-in-the-loop As foundation models grow ever more sophisticated, human-level quality judgments will become reality, giving rise to possible vicious (or, from the AI's point of view, virtuous) circles.

Benchmarking Research fields advance with their benchmarks. Data quality assessment benchmarks, including (preferably real-world) datasets of known quality and benchmark metrics for assessment methods, promise measurable research outcomes.

These challenges represent both opportunities and necessities as our field moves toward AI-driven data ecosystems. Addressing them will require interdisciplinary collaboration across computer science, statistics, domain expertise, and human-computer interaction research.

Assessing data quality is only a first step – we need to act upon the outcomes. The results of data quality assessment can be used in various ways. First, they can create concrete input for data cleaning algorithms [7]. Low-quality data can also be enhanced by collecting more data, or synthetically augmenting them, for instance to increase completeness or diversity. Ultimately, comprehensive quality assessment enables informed decision-making about data's *fitness for use*, ensuring that AI systems are built on data foundations that meet their intended needs.

Acknowledgement. This article accompanies an invited talk at the ADBIS 2025 conference. It is based on research together with Divya Bhadauria, Carolina Cortes, Lorena Etcheverry, and Philipp Hildebrandt in the context of the Metis project: www.metisdq.org

References

1. Ailamaki, A., et al.: The Cambridge report on database research. CoRR (2025). https://doi.org/10.48550/arxiv.2504.11259
2. Batini, C., Cappiello, C., Francalanci, C., Maurino, A.: Methodologies for data quality assessment and improvement. ACM Comput. Surv. **41**(3), 1–52 (2009). https://doi.org/10.1145/1541880.1541883
3. Batini, C., Scannapieco, M.: Data and Information Quality: Dimensions. Principles and Techniques. Springer, Berlin Heidelberg (2016)
4. Bleifuß, T., Bornemann, L., Johnson, T., Kalashnikov, D.V., Naumann, F., Srivastava, D.: Exploring change - a new dimension of data analytics. PVLDB **12**(2), 85–98 (2018)
5. Ehrlinger, L., Werth, B., Wöß, W.: Automated continuous data quality Measurement with QualIe. Int. J. Adv. Softw **11**(3), 400–417 (2018)
6. European Commission: Artificial Intelligence Act (2024). https://eur-lex.europa.eu/legal-content/EN/TXT/?uri=CELEX:32024R1689
7. Ilyas, I.F., Chu, X.: Data Cleaning, ACM Books, vol. 28. ACM (2019). https://doi.org/10.1145/3310205
8. Mohammed, S., et al.: The effects of data quality on machine learning performance on tabular data. Inf. Syst. (IS) **132**, 102549 (2025). https://doi.org/10.1016/J.IS.2025.102549
9. Mohammed, S., Ehrlinger, L., Harmouch, H., Naumann, F., Srivastava, D.: The five facets of data quality assessment. SIGMOD Record **54**(2) (2025). https://sigmodrecord.org/2025/06/24/the-five-facets-of-data-quality-assessment/
10. Pipino, L., Lee, Y., Wang, R.: Data quality assessment. Commun. ACM **4**, 211–218 (2002)
11. Srivastava, D., Velegrakis, Y.: Intensional associations between data and metadata. In: Proceedings of the International Conference on Management of Data (SIGMOD), pp. 401–412. ACM (2007) https://doi.org/10.1145/1247480.1247526
12. Wang, Z., et al.: Data management for training large language models: a survey. CoRR (2024). https://doi.org/10.48550/arxiv.2312.01700

Query Optimization

Independence Rules: Analysis of Nine Simple Cardinality Estimators and Their Impact on Plan Quality

Daniel Flachs(✉) and Guido Moerkotte

University of Mannheim, 68131 Mannheim, Germany
{flachs,moerkotte}@uni-mannheim.de

Abstract. We consider nine simple cardinality estimators, seven from previous work, and two new ones, and evaluate their performance with respect to plan quality using the Join Order Benchmark (JOB). Three estimators from the literature rely on the frequently used independence assumption and are prone to underestimation. In contrast, five estimators always overestimate and promote pessimistic plans. Our major finding is that all cardinality estimators produce relatively good plans. Surprisingly, the best contender uses the independence assumption under rule L.

Keywords: query optimization · cardinality estimation · cost model

1 Introduction

Lately, we have observed a tendency to simplify cardinality estimation [4,11]. An extreme example is SimpliSquared [4]: Instead of applying a complicated approach to cardinality estimation, only the cardinalities of (unfiltered) base relations are used. Further, no cost functions are required to generate a logical plan. Still, as their experiments indicate, the performance is quite acceptable [4]. The presumable reason is that it favors the execution of foreign-key/key joins (FKK-joins, 1:n-joins) over n:m-joins. Unfortunately, there is a downside to SimpliSquared [4] shared with UES [11]: both approaches only yield *logical plans*. In order to derive an executable *physical plan*, the produced logical plan is transformed to SQL, which is then given to PostgreSQL to derive the physical plan. This has two major disadvantages:

1. reliance on PostgreSQL's cardinality estimation, and
2. reliance on PostgreSQL's cost function.

Further, both the truly optimal plan and the worst plan remain unknown. Thus, the runtimes can only be judged relatively to other plans, but cannot be put into perspective by comparing them with the costs of the optimal and the worst plan. These problems could be solved if we had cardinality estimators and cost

The original version of the chapter has been revised. A correction to this chapter can be found at https://doi.org/10.1007/978-3-032-05281-0_20

P. K. Chrysanthis et al. (Eds.): ADBIS 2025, LNCS 16043, pp. 11–27, 2026.
https://doi.org/10.1007/978-3-032-05281-0_2

functions which allow for a direct evaluation of plan costs and thus enable the generation of physical plans.

One of the prominent questions in database engineering is how much effort should be put into the design of cardinality estimators to replace PostgreSQL's cardinality estimator. A stunning conclusion was made by Kumar and Stonebraker in their paper reporting on experiments showing a negligible influence of cardinality estimation errors on plan quality [15]:

> *We can conclude that data managers should make crude selectivity estimates and not spend a great deal of time attempting to provide accurate selectivity values.*

Similar in spirit is Sec. 4.2 entitled *Good Plans Despite Bad Cardinalities* in the paper by Leis et al. [16]. In this section, the authors observe that

> *[...] the performance of most queries is close to the one obtained using the true cardinalities. Given the bad quality of the cardinality estimates, we consider this to be a surprisingly positive result.*

Thus, we deem it interesting to design crude cardinality estimators, which also allow for the generation of physical plans (in conjunction with cost functions) and evaluate their performance using different plan generators.

Applying the independence assumption (IA) for join predicates is a well-known, simple approach to cardinality estimation. To estimate the result size of a given join, the estimator multiplies the input sizes with the product of the selectivities of all applicable join predicates. This is called the *multiplicative rule (M)* [26,30], and we denote this estimator by $\mathrm{CE}_{\text{IA-M}}$. It has obvious problems with join predicates derived from the transitivity of equality, which, however, are indispensable for the generation of good plans. To remedy this situation, Swami and Schiefer proposed two alternative cardinality estimators [30]: $\mathrm{CE}_{\text{IA-S}}$ applies the *smallest selectivity rule (S)*, and $\mathrm{CE}_{\text{IA-L}}$ the *largest selectivity rule (L)*. All three cardinality estimators have a tendency towards underestimations and require the same input: the selectivities of the join predicates. The cardinality estimators extracted from Hertzschuch et al. promote pessimistic plans, where cardinalities are always overestimated [11]. They take the maximum frequency of join attribute values as an additional input. Our four cardinality estimators take different inputs. The simplest estimator $\mathrm{CE}_{\text{base}}$ needs only the cardinalities of (unfiltered) base relations. CE_{sel} additionally receives the selectivities of single table selection predicates. $\mathrm{CE}_{\text{j-base}}$ requires cardinalities of joins between unfiltered base relations, and $\mathrm{CE}_{\text{j-sel}}$ additionally receives the selectivities of selection predicates. Having a broad coverage of possible inputs allows for answering some interesting questions like: What is worse, assuming independence for join predicates or independence for selection predicates?

In order to generate physical plans from logical plans, cost functions are required to decide, e.g., upon the join implementation to use for a specific logical join. Using PostgreSQL's cost function is not an option due to its unknown errors and its well-known flaws [33]. We construct two cost functions, CF_{tru} and CF_{est}

from measured experimental runtimes. The former cost function is tabulated and more precise, and will serve as a proxy for the true execution costs. The latter is based on q-optimal linear or exp-linear approximations of the measurements, and is less precise, allowing to evaluate the impact of cost function errors.

The last missing ingredient to plan generation is the join ordering algorithm. Here, we use DPccp [20] to generate optimal plans, as well as the two heuristics GooCost [18,27], and GooCard [6].

The rest of the paper is organized as follows. Section 2 serves as a primer on plan generation. Readers with a basic knowledge in query optimization can skim through it. Section 3 presents the different cardinality estimators. Section 4 contains our evaluation. Section 5 presents related work. Finally, Sect. 6 concludes the paper.

2 Preliminaries

This section summarizes basic notions from query optimization needed for the rest of the paper, and introduces our notation (see also Table 1).

Table 1. Notation Used Throughout the Paper.

$\mathcal{R} := \{R_1, \ldots, R_n\}$	set of relations to be joined	p_i	selection predicate on R_i
$\mathcal{A}(R_i) := \{A_{i,j}\}$	attributes of relation R_i ($j = 1, \ldots, m_i$)	JP	set of all join predicates
		JA	set of all join attributes
$(A_{i,k} = A_{j,l})$	join predicate between R_i, R_j	$p_{i,j}$	set of all join predicates between R_i and R_j
$\mathtt{JA}(R_i)$	$:= \mathcal{A}(R_i) \cap \mathtt{JA}$ join attributes of R_i	$\mathcal{F}(p)$	attributes references by some predicate p
S, S_1, S_2	plan classes		
$\mathrm{ccp}(S_1, S_2)$	csg-cmp pair	$\mathcal{A}(S)$	attributes provided by the relations in S
$\mathtt{JP}(S_1, S_2)$	join predicates between S_1, S_2		

2.1 Relations, Attributes, and Predicates

For a given query, we denote by $\mathcal{R} := \{R_1, \ldots, R_n\}$ the set of relations in its `from`-clause. The set of attributes of R_i is denoted by $\mathcal{A}(R_i) := \{A_{i,1}, \ldots, A_{i,m_i}\}$. The single-table selection predicates for R_i are denoted by p_i. If there is some relation R_j without a selection predicate, we assume a selection predicate $p_j \equiv$ true. We do not impose any restriction on selection predicates but assume that they are pushed down. The set of all conjunctively occurring equality-based join predicates between any two relations R_i and R_j of the form $A_{i,k} = A_{j,l}$ is denoted by JP. There may be multiple such predicates connecting two relations R_i and R_j. In particular, this occurs for composite keys. We abbreviate the set of all equality predicates connecting R_i and R_j by $p_{i,j}$ and assume that they are conjunctively connected. The set of attributes accessed by a selection or join predicate p is denoted by $\mathcal{F}(p)$. If $\mathcal{P}$ is a set of predicates, we define $\mathcal{F}(\mathcal{P}) := \bigcup_{p \in \mathcal{P}} \mathcal{F}(p)$. Thus, $\mathtt{JA} := \mathcal{F}(\mathtt{JP})$ denotes the set of join attributes.

2.2 Query Graph, Plan Class, and CCP

The *query graph (QG)* has the relations in $\mathcal{R}$ as nodes and an edge connecting R_i and R_j for every non-empty set of join predicates $p_{i,j} \neq \emptyset$. A *plan class* $S \subseteq \mathcal{R}$ is a subset of the relations inducing a connected subgraph of the query graph. In this case, we write $\mathrm{pc}(S)$. The set of attributes of a plan class S is defined as $\mathcal{A}(S) := \bigcup_{R_i \in S} \mathcal{A}(R_i)$. Two subsets $S_1, S_2 \subseteq \mathcal{R}$ form a *CCP* (connected subgraph complement pair) if $\mathrm{pc}(S_1)$, $\mathrm{pc}(S_2)$, $S_1 \cap S_2 = \emptyset$, and there is at least one join predicate ($p_{i,j} \neq \emptyset$) connecting R_i and R_j with $R_i \in S_1$ and $R_j \in S_2$ [20]. In this case, we write $\mathrm{ccp}(S_1, S_2)$.

Plan classes and CCPs play a predominant role in plan generation. CCPs span the complete search space of logical plans if cross products are to be avoided. As we will see in Sect. 2.5, where we discuss some more details about plan generation, any cardinality estimator must be able to provide a cardinality estimate for every plan class.

2.3 Uniqueness

For two of our cardinality estimators, we need to determine whether the join predicate connecting two plan classes S_1 and S_2 with $\mathrm{ccp}(S_1, S_2)$ implies uniqueness on S_1 and/or S_2.

The join predicates $\mathtt{JP}(S_1, S_2)$ for $\mathrm{ccp}(S_1, S_2)$ are the set of all join predicates $A_{i,k} = A_{j,l}$ such that $R_i \in S_1$ and $R_j \in S_2$. The set of join attributes of S_i ($i = 1, 2$) is then defined as $\mathtt{JA}(S_i, (S_1, S_2)) := \mathcal{F}(\mathtt{JP}(S_1, S_2)) \cap \mathcal{A}(S_i)$. We say that $\mathrm{ccp}(S_1, S_2)$ determines S_i uniquely if $\mathtt{JA}(S_i, (S_1, S_2))$ is a (super-) key of S_i. If $\mathtt{K}(S_i)$ denotes the set of keys of S_i, we can express this as

$$\mathfrak{U}(S_i, (S_1, S_2)) := \exists \kappa \in \mathtt{K}(S_i) : \kappa \subseteq \mathtt{JA}(S_i, (S_1, S_2)).$$

In order to derive this uniqueness, we need to determine the keys for plan classes S. We start by assuming that for every relation R_i, the set of primary and secondary keys is given as $\mathtt{K}(\{R_i\})$. For a $\mathrm{ccp}(S_1, S_2)$, we can determine the set of keys $\mathtt{K}(S_1 \cup S_2)$ as follows. If there exists a key $\kappa \in \mathtt{K}(S_i)$ such that $\kappa \subseteq \mathtt{JA}(S_i, (S_1, S_2))$, then $\mathtt{K}(S_{3-i}) \subseteq \mathtt{K}(S_1 \cup S_2)$. If neither is true, then $\mathtt{K}(S_1 \cup S_2)$ contains $\kappa_1 \cup \kappa_2$ for all $\kappa_1 \in \mathtt{K}(S_1)$ and for all $\kappa_2 \in \mathtt{K}(S_2)$.

2.4 Error Measures

To measure the error between a true value and an estimate, we use the *q-error* [21]. Let $x \geq 0$ be some true value and $\hat{x} \geq 0$ be its estimate, then

$$\mathrm{qerr}(\hat{x}, x) := \begin{cases} 1 & \text{if } \hat{x} = x \\ \infty & \text{if } (\hat{x} = 0 \vee x = 0) \wedge \hat{x} \neq x \\ \hat{x}/x & \text{if } \hat{x} > x > 0 \\ x/\hat{x} & \text{if } x > \hat{x} > 0 \end{cases}$$

The q-error is a relative error measure that is symmetric for both under- and overestimates, and has a minimum of 1. Since the q-error does not differentiate between over- or an underestimates, we also use the *p-error* [21]:

$$\operatorname{perr}(\hat{x}, x) := \begin{cases} -(\operatorname{qerr}(\hat{x}, x) - 1) & \text{if } \hat{x} < x \\ (\operatorname{qerr}(\hat{x}, x) - 1) & \text{if } \hat{x} \geq x \end{cases}$$

Thus, positive p-errors indicate overestimates, negative p-errors underestimates.

2.5 Plan Generation

The task of finding a join tree for a given query graph can be split into two distinct problems: the *join ordering* that decides which relations and subtrees to join next, and the *operator selection* that chooses the most suitable physical operator and argument order to join two subtrees T_1, T_2. The former problem is tackled by optimal join order ordering algorithms like DPccp [20], or heuristics, like GooCost [18,27] and GooCard [6].

Let us consider DPccp as an example: It iterates over each CCP and stores the (cost-wise) optimal join tree for each plan class in an associative data structure mapping a plan class to its best plan. The operator selection problem is decided by a build-plan subroutine, which is called by the join ordering algorithm. Build-plan decides which physical join operators are considered, and which argument order is better, $T_1 \bowtie T_2$ or $T_2 \bowtie T_1$.

Both join ordering and build-plan apply cost functions to decide on the best alternative. Cost functions typically use cardinalities as their input, so cardinality estimators must provide estimates for each plan class enumerated by the plan generator. For a more detailed discussion of the interplay of join ordering algorithms and build-plan procedures, the reader is referred to previous work [8].

2.6 Join Implementations, Cost Functions, and the Loss Factor

In this work, we consider two physical main-memory hash join operators: the *chaining hash join (CH-join)* and the *3D hash join (3D-join)* [7]. Their main difference is the hash table data structure which is built and probed during the join. The CH-join uses a hash table that resolves collisions by collecting all colliding keys into one linked list for each hash table bucket. The 3D-join uses a 3D hash table that groups duplicate keys together in a hierarchical collision chain organization with main and sub nodes. Further, for both physical operators, we consider two variants for the physical design of collision chain nodes (unpacked and packed), and three prefetching variants (no prefetching, rolling prefetching, and AMAC). The setup is exactly the same as in previous work [8], so due to space constraints, we refrain from repeating the details.

For each join implementation variant considered, we construct cost functions in two levels of precision from runtime experiments: a more precise, but also more complex *tabulated* cost function, and a less precise, but less complex *functional* cost function. We use the more precise tabulated cost functions as $\mathrm{CF}_{\mathrm{tru}}$ and the simpler but less precise functional cost functions as CF_{est}. A more detailed description can be found in previous work [8]. Note that both CF_{tru} and CF_{est} exhibit errors with regard to the true execution cost of the respective join. We assume CF_{tru} to be the true execution cost to have a notion of 'true optimality'. For our evaluation, only the relative errors between CF_{tru} and CF_{est} matter.

Finally, we define the *loss factor* of a plan, where a plan is a physical query evaluation plan which provides all the information necessary for its execution. Among other things, it specifies the operator implementations and argument order. If we consider, for instance, a hash join, we must decide on build and probe sides. An operator tree in the relational algebra without this physical information is called a *logical plan* (see also Sect. 2.5). Given a plan class S and a plan P for S, we define the *loss factor* of P as the true cost of P divided by the true cost of the overall best plan for S.

3 Join Cardinality Estimators

As became clear in Sect. 2.5, calculating the costs for all join implementations for each CCP requires the cardinality of each plan class as input. The true cardinality of a plan class S is given by

$$CE_{tru}(S) := |\bowtie_{R_i \in S} \sigma_{p_i}(R_i)|,$$

where we left out the join predicates for ease of notation. Using the true cardinality in a plan generator leads to the notion of *exact cardinality query optimization* (ECQO) [2,32]. All join cardinality estimators must be capable to provide an estimate for $CE_{tru}(S)$ for all plan classes S.

Table 2. The inputs of the different join cardinality estimators. $CE_{IA\text{-}X}$ refers to all IA-based estimators from Sect. 3.1.

	CE_{tru}	CE_{base}	CE_{sel}	$CE_{up\text{-}base}$	$CE_{up\text{-}sel}$	$CE_{IA\text{-}X}$	$CE_{j\text{-}base}$	$CE_{j\text{-}sel}$
$\|R_i\|$		+	+	+	+	+	+	+
$\|\sigma_{p_i}(R_i)\|$			+		+	+		+
$\|R_i \bowtie_{p_{i,j}} R_j\|$						+	+	+
$\|\sigma_{p_i}(R_i) \bowtie_{p_{i,j}} \sigma_{p_j}(R_j)\|$						+		
$\|\bowtie_{R_i \in S} R_i\|$							+	+
$MF(A_i)$ $(A_i \in \mathtt{JA})$				+	+			
$\|\bowtie_{R_i \in S} \sigma_{p_i}(R_i)\|$	+							

Except maybe for cardinalities of base relations, all other estimates, e.g., for $|\sigma_{p_i}(R_i)|$, will be prone to errors. Even worse, errors propagate exponentially from the bottom to the top of a plan [12,21]. In order to clearly identify the errors inherent to the different join cardinality estimators, we provide them with precise inputs. Table 2 indicates which estimator receives which input precisely.

We skim through this table to motivate the different estimators. The next subsections describe each of them in more detail. The first column of Table 2 shows the cardinalities of different algebraic expressions used by the different estimators. The last row contains the cardinalities for which estimates must be produced. The estimator CE_{tru} gets the true cardinalities as input and thus realizes the ECQO scenario. We will evaluate three cardinality estimators ($CE_{IA\text{-}M}$, $CE_{IA\text{-}S}$, and $CE_{IA\text{-}L}$), all of which use the independence assumption for join predicates. Since they all receive the same input, they are summarized under the column $CE_{IA\text{-}X}$. From the base and filtered cardinalities of base relations (rows 1, 2), they can derive the selectivity of selection predicates p_i correctly. Further, from the unfiltered and filtered results of binary joins (rows 3, 4), they derive the join selectivities. These are also given precisely. Next, let us turn to column CE_{base}. Here, only the sizes of (unfiltered) base relations are given. This estimator's purpose is to emulate the approach by Datta et al., whose join ordering heuristic SimpliSquared also only relies on base cardinalities [4]. As we will see, this estimator heavily punishes n:m-joins. The estimator CE_{sel} additionally receives the cardinalities of unfiltered and filtered base relations to calculate and use the selectivity of selection predicates. Thus, we can evaluate their usefulness. The estimator $CE_{j\text{-}base}$ receives the correct cardinalities of all plan classes S if only base relations are joined. Thus, it also completely ignores any selections, but, opposed to CE_{base}, does not have to make crude guesses about join result sizes. $CE_{j\text{-}sel}$ extends this estimator by information about filtered base relations. It will apply the independence assumption for selection predicates, and only for these. Thus, it ignores potential join-crossing correlations [17]. We can then compare this to $CE_{IA\text{-}X}$ to see what is worse, independence assumption for join predicates or selection predicates. The two upper bound-based cardinality estimators $CE_{up\text{-}base}$ and $CE_{up\text{-}sel}$ are special in terms of their inputs: In addition to the base cardinalities, they also get the maximum frequency $MF(A_i)$ for all join attribute values $A_i \in \mathtt{JA}$. We now define these cardinality estimators and additionally motivate their use.

3.1 Join Cardinality Estimation Under Independence Assumption

We briefly review three existing cardinality estimators based on the independence assumption (IA), namely $CE_{IA\text{-}M}$, $CE_{IA\text{-}S}$, and $CE_{IA\text{-}L}$. We suspect that the one most commonly used is $CE_{IA\text{-}M}$, which applies the *multiplicative rule (rule M)*: It takes the size of the cross product of all relations contained in some plan class S and multiplies it with the product of the selectivities of all applicable join predicates [26].

While $CE_{IA\text{-}M}$ simply multiplies all join predicate selectivities, $CE_{IA\text{-}S}$ and $CE_{IA\text{-}L}$ use only a certain subset of selectivities. In particular, they use one 'representative' selectivity for each equivalence class from the equivalence relation induced by the equality of join attributes: $CE_{IA\text{-}S}$ *(rule S)* picks the smallest, $CE_{IA\text{-}L}$ *(rule L)* the largest selectivity, respectively. For details on these estimators, the reader is referred to the literature [30].

Note that all three estimators receive the true filtered cardinalities as well as the true join selectivities as input. Thus, for plan classes with at most two relations, the estimates are precise. Errors can occur only in larger plan classes solely because independence is assumed for join predicates. This allows us to evaluate the inherent properties of these estimators, since their errors are never caused by incorrect inputs, e.g., wrong estimates for the filtered base relations.

Swami and Schiefer argue that $\text{CE}_{\text{IA-L}}$ is the only correct way to handle join cardinality estimation via the independence assumption [30]. We thus expect that this estimator is the most precise one and will ultimately lead to the best plans compared to $\text{CE}_{\text{IA-M}}$ and $\text{CE}_{\text{IA-S}}$.

3.2 Join Cardinality Estimation Based on Upper Bounds

The UES approach by Hertzschuch et al. [11] computes upper bounds for the intermediate cardinalities. It uses base table cardinalities and maximum frequencies of join attribute values as input. In their paper, cardinality estimation and plan generation are tightly integrated. Therefore, we extracted only the parts relevant for computing the upper bounds, making their approach a standalone cardinality estimator. Further, we resolved some imprecision regarding the cardinality computation for arbitrary plan classes and for updating maximum frequency values. We provide two variants of this estimator: $\text{CE}_{\text{up-base}}$ gets the unfiltered base table cardinalities as input and ignores single-table selection predicates; $\text{CE}_{\text{up-sel}}$ additionally gets the true filtered base table cardinalities.

3.3 Our Four Join Cardinality Estimators

After reviewing existing join cardinality estimators, we will now describe our four estimators.

The Crude Estimators. The first two estimators (CE_{base} and CE_{sel}) were proposed in previous work [8]. We repeat their definition for convenience. We first define both estimators for plan classes containing a single relation $R_i \in \mathcal{R}$:

$$\text{CE}_{\text{X}}(\{R_i\}) := \begin{cases} |R_i| & \text{if X = `base'} \\ |\sigma_{p_i}(R_i)| & \text{if X = `sel'} \end{cases} \tag{1}$$

Then, for general plan classes, we define

$$\text{CE}_{\text{X}}(S) := \min_{\text{ccp}(S_1,S_2):\ S=S_1\cup S_2} \text{CE}_{\text{X}}(S,(S_1,S_2)), \tag{2}$$

where, with $\mathfrak{U}(S_i)$ abbreviating $\mathfrak{U}(S_i,(S_1,S_2))$,

$$\text{CE}_{\text{X}}(S,(S_1,S_2)) := \begin{cases} \min(\text{CE}_{\text{X}}(S_1),\text{CE}_{\text{X}}(S_2)) & \text{if } \ \mathfrak{U}(S_1) \wedge \ \mathfrak{U}(S_2) \\ \text{CE}_{\text{X}}(S_1) & \text{if } \neg\mathfrak{U}(S_1) \wedge \ \mathfrak{U}(S_2) \\ \text{CE}_{\text{X}}(S_2) & \text{if } \ \mathfrak{U}(S_1) \wedge \neg\mathfrak{U}(S_2) \\ \text{CE}_{\text{X}}(S_1)\cdot\text{CE}_{\text{X}}(S_2) & \text{if } \neg\mathfrak{U}(S_1) \wedge \neg\mathfrak{U}(S_2). \end{cases} \tag{3}$$

The idea is to use the smaller cardinality if both arguments are unique, the cardinality of the cross product if neither argument is unique, and the cardinality of the non-unique side if one of the argument relations is unique but not the other. The latter case relates to foreign-key/key (FKK) joins where the non-unique side contains the foreign keys and determines the result size.

Observe that $\mathrm{CE}_{\text{base}}$ ignores single-table selection predicates and will always produce an overestimate. Further, its only input are the cardinalities of the unfiltered base relations. Finally, by taking the size of the cross product if neither side is unique, n:m-joins are punished heavily. This steers any plan generator to avoid them and instead prefer FKK-joins. This idea has often been applied to devise plan generator heuristics [4,11]. Embedding this preference for FKK-joins into the cardinality estimator allows to advise *any* plan generator to prefer FKK-joins. Further, no third-party cardinality estimator is necessary to produce the actual physical plans, as is the case for SimpliSquared [4] and UES [11]. Both use PostgreSQL to convert their logical plans into physical plans. In particular, they use the PostgreSQL cardinality estimator and the PostgreSQL cost function for the build-plan procedure. The tuning (e.g., number of allocated statistics slots), errors, and other properties of both have not been reported and may change from version to version. In contrast, our cardinality estimators, whose errors we report, can readily be used by build-plan, and their definition is stable.

The J-Estimators. Now, we introduce the next two, new cardinality estimators $\mathrm{CE}_{\text{j-base}}$ and $\mathrm{CE}_{\text{j-sel}}$. Denote by $\mathrm{sel}(p_i) := |\sigma_{p_i}(R_i)|/|R_i|$ the selectivity of a selection predicate p_i on R_i [26]. Then, we defined them as follows:

$$\mathrm{CE}_{\text{j-base}}(S) := |\bowtie_{R_i \in S} R_i| \tag{4}$$

$$\mathrm{CE}_{\text{j-sel}}(S) := \mathrm{CE}_{\text{j-base}}(S) \cdot \prod_{R_i \in S} \mathrm{sel}(p_i) \tag{5}$$

Here, the cardinalities of unfiltered base relations as well as the cardinalities for joining them are given precisely. For $\mathrm{CE}_{\text{j-base}}$, this is the only input. $\mathrm{CE}_{\text{j-base}}$ never underestimates the true cardinality. It is thus an upper bound of the true cardinality. Further, since $\mathrm{CE}_{\text{j-base}}$ ignores selection predicates, the cardinalities $\mathrm{CE}_{\text{j-base}}(S)$ can be precomputed and readily reused among different queries. Last but not least, $\mathrm{CE}_{\text{j-base}}$ gives rise to the notion of pessimistic plans, a term used, but not defined in the literature [11]. Here, we can define a *pessimistic plan* as a plan that is optimal under $\mathrm{CE}_{\text{j-base}}$. Hence, it is a plan for which the costs will only decrease, regardless of the true selectivities of the selection predicates. This is not true for what has been called an *optimistic plan*, where selectivities are possibly underestimated [11].

$\mathrm{CE}_{\text{j-sel}}$ takes the estimate produced by $\mathrm{CE}_{\text{j-base}}$ and multiplies it by the selectivities of all applicable single-table selection predicates. Thus, it assumes independence for selection predicates, thereby ignoring join crossing correlations [17]. This differs from the IA-based estimators from Sect. 3.1, which assume independence for join predicates. Separating these two allows us to answer interesting questions like which of these alternatives is worse.

4 Evaluation

This section describes our evaluation. We first present our experimental setup. Afterwards, we analyze the errors of the different cardinality estimators in isolation before examining their impact on plan quality.

4.1 Experimental Framework

For the evaluation, we use the Join Order Benchmark (JOB) [16,17]. It consists of 113 queries from 33 query templates on the Internet Movie Database (IMDb) schema and dataset. It contains challenging select-project-join queries with four to 17 relations on real-world data, which is why it is frequently used in the literature to evaluate the quality of plan generators [4,11,19,28,32].

Cardinality estimators (CE), cost functions (CF), join ordering algorithms (JOA), and the build-plan procedure (BP) are orthogonal components involved in plan generation. In order to systematically evaluate and compare the performance of our newly proposed cardinality estimators, we implemented a framework for plan generator benchmarking, which we call *PgBench*. In the following, we report on the performance of the existing cardinality estimators $CE_{\text{IA-M}}$, $CE_{\text{IA-S}}$, $CE_{\text{IA-L}}$, $CE_{\text{up-base}}$, and $CE_{\text{up-sel}}$ in comparison to our estimators CE_{base}, CE_{sel}, $CE_{\text{j-base}}$, and $CE_{\text{j-sel}}$. Further, we consider the join ordering algorithms DPccp, GooCost, and GooCard. As cost functions, we use CF_{tru} and CF_{est} from Sect. 2.6. The build-plan procedure used exhaustively evaluates the costs of all available join implementation variants with both possible join argument orders and picks the cheapest one (see also Sect. 2.6).

4.2 Cardinality Estimation in Isolation

We first analyze the errors of the cardinality estimators in isolation. Figure 1 shows the errors of the three IA-based estimators, the pessimistic upper-bound estimators, and our four estimators. Each plot illustrates the error distribution of the respective estimator. Included are all cardinality estimates for all plan classes enumerated during plan generation, i.e., all connected subgraphs of the query graph. The plan class size is shown on the x-axis. Plan classes of size 1 represent base relations (potentially with filters), plan classes of size 2 contain single joins of two (filtered) base relations, and so on. The y-axis is log scale and shows the p-error, where errors above 0 are over- and errors below 0 are underestimates. Boxes range from the 25th to the 75th percentile and are split by the median. Whiskers range from the 5th to the 95th percentile. Values outside the whiskers are shown as dots. Note that the y-axis is not the same across different plots.

Let us first turn our attention to the IA-reliant estimators in Figs. 1a–c. As one would expect, all three estimators produce mostly underestimates due to the positive correlation between predicates. By design, they return perfect estimates for cardinalities of base tables and joins of two base tables, since they receive their true cardinalities as input. While $CE_{\text{IA-M}}$ multiplies all applicable

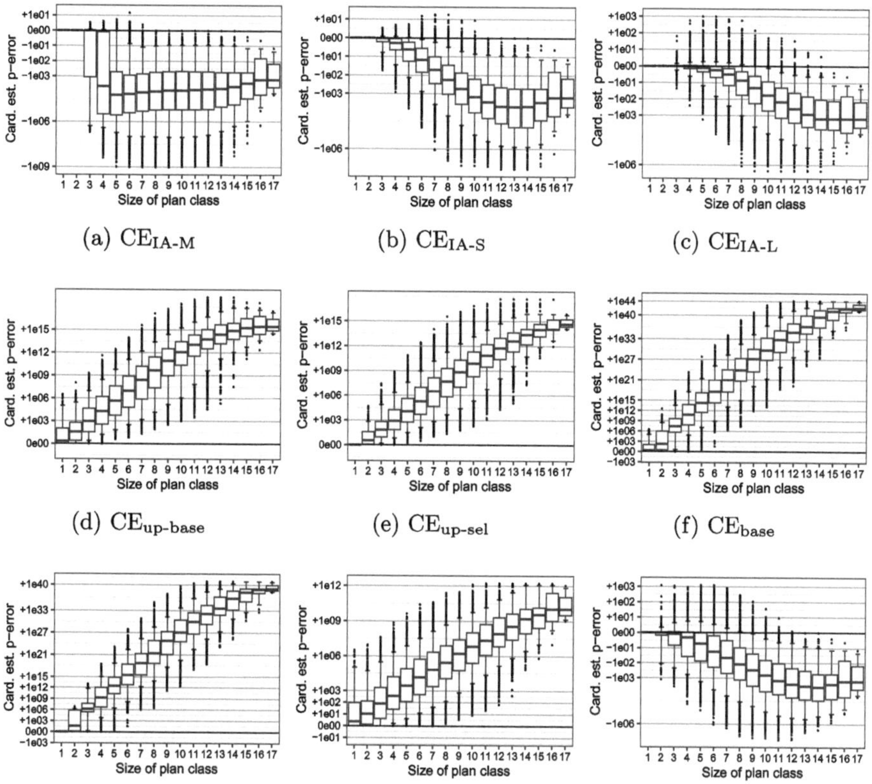

(a) $CE_{IA\text{-}M}$ (b) $CE_{IA\text{-}S}$ (c) $CE_{IA\text{-}L}$

(d) $CE_{up\text{-}base}$ (e) $CE_{up\text{-}sel}$ (f) CE_{base}

(g) CE_{sel} (h) $CE_{j\text{-}base}$ (i) $CE_{j\text{-}sel}$

Fig. 1. Errors of all nine cardinality estimators. The x-axis shows the plan class size (number of relations). The y-axis is log scale and shows the p-error. Boxes range from the 25th to the 75th percentile and are split by the median. Whiskers range from the 5th to the 95th percentile.

join selectivities for the given plan class, $CE_{IA\text{-}S}$ and $CE_{IA\text{-}L}$ only pick a 'representative' selectivity for each equivalence class. Differently put, they multiply fewer selectivities, which explains why they do not underestimate as severely as $CE_{IA\text{-}M}$. $CE_{IA\text{-}S}$ picks the smallest selectivity per equivalence class, $CE_{IA\text{-}L}$ the largest, but they both pick the same number of selectivities to multiply. As smaller selectivities reduce the estimated cardinality more, $CE_{IA\text{-}L}$ underestimates a little less severely than $CE_{IA\text{-}S}$, and produces more overestimates.

Next, we examine the errors of the pessimistic cardinality estimators from UES in Figs. 1d and 1e. As expected and in contrast to the IA estimators, they only overestimate. By design, $CE_{up\text{-}sel}$ yields perfect estimates for filtered base tables. The error and error variability for $CE_{up\text{-}sel}$ is smaller than for $CE_{up\text{-}base}$.

Table 3. Maximum and average loss factors over all plans for JOB.

CF	tru	IA-M	IA-S	IA-L	up-base	up-sel	base	sel	j-base	j-sel
					DPccp					
tru	1.00	10.34	1.99	2.31	12.42	3.16	16.12	7.23	6.80	2.30
	1.00	1.32	1.08	1.03	2.35	1.56	2.56	1.95	2.26	1.25
est	3.79	17.21	3.79	3.79	13.41	4.64	6.44	4.95	7.69	3.81
	1.26	1.49	1.29	1.27	2.64	1.85	2.55	2.28	2.42	1.33
					GooCost					
tru	1.45	2.53	1.95	1.50	7.32	5.10	7.32	8.35	6.64	2.40
	1.10	1.29	1.16	1.11	2.29	1.78	2.30	2.15	2.13	1.25
est	2.45	2.45	2.45	2.45	7.96	8.04	7.96	7.37	7.61	2.45
	1.27	1.27	1.27	1.27	2.27	2.03	2.31	2.25	2.24	1.28
					GooCard					
tru	1.43	9.62	2.16	1.71	7.27	6.03	7.27	12.74	6.59	1.75
	1.06	1.45	1.15	1.10	2.24	1.72	2.30	2.26	2.10	1.16
est	1.49	9.53	2.02	1.63	7.94	7.73	7.94	7.22	7.61	1.96
	1.12	1.42	1.18	1.16	2.34	1.99	2.40	2.30	2.20	1.15

Let us now consider CE_{base} and CE_{sel} in Figs. 1f and 1g. As expected, the estimators only overestimate. This is due to the fact that they ignore join selectivities and compute the join cardinalities based on the uniqueness property instead. CE_{sel} is slightly better than CE_{base}. While CE_{base} disregards single-table selections, CE_{sel} takes them into account. The resulting reduction in base table cardinality is then propagated through the joins. CE_{sel} receives the true filtered base table cardinalities as input, resulting in perfect estimates for plan classes of size 1.

$CE_{j\text{-}base}$ also overestimates, but to a much more limited degree. For joins of unfiltered base tables, it returns the correct cardinality. Hence, the estimation errors shown in Fig. 1h result solely from disregarding selections. As already mentioned in Sect. 3.3, $CE_{j\text{-}base}$ gives an upper bound for the true cardinalities, and is related to pessimistic plans. For $CE_{j\text{-}sel}$, the picture changes, as it produces both over- and underestimates (see Fig. 1i). However, underestimates make up at least 75% of estimates across all plan classes of size 2 or above. The underlying assumption leading to underestimates is the independence assumption on single-table selection predicates. In other words, $CE_{j\text{-}sel}$ highlights the influence of join-crossing correlations of selection predicates.

4.3 Plan Quality

In this section, we turn to plan quality in the big picture, and how it is influenced by the choice of join ordering algorithm, cost function, and cardinality estimator.

Table 3 summarizes all of our results. One block of numbers always belongs to the respective join ordering algorithm (JOA). The first table column indicates which cost function (CF) was applied by the build-plan procedure and the join ordering algorithm. For every combination of JOA and CF, there exist two lines. The first (second) line contains the maximum (average) loss factor taken over all JOB queries. The ten numbers in each row correspond to the cardinality estimator (CE) applied, indicated in the header line. The four rightmost columns refer to our cardinality estimators. For GooCard, the CF column refers only to build-plan, as GooCard makes its decision solely based on cardinalities.

Before discussing individual results, let us first set the numbers in Table 3 into perspective. By definition, the best possible loss factor is 1. Modifying DPccp to always pick the worst instead of the best possible plan allows us to determine that the maximum and average loss factor of the worst possible plans across all JOB queries using true cardinalities is 35 636 and 1160, respectively. As we will see, our results are orders of magnitude away from this worst case.

Table 3 allows for all sorts of comparisons. In the following, we will highlight some interesting aspects.

Looking only at column 'tru' (for CE_{tru}) lets us analyze the loss factor of different plan generators caused by inaccuracies in the cost function alone. By definition, DPccp with CE_{tru} and CF_{tru} yields optimal plans in all cases, which is why the maximum and average loss factors are both 1. Going from a precise to a less precise cost function (i.e., from CF_{tru} to CF_{est}) incurs a loss factor of 1.26 on average and of at most 3.79. Interestingly, GooCard performs surprisingly well, with a maximum (average) loss factor of 1.43 (1.06), with CF_{tru} for build-plan.

Consider the first two lines for DPccp, which shows the performance under CF_{tru} and different cardinality estimators. The maximum (average) loss factors for the IA-based cardinality estimators are 10.34 (1.32), 1.99 (1.08), and 2.31 (1.03). Hence, there is quite a big gap in the worst case between $CE_{\text{IA-M}}$ and the other two. Surprisingly, the loss factor on average seems quite acceptable, despite the fact that the underlying independence assumption is usually heavily criticized in the literature. For the pessimistic estimators $CE_{\text{up-base}}$ and $CE_{\text{up-sel}}$, the maximum (average) loss factors are 12.42 (2.35) and 3.16 (1.56), respectively, which is (mostly) worse than the IA-based estimators. Our crude estimators yield maximum (average) loss factors of 16.12 (2.56) and 7.23 (1.95); the J-estimators yield 6.8 (2.26) and 2.3 (1.25). Going to the less precise cost function CF_{est}, the average loss factors across all estimators become worse or stay about the same (for CE_{base}). However, the maximum loss factor improves for CE_{base} (6.44) and CE_{sel} (4.95). This indicates that errors in the cost function and the estimators partially cancel each other out.

Next, let us consider GooCost. There is a visible difference in performance between the more underestimation-prone cardinality estimators ($CE_{\text{IA-X}}$ and $CE_{\text{j-sel}}$) and those that tend to rather overestimate. GooCost seems to work better when supplied with cardinalities that are too small. Overall, its performance is surprisingly good for a heuristic approach, having maximum loss factors that do not exceed 2.53 for all underestimating CEs under CF_{tru}. With CF_{est}, the maximum loss is even slightly better. Further, if true cardinalities and true costs are not available, one can still achieve an average loss of only 27% using $CE_{IA\text{-}M}$.

Finally, in agreement with Swami and Schiefer [30], we can conclude that for IA-based cardinality estimation, rule M ($CE_{IA\text{-}M}$) and rule S ($CE_{IA\text{-}S}$) are inferior to rule L ($CE_{IA\text{-}L}$) in almost all cases in terms of loss.

5 Related Work

Although this paper touches upon join ordering algorithms, join algorithms, cost functions, and cardinality estimation, we focus on related work for the latter.

An excellent overview of cardinality estimators for selection predicates on single tables is provided by Cormode et al. [3].

To estimate the result size of joins, simple statistics can be used [5,9,26]. The prevailing approach seems to be sampling [3]. This may not come as a surprise, as Kaushik et al. show that sampling comes close to the optimum from a space-complexity perspective [13]. Alternatives are sketches (see the excellent overview by Rusu and Dobra [25]), kernel density-based estimators [14], bloom filters [23], and histograms [1,22]. Finally, there exist learning approaches to predict join cardinalities. Although well over 100 papers on learning cardinality estimates [29,34] exist, such approaches rarely made it into real systems. One notable exception is DB2, which applies a special lightweight learning approach with very restrictive time and space constraints [31].

Most of these approaches work well for joins of two base relations. Except for sketches, other approaches allow for selection predicates on the relations to be joined. Thus, these approaches can be used to provide the inputs $R_i \bowtie_{p_{i,j}} R_j$ and $\sigma_{p_i}(R_i) \bowtie_{p_{i,j}} \sigma_{p_j}(R_j)$ required by our join cardinality estimators. Some of the approaches also allow for the estimation of multiple joins and can thus be viewed as an alternative to our estimators [9].

6 Conclusion

Given that the worst case loss factor for any cardinality estimator in any of the tested scenarios is about 17 and the maximum loss factor of the worst plans is 35 636, we can conclude that for JOB, any of the tested cardinality estimators work quite well, although they all exhibit huge errors in their estimates. Knowing about the sensitivity of plan generators to selectivities [10,24] only adds to the wonder of working query optimizers. Even the simplest of our cardinality estimators, CE_{base}, which only relies on the cardinalities of base relations, has a maximum loss factor of only about 7.3 for GooCard and GooCost under CF_{tru}. $CE_{IA\text{-}L}$ performs surprisingly well, especially with the heuristic JOAs. Therefore, it can be worthwhile for database system developers to use these simple, fast to implement approaches in their optimizer in earlier stages of development.

Our findings, however, are currently limited to a single benchmark and might not generalize to arbitrary workloads. For instance, JOB does not contain cases where $|p_{i,j}| > 1$. We therefore aim to test the estimators for more benchmarks in future work. Further, CF_{tru} is a precise, but still error-prone cost function. It is hence desirable to measure the true execution time of the generated plans, for which we plan to extend our evaluation framework to include code generation.

The next big challenge in query optimization will be to find the intrinsic properties of real data, queries, and cardinality estimators that allow even highly erroneous cardinality estimators to result in acceptable plans. Once this problem has been solved, we can take a deeper look into those cases where these properties do not hold and hopefully come up with better estimators for exactly these cases.

Acknowledgement. We would like to thank Simone Kehrberg for proofreading the paper, Nazanin Rashedi for helpful comments and discussions, and Manuel Krakowski for extracting the standalone cardinality estimators from UES. We would also like to thank the anonymous reviewers for their suggestions to improve the paper.

References

1. Alway, K., Nica, A.: Constructing join histograms from histograms with q-error guarantees. In: Proc. of the ACM SIGMOD Conf. on Management of Data, pp. 2245–2246 (2016)
2. Chaudhuri, S., Narasayya, V., Ramamurthy, R.: Exact cardinality query optimization for optimizer testing. In: Proc. Int. Conf. on Very Large Data Bases (VLDB), pp. 994–1005 (2009)
3. Cormode, G., Garofalakis, M., Haas, P., Jermaine, C.: Synopses for Massive Data: Samples, Histograms, Wavelets, Sketches. NOW Press (2012)
4. Datta, A., Tsan, B., Izenov, Y., Rusu, F.: Simpli-squared: optimizing without cardinality estimates. In: SiMoD '24: Proceedings of the 2nd Workshop on Simplicity in Management of Data, pp. 1–10 (2024)
5. Dell'Aquila, C., Lefons, E., Tangorra, F.: Estimation of database unique values. WEAS Trans. Inf. Sci. Appl. **1**(1), 280–285 (2004)
6. Fegaras, L.: A new heuristic for optimizing large queries. In: Quirchmayr, G., Schweighofer, E., Bench-Capon, T.J.M. (eds.) DEXA 1998. LNCS, vol. 1460, pp. 726–735. Springer, Heidelberg (1998).
7. Flachs, D., Müller, M., Moerkotte, G.: The 3D hash join: building on non-unique join attributes. In: Proc. Conference on Innovative Data Systems Research (CIDR) (2022)
8. Flachs, D., Moerkotte, G.: On the performance of query optimization without cost functions and very simple cardinality estimation. In: DBKDA 2025, The Seventeenth International Conference on Advances in Databases, Knowledge, and Data Applications, pp. 58–64 (2025)
9. Gelder, A.V.: Multiple join size estimation by virtual domains. In: Proc. ACM SIGMOD/SIGACT Conf. on Princ. of Database Syst. (PODS), pp. 180–189 (1993)
10. Haritsa, J.: The Picasso database query optimizer visualizer. Proc. VLDB Endowment (PVLDB) **3**(2), 1517–1520 (2010)
11. Hertzschuch, A., Hartmann, C., Habich, D., Lehner, W.: Simplicity done right for join ordering. In: Proc. Conference on Innovative Data Systems Research (CIDR) (2021)

12. Ioannidis, Y.E., Christodoulakis, S.: On the propagation of errors in the size of join results. In: Proc. of the ACM SIGMOD Conf. on Management of Data, pp. 268–277 (1991)
13. Kaushik, R., Naughton, J., Ramakrishnan, R.: Synopses for query optimization: a space-complexity perspective. ACM Trans. Database Syst. **30**(4), 1102–1127 (2005)
14. Kiefer, M., Heimel, M., Breß, S., Markl, V.: Estimating join selectivities using bandwidth-optimized kernel density models. Proc. VLDB Endowment (PVLDB) **10**(13), 2085–2096 (2017)
15. Kumar, A., Stonebraker, M.: The effect of join selectivities on optimal nesting order. SIGMOD Rec. **16**(1), 28–41 (1987)
16. Leis, V., Gubichev, A., Mirchev, A., Boncz, P., Kemper, A., Neumann, T.: How good are query optimizers, really? Proc. VLDB Endowment (PVLDB) **9**(3) (2015)
17. Leis, V., et al.: Query optimization through the looking glass, and what we found running the join order benchmark. VLDB J. **27**, 643–668 (2018)
18. Lohman, G.: Heuristic method for joining relational database tables. IBM Tech. Discl. Bull. **30**(9), 8–10 (1988)
19. Marcus, R., et al.: Neo: a learned query optimizer. Proc. VLDB Endow. **12**(11), 1705–1718 (2019)
20. Moerkotte, G., Neumann, T.: Analysis of two existing and one new dynamic programming algorithm for the generation of optimal bushy trees without cross products. In: Proc. Int. Conf. on Very Large Data Bases (VLDB), pp. 930–941 (2006)
21. Moerkotte, G., Neumann, T., Steidl, G.: Preventing bad plans by bounding the impact of cardinality estimation errors. Proc. VLDB Endowment (PVLDB) **2**(1), 982–993 (2009)
22. Müller, M., Flachs, D., Moerkotte, G.: Memory-efficient key/foreign-key join size estimation via multiplicity and intersection size. In: Proc. IEEE Conference on Data Engineering, pp. 984–995 (2021)
23. Mullin, J.: Estimating the size of a relational join. Inf. Syst. **18**(3), 189–196 (1993)
24. Reddy, N., Haritsa, J.: Analyzing plan diagrams of database query optimizers. In: Proc. Int. Conf. on Very Large Data Bases (VLDB), pp. 1228–1240 (2005)
25. Rusu, F., Dobra, A.: Sketches for size of join estimation. ACM Trans. Database Syst. **33**(3), 15 (2008)
26. Selinger, P., Astrahan, M., Chamberlin, D., Lorie, R., Price, T.: Access path selection in a relational database management system. In: Proc. of the ACM SIGMOD Conf. on Management of Data, pp. 23–34 (1979)
27. Shekita, E., Tan, K.L., Young, H.: Multi-join optimization for symmetric multiprocessors. In: Proc. Int. Conf. on Very Large Data Bases (VLDB), pp. 479–492 (1993)
28. Sun, J., Li, G.: An end-to-end learning-based cost estimator. Proc. VLDB Endow. **13**(3), 307–319 (2019)
29. Sun, L.: Papers for database systems powered by artificial intelligence (machine learning for database). https://github.com/LumingSun/ML4DB-paper-list. Accessed Nov 2024
30. Swami, A., Schiefer, K.B.: On the estimation of join result sizes. In: Jarke, M., Bubenko, J., Jeffery, K. (eds.) EDBT 1994. LNCS, vol. 779, pp. 287–300. Springer, Heidelberg (1994).
31. Tan, W., et al.: Query predicate selectivity using machine learning in DB2. In: CASCON '21: Proc. 31st Annual Int. Conf. on Computer Science and Software Engineering (2021)
32. Trummer, I.: Exact cardinality query optimization with bounded execution cost. In: Proc. of the ACM SIGMOD Conf. on Management of Data (2019)

33. Wu, W., Chi, Y., Zhu, S., Tatemura, J., Hacigümüs, H., Naughton, J.: Predicting query execution time: Are optimizer cost models really unusable? In: Proc. IEEE Conference on Data Engineering, pp. 1081–1092 (2013)
34. Zhou, X., Chai, C., Li, G., Sun, J.: Database meets artificial intelligence: a survey. IEEE Trans. Knowl. Data Eng. **34**(3), 1096–1116 (2022)

Vector Database Benchmarking: A Face Retrieval Use Case

Teodor Sakal Francišković(✉), Nikola Todorović, Jelena Hrnjak, and Vladimir Dimitrieski

Faculty of Technical Sciences, University of Novi Sad, Novi Sad, Serbia
teodor.sakal_franciskovic@uns.ac.rs

Abstract. Recently, several database management systems shave been developed to support the efficient handling of high-dimensional vector data and fast and accurate similarity searches. These systems differ significantly in their overall architecture and implementation of indexing and querying techniques, which may impact their performance in real-world scenarios. However, no official performance comparison exists to guide users in selecting the right system. We present a controlled experiment evaluating six widely used vector database management systems in a real-world face retrieval scenario to fill this gap. Although the use case involves face retrieval, the results are broadly applicable, as the core evaluation focuses on database performance over vector embeddings, regardless of their origin. The scenario utilises a dataset of 1TB of images from which a total of 1,730,000 face vector embeddings were identified and extracted; the extraction was performed using two different vectorisation techniques, resulting in vectors of two different dimensions, i.e., sizes. The evaluation is conducted for both vector sizes on the dataset divided into small, medium, and large subsets to investigate the impact of different data processing loads on system performance. The evaluated performance aspects include insertion and deletion times, initialisation overhead, and search latency. In addition to performance metrics, face retrieval quality is assessed using precision, F1, and specificity.

Keywords: Benchmarking · Vector Databases · Face Retrieval · Similarity Search · HNSW · High-Dimensional Data

1 Introduction

In recent years, many real-world applications have come to rely on high-dimensional data such as images, text, and audio. Systems like facial recognition, recommendation engines, voice assistants, and biometric platforms convert this data into numerical vector representations to enable efficient processing. For example, facial retrieval systems generate face embeddings, search engines embed queries and documents into a shared vector space, and recommendation systems model user behavior and product features as vectors. As these applications grow in scale and complexity, so does the demand for efficient storage, search, and management of large vector datasets.

P. K. Chrysanthis et al. (Eds.): ADBIS 2025, LNCS 16043, pp. 28–43, 2026.
https://doi.org/10.1007/978-3-032-05281-0_3

To meet this need, numerous specialised vector database management systems (VDBMS) have emerged [14–17,19,20]. These systems differ in indexing, querying, and architecture, which affects performance across various data types, use cases, and deployment setups. While vendor blogs and talks often highlight capabilities [22,24], most lack rigorous, controlled evaluation. This creates a gap in the literature and motivates the need for systematic benchmarking to better understand each system's strengths and limitations.

This paper aims to evaluate several modern VDBMS in the context of a face retrieval scenario, which is a common real-world application found in domains such as security systems and photo organisation tools. The goal is to assess the performance of these VDBMS under conditions that reflect realistic usage patterns. The evaluation covers key performance aspects such as insertion and deletion times, initialisation overhead, and search latency. In addition to database-related metrics, the quality of face retrieval is also assessed using measures such as precision, F1, specificity, and other relevant indicators, as they provide a more comprehensive evaluation of how accurately and reliably the system retrieves the correct faces. The dataset consists of 1TB of images, divided into small, medium, and large subsets, from which approximately 1,730,000 face embeddings were identified and extracted. Two different vectorisation techniques were employed to explore the impact of vector size on performance: ArcFace and DINO, which produce 512- and 768-dimensional vectors, respectively. This diversified setup was designed to quantify VDBMS performance across different workload scales systematically.

This paper is organised as follows. Section 2 provides the theoretical background necessary to understand the key concepts behind vector representations, similarity search, and face embeddings. Section 4 reviews related work on database benchmarking, face retrieval systems, and the evaluation of approximate nearest neighbour (ANN) algorithms. It also outlines the existing research gaps that this study aims to address. Section 5 presents the methodology. Section 6 showcases and discusses the results, while also outlining the threats to validity. Finally, Sect. 7 concludes the paper.

2 Theoretical Background

The increasing volume of high-dimensional data in AI applications has driven the adoption of VDBMS for efficient similarity searches. Efficient management and querying of face embeddings—stored as vectors—remains a fundamental challenge in fields such as facial retrieval, image retrieval, and recommendation systems [1]. ANN techniques have been widely studied to enhance search efficiency, leading to the development of specialised VDBMS.

VDBMS are specialised systems designed to efficiently store, index, and query high-dimensional vector data, which are numerical representations capturing the semantic essence of complex data types such as text, images, audio, and video. Unlike traditional relational DBMS that manage structured data in rows and columns, VDBMS handle unstructured data by representing it as dense vectors in a vector space. This allows for similarity searches based on semantic meaning rather than exact matches. To enable scalable similarity searches, VDBMS often employ ANN algorithms, which are crucial for maintaining both speed and accuracy when searching across massive datasets. One of the most effective approaches to ANN search is the use of Hierarchical Navigable Small World (HNSW) graphs. Many VDBMS employ HNSW as the default algorithm for similarity search due to its scalability and efficiency in handling high-dimensional data [1].

HNSW is a graph-based ANN search algorithm first introduced in [12]. It enhances the Navigable Small World (NSW) approach by organising nodes into multiple hierarchical layers, similar to skip lists, where elements are arranged into levels for faster search by skipping over multiple elements at once. In skip lists, elements are organised in layers, with each higher layer acting as an "express lane" that allows for faster traversal by skipping over multiple elements [13]. In the NSW approach, nodes exist in a single layer, forming proximity-based edges, which allows efficient navigation but can be suboptimal for large datasets [12].

As shown in Fig. 1, HNSW consists of multiple layers:

- **Bottom Layer:** Represents the original NSW graph where nodes are fully connected based on their nearest neighbors, meaning that each node is directly linked to the most similar nodes present.
- **Higher Layers:** Act as progressively sparser versions of the graph, where only a subset of nodes exist, forming long-range connections that allow faster navigation through the structure.

The search begins at the uppermost layer, where the sparse nodes allow for rapid navigation. The search process moves down through progressively denser layers, refining the search until the closest neighbour is identified. This hierarchical structure substantially speeds up the search process in comparison to conventional nearest-neighbor search techniques [12]. The search performance relies on several hyperparameters:

- **M:** Defines the number of bidirectional links (neighbors) each node can have. A higher value improves recall but increases memory consumption.
- **efConstruction:** Determines the size of the dynamic candidate list during graph construction. A higher value leads to a more optimal graph, but increases the construction time. Furthermore, this hyperparameter impacts both search efficiency and accuracy.

Despite its efficiency, the HNSW algorithm has some downsides. The construction of the multilayered graph can be computationally expensive, especially for large datasets, as it requires a considerable amount of time to build the hierarchical structure. Additionally, memory consumption can grow significantly with higher values of the hyperparameters mentioned above, making it less suitable for resource-constrained environments. Furthermore, while the hierarchical structure improves search speed, performance may be degraded if the dataset is highly dynamic, requiring frequent updates to the graph [12].

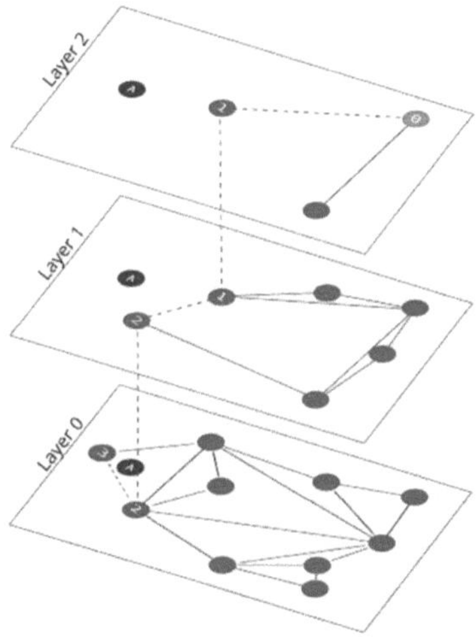

Fig. 1. Illustration of the hierarchical structure of the HNSW algorithm [28]

Efficient similarity search often relies on methods like HNSW to find the most similar embeddings in large datasets. This process involves comparing vector representations—such as face embeddings—using distance metrics like Euclidean or cosine similarity. When a query embedding is generated (e.g., from a new face), it is compared to stored embeddings, and the closest matches are retrieved [11]. For accurate embedding extraction, faces must first be detected using models like RetinaFace [5]. Deep learning models such as ArcFace [6] are commonly used to generate embeddings, while newer models like DINO [7] offer improved quality and reduced reliance on labeled data.

3 State-of-the-Art VDBMS

This subsection provides an overview of the VDBMS selected for this study, chosen based on their suitability in the real world for various applications. It considers their brief history, primary use cases, unique features, supported data structures, scalability, and compatibility with various programming languages and frameworks. It should be noted that while all evaluated VDBMS support the HNSW algorithm, Milvus is the only one that has made its HNSW implementation publicly available.

Milvus. [2,14] is an open-source VDBMS developed by Zilliz, optimised for high-performance similarity search over large-scale datasets. Originally built on FAISS, Milvus supports multiple ANN indexing methods, including IVF, HNSW, and ANNOY.

Data in Milvus is organised into collections, partitions, and segments, allowing efficient storage and retrieval at scale. It supports dense, binary, and sparse vectors and provides SDKs for Python, Java, Go, Node.js, and C#, enabling integration into face retrieval pipelines. Milvus supports horizontal scalability and can leverage GPU acceleration to improve search performance on large face datasets.

Weaviate. [15] is an open-source VDBMS developed by SeMI Technologies, built to handle high-dimensional vector data with efficient ANN search. Its design makes it suitable for face retrieval tasks, where fast and accurate similarity search over facial embeddings is essential.

Data is organised into "classes" with properties and associated vector embeddings, enabling both similarity-based search and structured filtering. Weaviate supports dense vectors and multiple similarity metrics such as cosine, dot product, and Euclidean distance. It provides client libraries for Python, JavaScript (and TypeScript), Go, and Java, and supports horizontal scaling for large-scale face embedding datasets.

Pgvector. [17] is an extension of a popular and well-established PostgreSQL open-source relational DBMS [18]. It utilises the existing PostgreSQL architecture to store vector data alongside already-existing relational data. It also introduces advanced mechanisms and indexing techniques for vector similarity search. Using pgvector is especially interesting for developers who want to avoid introducing additional components to an existing solution solely for handling vector data. However, relying on the existing PostgreSQL architecture might also introduce performance latency—the solution might be sub-optimal since it was not explicitly built to support vector data. Therefore, pgvector is a particularly significant candidate for the evaluation.

Qdrant is a popular VDBMS and a vector similarity search engine. Qdrant exposes a RESTful API and provides official client libraries for multiple popular programming languages, making it suitable for integration into face retrieval pipelines of various production environments.

Data in Qdrant is stored in collections comprising a named set of points, i.e., vectors with an optional additional payload. All data within one collection is divided into segments. Each segment has its independent vector and payload storage as well as indexes. It is also worth noting that Qdrant also supports horizontal scalability based on collection sharding across multiple cluster nodes.

Elasticsearch. [19] is an open-source, distributed search and analytics engine built on top of Apache Lucene. It is designed for high-throughput, low-latency workloads and supports horizontal scaling through a cluster-based architecture. Elasticsearch offers a RESTful API and official client libraries for several programming languages, making it suitable for integration into various production environments.

In the context of face retrieval, Elasticsearch supports dense vector fields, allowing for the storage and efficient similarity search of high-dimensional embeddings. It enables vector-based queries using ANN algorithms. It's distributed design ensures scalability, while internal mechanisms manage data replication and query distribution across nodes.

Chroma. [16] is an open-source VDBMS designed for storing and retrieving embeddings with associated metadata. It supports metadata-filtered vector searches, combining vector similarity with structured queries, making it suitable for face-retrieval tasks.

Chroma primarily supports dense vector embeddings and integrates easily with popular embedding providers. While it efficiently handles moderate-scale datasets, distributed scaling requires external orchestration. Chroma offers a RESTful API and client support for languages such as Python and JavaScript.

4 Related Work

Prior research has extensively explored the use of ANN techniques, comparing their performance across various datasets and retrieval strategies. One of the studies evaluated a variety of ANN algorithms—including Locality Sensitive Hashing (LSH), L2H, quantisation-based methods, tree-based techniques, and graph-based methods across 18 datasets with dimensions ranging from 100 to 4,096. The datasets included real-world image, text, video, and audio collections, as well as synthetic data. The evaluation focused on query latency and the quality of result sets, utilising metrics such as precision and recall. The findings indicated that graph-based algorithms, particularly Diversified Proximity Graph (DPG) and HNSW, offered superior search performance, demonstrating robustness across various datasets in terms of result quality, search time, and scalability [3].

To facilitate systematic benchmarking of ANN methods, the ANN Benchmarks tool was introduced [21], providing a standardised evaluation framework with automated testing, multiple quality metrics, and visualisation capabilities. This tool enables researchers and practitioners to compare algorithms effectively and optimise their selection based on dataset characteristics. The evaluation revealed that graph-based algorithms, particularly HNSW and Optimized Nearest Neighbor Graph (ONNG), achieved the highest query rates across most datasets. ONNG demonstrated increased robustness in the absence of global dataset structures. However, a notable drawback of graph-based approaches is the high preprocessing time required to construct data structures, making them less suitable for frequently changing datasets [4].

Beyond benchmarking, ANN algorithms are extensively utilised in VDBMS to enhance the efficiency of similarity searches. These VDBMS store high-dimensional data and leverage ANN techniques to perform approximate similarity matching, which is crucial for applications such as image retrieval, recommendation systems, and natural language processing. All of the popular VDBMS, including the ones from this study, rely on ANN algorithms to provide fast and scalable similarity search capabilities. However, ANN is not the only factor that influences the efficiency and usability of VDBMS. Other parameters, such as indexing strategies, hardware optimisation, and query processing techniques, also play a significant role in determining the overall performance of vector search systems. Recent research has explored various aspects of VDBMS, including their architectures, indexing techniques, and use cases. Such surveys [1,8] provide comprehensive overviews of existing solutions, detailing their underlying data structures, query optimisation techniques, and integration with ANN-based search algorithms. These studies have contributed to a better understanding of the design principles and trade-offs involved in the development of VDBMS, while also highlighting current challenges regarding VDBMS usage. However, these studies are primarily descriptive and do not perform quantitative evaluations, which are crucial for making informed decisions when selecting an appropriate VDBMS for a specific application.

Despite these advancements, there remains a notable gap in evaluating the real-world performance of VDBMS. While vendor-specific benchmarks exist [22] [23–27], they often lack objectivity and may not provide a fair comparison of different systems under standardised conditions. Unlike traditional DBMS, where extensive research has been conducted comparing relational and NoSQL systems across various performance metrics [9], VDBMS evaluations remain limited to proprietary studies or controlled experimental setups that do not fully reflect real-world workloads.

To fill this gap, our study will adopt established DBMS performance evaluation methods that are commonly used in comparisons between various DBMS types and image retrieval tasks [9,10]. By adopting the established metrics, we aim to provide a systematic and objective assessment of VDBMS in practical applications. This approach will help bridge the gap between theoretical performance claims and real-world deployment scenarios, providing valuable insights into the efficiency and reliability of various VDBMS solutions when handling datasets of varying scales.

5 Methodology

This study's use case is facial image retrieval, to identify and retrieve all occurrences of a specific individual across an image dataset. For this purpose, pictures from numerous large gatherings of people were collected. Subsequently, operations such as initialisation, insertion, search, and deletion were performed to evaluate the performance of various VDBMS.

5.1 Data Preparation

The initial dataset comprised 1.5 TB, which consisted of various file types, including images, videos, compressed archives, manually created print files, and system files. A manual data-cleaning process was performed, during which all non-picture type files were removed. After filtering out these files, the dataset was reduced to 0.99 TB (348,000 images). To facilitate analysis and experimentation, the dataset was further divided into three subsets:

- **Small Dataset:** ~33,000 images, from which ~192,500 faces (vectors) were extracted (ArcFace dataset size: 0.38 GB, Dino dataset size: 0.56 GB, around 10% of the whole dataset);
- **Medium Dataset:** ~170,000 images, from which ~815,500 faces (vectors) were extracted (ArcFace dataset size: 1.55 GB, Dino dataset size: 2.35 GB, around 50% of the whole dataset);
- **Large Dataset:** ~348,000 images, from which ~1,730,000 faces were extracted (ArcFace dataset size: 3.3 GB, Dino dataset size: 5.03 GB, 100% of the dataset).

Manual labelling was performed on a subset of 10 GB images for annotation. The labelling process involved determining whether a person was present in each image, specifically focusing on two individuals who had a significant presence at one of the events, meaning that they appeared in a large number of images. This was necessary to assess facial retrieval quality and ensure accurate identification within the dataset.

5.2 Experimental Setup

Each image in the dataset was first processed using RetinaFace for face detection. The detected faces were then transformed into high-dimensional vector embeddings using two models: DINO, which generates a 768-dimensional embedding, and ArcFace, which produces a 512-dimensional embedding.

All experiments were conducted on a machine equipped with an Intel Core i7-8665U processor featuring four cores and a base clock speed of 2.11 GHz. The system had 32 GB of RAM, and a 475 GB SSD provided storage, meaning all the datasets could be stored in RAM. Network performance was not a limiting factor, as all components were containerised and ran on the same host machine in a single-node, fully containerised setup using Docker.

The host machine operated on Windows 11, and all services, including the project code developed in Python 3.12 for face embedding generation and experimental execution, were deployed as separate containers. All containers were based on Linux images and were interconnected through a shared Docker network, facilitating communication during testing. Six VDBMS were evaluated:

- Milvus (milvusdb/milvus:v2.4.17, Ubuntu 22.04),
- Elasticsearch (elasticsearch:8.17.1, Ubuntu 20.04),
- Weaviate (cr.weaviate.io/semitechnologies/weaviate:1.28.2, Alpine-based),
- Chroma (ghcr.io/chroma-core/chroma:0.6.3, Debian-based),
- Qdrant (qdrant/qdrant:v1.13.2, Debian-based), and
- pgvector (pgvector/pgvector:0.8.0-pg17, Debian-based).

Docker Compose was used to manage and orchestrate the environment, and Python dependencies were installed separately inside each container using pip. All project code, including configuration files and experiment scripts, is publicly available on Github[1].

The benchmarking involved four core operations: initialisation, insertion, deletion, and search. These operations were selected to evaluate the performance and responsiveness of each VDBMS across key lifecycle stages. Initialisation refers to the creation of the VDBMS collection, including the setup of the necessary index structure for fast querying—in other words, creating a new collection in the vector database and configuring the indexing structure to enable efficient similarity search. Insertion involves adding data to the collection in batches of 500 rows to optimize memory usage and mitigate potential system bottlenecks—that is, adding batches of vector data to the collection while ensuring the data is properly written to disk. All reported insertion times include writing the data to disk. Deletion refers to removing the entire collection from the DBMS, effectively erasing all data stored in it, including permanently removing all stored vectors and associated metadata from the database. Insertion and deletion operations were performed three times for every combination of VDBMS, dataset, and embedding algorithm to ensure statistical reliability. Finally, search involves querying the DBMS to find the most similar faces to a provided query image for two different individuals, utilising the precomputed embeddings for similarity comparison—in essence, finding the most similar vectors in the database to a given query vector based on a defined similarity metric. The search operation was performed 100 times in parallel to evaluate the system's responsiveness and ensure statistical reliability. The number of returned faces was limited to 10,000 per query, with a cosine similarity threshold of 0.6 used to filter out less relevant matches. It is important to note that the implementation code for each of these four operations varied depending on the specific VDBMS being used.

The performance-oriented metrics included initialisation time, insertion time, deletion time, and search time, all measured in seconds, with lower times corresponding to faster performance. Additionally, mean, standard deviation (std), and 90th percentile (p90) values were also measured, where the 90th percentile represents the value below which 90% of the measured times fall. Requests Per Second (RPS) assessed the system's responsiveness during search operations under concurrent load, which was measured in successful queries per second, with higher values indicating better performance.

To assess the effectiveness of the HNSW algorithm within a particular VDBMS for facial retrieval, a set of classification metrics was employed—Precision, Recall, F1-score, Specificity, False Acceptance Rate (FAR), and False Rejection Rate (FRR)—as these metrics capture various dimensions of retrieval accuracy and system dependability. Notably, these metrics are influenced not only by the choice of VDBMS and the HNSW algorithm itself, but also by the quality of the vector representations generated by the underlying embedding model. Precision and recall are expressed as ratios (ranging from 0 to 1), where

[1] https://github.com/codesquaredtech/vector-db-benchmarking.

higher values are better. Precision measures how accurately the system identifies true matches, while recall evaluates how well the system captures all actual positive instances. The F1-score provides a balanced view by combining both, with higher values indicating better performance. Specificity is expressed as a ratio (ranging from 0 to 1), where higher values are better. It evaluates the system's ability to correctly reject non-matching instances, thereby reducing the likelihood of false positives [10]. FAR and FRR are expressed as ratios (ranging from 0 to 1), where lower values are better. FAR indicates how often incorrect matches are accepted, while FRR shows how often true matches are mistakenly rejected, offering insights into the system's practical reliability [10].

6 Results and Discussion

This section will provide the acquired experimental results of the evaluated VDBMS' initialisation, insertion, deletion, and search operations. Detailed raw results are available in the provided files on Github[2].

All of the benchmarked tables performed well in terms of **initialisation time**, with all **mean results** recorded below one second, as shown in Table 1. Milvus was the slowest, with a **mean initialisation time** of 0.6 s, while pgvector and Chroma were the fastest, recording **mean times** of 0.02 and 0.04 s, respectively, making them approximately 30 and 15 times faster in comparison to Milvus.

Table 1. Comparison of initialisation times (in seconds).

	Mean	Std	P90
Milvus	0.600	0.010	0.617
Weaviate	0.180	0.010	0.194
pgvector	**0.020**	**0.007**	**0.020**
Qdrant	0.220	0.020	0.237
Elasticsearch	0.470	0.360	0.830
Chroma	0.040	0.010	0.054

The results present in Fig. 2 show how **insertion** performance changes as more data accumulates over time, with batch size held constant. Milvus and Qdrant maintain consistently low and stable insert times, indicating good scalability. Elasticsearch also performs well, though it shows occasional spikes likely due to internal processes. In contrast, Chroma's insert times steadily increase, suggesting it handles growing data volumes less efficiently. Pgvecto.rs and Weaviate are mostly stable but show occasional performance drops, likely tied to memory or system events. Overall, Milvus and Qdrant offer the most reliable performance under sustained data ingestion.

[2] https://github.com/codesquaredtech/vector-db-benchmarking.

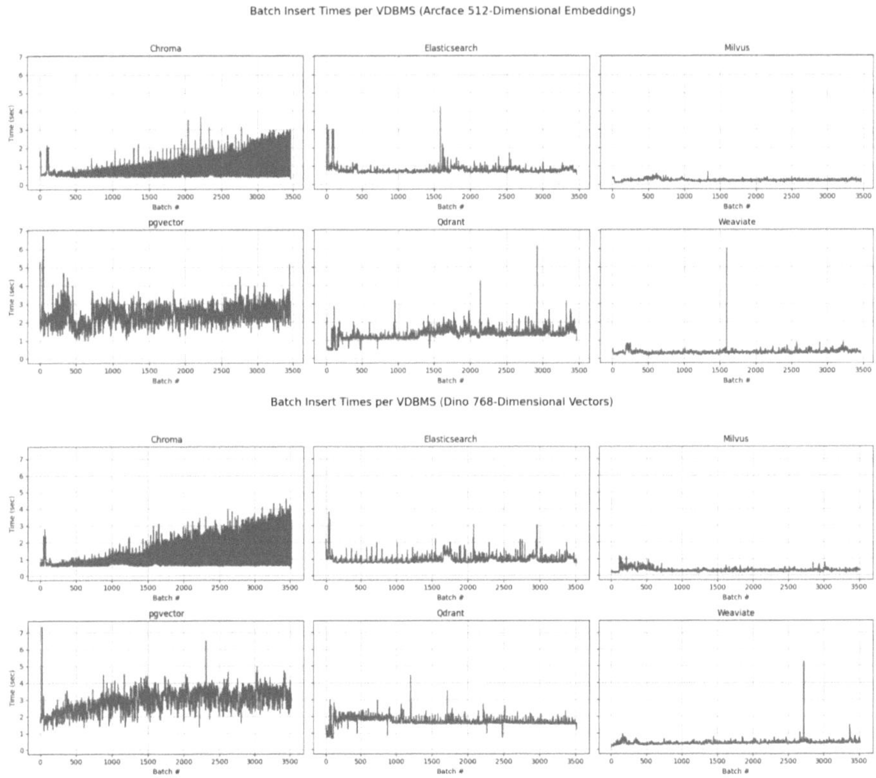

Fig. 2. Comparison of batch insertion times across different VDBMS for both 512- and 768-dimensional vectors.

As presented in Table 2, Milvus achieves consistently fast **mean** and **90th percentile (P90) deletion times** across all dataset sizes and embedding dimensions, highlighting its strong optimisation for deletion tasks. Chroma also performs well, often closely matching Milvus in speed while maintaining the most stable **deletion times**, as indicated by its low **variability**. Elasticsearch shows solid **deletion** performance, generally close to Chroma, though with somewhat higher **variability** across different conditions. Qdrant and Weaviate perform reasonably well on smaller datasets but experience noticeable slowdowns as dataset size increases. Pgvector consistently exhibits the slowest **mean** and **P90 deletion times** under all tested conditions. Unlike **insertion operations**, **deletion performance** appears less sensitive to changes in embedding dimensionality, with 512- and 768-dimensional embeddings resulting in broadly similar timings across most VDBMS.

Table 2. Comparison of deletion times (in seconds) across different dataset sizes and embeddings.

ArcFace (512-dimensional embeddings)									
	Small Dataset (~192,500 vectors)			Medium Dataset (~815,500 vectors)			Large Dataset (~1,730,000 vectors)		
	Mean	Std	P90	Mean	Std	P90	Mean	Std	P90
Milvus	**0.014**	**0.001**	**0.015**	**0.020**	0.010	0.033	**0.026**	0.008	**0.034**
Weaviate	0.470	0.190	0.660	0.700	0.030	0.726	0.740	0.100	0.850
pgvector	1.730	0.130	1.860	8.830	0.450	9.290	46.130	2.600	48.620
Qdrant	0.230	0.010	0.236	1.090	0.003	1.092	2.050	0.140	2.180
Elasticsearch	0.030	0.004	0.033	0.050	0.007	0.057	0.063	0.030	0.093
Chroma	0.030	0.001	0.028	0.022	**0.001**	**0.024**	0.056	**0.007**	0.063
Dino (768-dimensional embeddings)									
	Small Dataset (~192,500 vectors)			Medium Dataset (~815,500 vectors)			Large Dataset (~1,730,000 vectors)		
	Mean	Std	P90	Mean	Std	P90	Mean	Std	P90
Milvus	**0.017**	0.008	**0.024**	**0.024**	0.003	0.027	**0.026**	0.008	**0.034**
Weaviate	0.330	0.080	0.410	0.850	0.065	0.910	0.740	0.100	0.850
pgvector	2.420	0.100	2.520	12.830	0.080	12.890	46.120	2.600	48.620
Qdrant	0.226	0.009	0.234	1.540	0.114	1.660	2.050	0.136	2.175
Elasticsearch	0.028	0.005	0.033	0.060	0.030	0.090	0.063	0.030	0.093
Chroma	0.033	**0.004**	0.037	**0.024**	**0.002**	**0.025**	0.056	**0.007**	0.063

The **search performance** evaluation, summarised in Table 3, shows that Milvus generally offers the fastest **search performance**, consistently achieving the lowest **latency** and highest **throughput**. However, its **retrieval accuracy**, particularly the **F1-score**, ranks among the worst of the evaluated VDBMS. Weaviate provides a strong overall balance, maintaining fast **search speeds** and relatively stable **retrieval accuracy** across both embedding types. Chroma exhibits the slowest and least consistent **search performance**, coupled with higher **False Acceptance Rates (FAR)** and lower **recall** as dataset size increases, although its overall **F1-scores** remain among the best. Qdrant and pgvector deliver moderate results in both **speed** and **accuracy**, while Elasticsearch shows similarly average **search performance** but tends to suffer from higher error rates on larger datasets. Notably, higher **search speed** does not necessarily guarantee better **retrieval accuracy**, as illustrated by Milvus. Additionally, while all VDBMS experience increased **latency** with larger datasets, Chroma's **performance** degrades more sharply compared to Milvus and Weaviate. Overall, the choice between ArcFace and Dino embeddings significantly affects **face retrieval performance**, with ArcFace outperforming Dino due to superior embedding quality and feature distinctiveness in the context of face-related embeddings.

In summary, the benchmarking results highlight that important differences emerge depending on the workload type, dataset size, and embedding choice. Milvus consistently leads in **raw speed** across **insertion**, **deletion**, and **search tasks** but often sacrifices **retrieval accuracy**. Weaviate offers a strong overall balance between **performance** and **retrieval quality**, making it a versatile

Table 3. Comparison of search-related metrics across different dataset sizes and embedding types, for two individuals.

Person 1

ArcFace (512-dimensional embeddings)

	Small Dataset (~192,500 vectors)										Medium Dataset (~815,500 vectors)										Large Dataset (~1,730,000 vectors)									
	Search Performance				Face Retrieval Metrics						Search Performance				Face Retrieval Metrics						Search Performance				Face Retrieval Metrics					
	Mean	Std	P90	RPS	Prec.	Rec.	F1	Spec.	FAR	FRR	Mean	Std	P90	RPS	Prec.	Rec.	F1	Spec.	FAR	FRR	Mean	Std	P90	RPS	Prec.	Rec.	F1	Spec.	FAR	FRR
Milvus	1.58	0.86	4.14	0.64	0.997	0.51	0.67	0.999	0.001	0.49	2.46	1.01	3.91	0.19	0.997	0.51	0.67	0.999	0.001	0.49	5.82	1.40	8.13	0.10	0.997	0.51	0.67	0.999	0.0001	0.49
Weaviate	1.21	0.66	2.08	0.51	0.870	0.90	0.88	0.980	0.020	0.10	1.61	0.80	2.45	0.15	0.700	0.91	0.79	0.980	0.020	0.09	1.60	0.94	3.25	0.07	0.480	0.92	0.63	0.970	0.0300	0.08
pgvector	4.52	0.52	4.93	0.73	0.997	0.51	0.67	0.999	0.001	0.49	18.10	1.10	19.20	0.23	0.997	0.51	0.67	0.999	0.001	0.49	33.50	9.90	39.90	0.13	0.997	0.51	0.67	0.999	0.0001	0.49
Qdrant	0.72	0.33	1.10	0.38	0.997	0.51	0.67	0.999	0.001	0.49	1.07	0.37	1.37	0.13	0.997	0.51	0.67	0.999	0.001	0.49	1.36	0.62	1.92	0.08	0.997	0.51	0.67	0.999	0.0001	0.49
Elasticsearch	2.22	0.25	2.56	0.49	0.870	0.90	0.88	0.980	0.020	0.10	3.07	0.46	3.83	0.14	0.700	0.91	0.79	0.980	0.020	0.09	4.76	1.56	5.88	0.07	0.470	0.92	0.62	0.970	0.0300	0.08
Chroma	37.40	20.30	52.7	0.21	0.990	0.76	0.86	0.999	0.001	0.24	29.3	20.50	49.90	0.16	0.994	0.66	0.79	0.999	0.001	0.34	15.50	19.60	48.70	0.14	0.994	0.66	0.79	0.999	0.0001	0.34

Dino (768-dimensional embeddings)

	Small Dataset (~192,500 vectors)										Medium Dataset (~815,500 vectors)										Large Dataset (~1,730,000 vectors)									
	Search Performance				Face Retrieval Metrics						Search Performance				Face Retrieval Metrics						Search Performance				Face Retrieval Metrics					
	Mean	Std	P90	RPS	Prec.	Rec.	F1	Spec.	FAR	FRR	Mean	Std	P90	RPS	Prec.	Rec.	F1	Spec.	FAR	FRR	Mean	Std	P90	RPS	Prec.	Rec.	F1	Spec.	FAR	FRR
Milvus	1.1	0.5	1.97	0.25	0.21	0.7	0.32	0.64	0.36	0.34	1.2	0.27	1.54	0.06	0.11	0.51	0.18	0.81	0.19	0.49	1.94	0.45	2.4	0.035	0.085	0.42	0.14	0.88	0.12	0.62
Weaviate	1.7	0.9	3.13	0.30	0.22	0.6	0.32	0.69	0.31	0.40	2.8	1.53	4.64	0.09	0.12	0.37	0.18	0.87	0.13	0.63	2.03	1.23	4.0	0.060	0.090	0.30	0.14	0.92	0.08	0.70
pgvector	6.2	1.1	8.06	0.22	0.22	0.6	0.32	0.69	0.31	0.40	22.7	8.30	31.90	0.07	0.12	0.37	0.18	0.87	0.13	0.63	34.60	10.70	45.8	0.050	0.090	0.30	0.14	0.92	0.08	0.70
Qdrant	2.1	0.9	2.97	0.11	0.22	0.6	0.32	0.69	0.31	0.40	2.9	0.74	3.71	0.03	0.12	0.37	0.18	0.87	0.13	0.63	3.00	1.17	4.2	0.020	0.090	0.30	0.14	0.92	0.08	0.70
Elasticsearch	6.5	0.6	7.45	0.25	0.22	0.6	0.32	0.69	0.31	0.40	7.3	1.63	8.80	0.09	0.11	0.37	0.18	0.87	0.13	0.63	7.72	2.32	10.1	0.054	0.092	0.30	0.14	0.92	0.08	0.70
Chroma	15.0	19.6	48.0	0.23	0.22	0.6	0.32	0.69	0.31	0.40	11.6	18.80	47.50	0.09	0.12	0.37	0.18	0.87	0.13	0.63	23.80	25.10	54.4	0.048	0.091	0.30	0.14	0.92	0.08	0.80

Person 2

ArcFace (512-dimensional embeddings)

	Small Dataset (~192,500 vectors)										Medium Dataset (~815,500 vectors)										Large Dataset (~1,730,000 vectors)									
	Search Performance				Face Retrieval Metrics						Search Performance				Face Retrieval Metrics						Search Performance				Face Retrieval Metrics					
	Mean	Std	P90	RPS	Prec.	Rec.	F1	Spec.	FAR	FRR	Mean	Std	P90	RPS	Prec.	Rec.	F1	Spec.	FAR	FRR	Mean	Std	P90	RPS	Prec.	Rec.	F1	Spec.	FAR	FRR
Milvus	2.87	4.76	3.35	0.61	0.99	0.66	0.80	0.999	0.001	0.34	6.19	6.72	8.61	0.12	0.99	0.66	0.79	0.999	0.001	0.34	12.10	17.40	19.30	0.08	0.99	0.66	0.79	0.9998	0.0002	0.34
Weaviate	1.06	0.58	1.98	0.48	0.78	0.97	0.87	0.970	0.030	0.03	1.81	0.98	3.33	0.14	0.59	0.97	0.74	0.970	0.030	0.03	1.71	0.91	3.12	0.07	0.45	0.97	0.62	0.9700	0.0300	0.03
pgvector	4.40	0.41	4.83	0.67	0.99	0.66	0.80	0.999	0.001	0.34	16.90	4.08	20.00	0.21	0.99	0.66	0.79	0.999	0.001	0.34	36.30	8.90	41.20	0.11	0.99	0.66	0.79	0.9998	0.0002	0.34
Qdrant	0.80	0.26	1.18	0.32	0.99	0.66	0.80	0.999	0.001	0.34	1.30	0.61	1.63	0.11	0.99	0.66	0.79	0.999	0.001	0.34	1.46	0.68	2.01	0.07	0.99	0.66	0.79	0.9998	0.0002	0.34
Elasticsearch	2.96	1.03	3.63	0.44	0.75	0.97	0.85	0.960	0.040	0.03	4.39	1.76	5.45	0.12	0.52	0.97	0.68	0.960	0.040	0.03	6.66	4.81	15.80	0.06	0.38	0.98	0.55	0.9600	0.0400	0.03
Chroma	41.00	17.60	52.10	0.18	0.98	0.91	0.94	0.997	0.003	0.09	19.80	22.3	49.70	0.18	0.98	0.81	0.88	0.999	0.001	0.19	12.00	18.40	46.40	0.14	0.98	0.81	0.89	0.9997	0.0003	0.19

Dino (768-dimensional embeddings)

	Small Dataset (~192,500 vectors)										Medium Dataset (~815,500 vectors)										Large Dataset (~1,730,000 vectors)									
	Search Performance				Face Retrieval Metrics						Search Performance				Face Retrieval Metrics						Search Performance				Face Retrieval Metrics					
	Mean	Std	P90	RPS	Prec.	Rec.	F1	Spec.	FAR	FRR	Mean	Std	P90	RPS	Prec.	Rec.	F1	Spec.	FAR	FRR	Mean	Std	P90	RPS	Prec.	Rec.	F1	Spec.	FAR	FRR
Milvus	2.41	3.38	3.20	0.91	0.23	0.11	0.15	0.950	0.050	0.88	5.66	10.20	8.77	0.07	0.09	0.26	0.13	0.900	0.10	0.74	8.64	20.1	11.20	0.02	0.064	0.33	0.107	0.885	0.115	0.672
Weaviate	1.74	1.05	2.87	0.30	0.19	0.57	0.28	0.686	0.314	0.43	2.06	1.18	3.95	0.09	0.09	0.33	0.14	0.867	0.13	0.67	3.05	1.8	5.03	0.07	0.064	0.25	0.102	0.918	0.082	0.766
pgvector	5.58	1.07	6.57	0.75	0.23	0.11	0.15	0.950	0.050	0.88	29.00	10.00	38.20	0.06	0.09	0.26	0.13	0.900	0.10	0.74	37.00	12.5	45.50	0.05	0.064	0.23	0.101	0.918	0.082	0.766
Qdrant	0.75	0.25	1.10	0.48	0.23	0.11	0.15	0.950	0.050	0.88	2.60	0.92	3.47	0.04	0.09	0.26	0.13	0.900	0.10	0.74	4.08	2.8	6.20	0.02	0.064	0.23	0.101	0.918	0.082	0.766
Elasticsearch	7.35	2.34	10.20	0.25	0.19	0.57	0.28	0.686	0.314	0.43	10.50	6.30	17.30	0.07	0.09	0.33	0.14	0.867	0.13	0.67	8.53	4.0	11.90	0.05	0.064	0.23	0.101	0.918	0.082	0.766
Chroma	19.50	21.70	48.00	0.21	0.19	0.57	0.28	0.686	0.314	0.43	10.10	15.60	35.00	0.09	0.09	0.33	0.14	0.867	0.13	0.67	13.70	22.1	55.40	0.05	0.064	0.24	0.101	0.918	0.082	0.762

choice for face retrieval applications. Chroma, despite slower and more variable **search performance**, achieves some of the highest **F1-scores**, indicating strong **retrieval effectiveness**, albeit with higher **error rates** and poorer **scalability** under heavy loads. Pgvector, Qdrant, and Elasticsearch demonstrate moderate but reliable **performance**, though they generally lag behind the top performers in either **speed** or **retrieval accuracy**, particularly on larger datasets. Importantly, the embedding model has a profound influence on VDBMS **performance**, significantly impacting **precision**, **recall**, **F1-score**, **specificity**, **FAR**, and **FRR** across all VDBMS.

The testing in this study was performed using the default settings for each VDBMS, which may not be the most optimal configuration for all cases. For instance, PostgreSQL could have benefited from the COPY command for faster data insertion, but this would have deviated from the approach used with other VDBMS. Additionally, a unified approach was applied across all VDBMS, which may not have been the most efficient method for each system's specific strengths. For example, the configurable HNSW parameters were kept the same for all VDBMS (efConstruct=200, M=16), even though each VDBMS may have different optimizations or recommended settings for HNSW indexing. The hardware specifications used could also represent a limitation, as varying requirements for CPU, memory, and I/O performance across different VDBMS may have influenced the results.

Several potential threats to the validity of our study should be considered. Firstly, all experiments were conducted on a single-node Docker environment, which may not fully represent other configurations or real-world deployment scenarios. Secondly, only a subset of the data, where faces were present, was labelled, which raises the possibility that faces from other events could affect the performance of face retrieval metrics. Lastly, the results could vary if parameters, such as those for the HNSW index, were optimised further.

7 Conclusion

In this study, we benchmarked six VDBMS—Milvus, Weaviate, pgvector, Qdrant, Elasticsearch, and Chroma—for a face retrieval use case, evaluating performance in tasks like initialization, insertion, deletion, and search. Our work addresses a literature gap by providing an objective, comparative analysis of VDBMS performance with real-world benchmarking. Milvus showed superior insertion and deletion speeds but had lower retrieval accuracy, while Weaviate balanced speed and accuracy well, making it a strong choice for face retrieval. Chroma, though slower in search, achieved high F1-scores, indicating strong effectiveness. Pgvector, Qdrant, and Elasticsearch offered moderate but consistent performance, especially on larger datasets. Additionally, we highlight the impact of embedding choices on retrieval accuracy, with ArcFace outperforming Dino, offering valuable guidance for selecting the right VDBMS and embedding combination.

Moving forward, we plan to expand this benchmark by incorporating a wider range of embeddings, adding additional performance metrics such as memory usage to see to which extent it influences the performance of VDBMS, and applying the findings to real-world applications. This will allow for a more detailed investigation of how VDBMS behave with varying embedding dimensionalities and a larger volume of images, ensuring that high-performance results can be leveraged in practical face retrieval applications. Additionally, the vector dataset size will intentionally be designed to exceed available RAM capacity, to assess system performance under realistic, memory-constrained conditions typical of large-scale deployments.

Acknowledgment. This research has been supported by Code^2 DOO Novi Sad(https://codesquared.io/); and by the Ministry of Science, Technological Development and Innovation (Contract No. 451-03-65/2024-03/ 200156) and the Faculty of Technical Sciences, University of Novi Sad through project "Scientific and Artistic Research Work of Researchers in Teaching and Associate Positions at the Faculty of Technical Sciences, University of Novi Sad" (No. 01-3394/1).

References

1. Pan, J.J., Wang, J., Li, G.: Survey of vector database management systems. VLDB J. **33**(5), 1–25 (2024). https://doi.org/10.1007/s00778-024-00864-x
2. Wang, J., et al.: Milvus: a purpose-built vector data management system. In: Proceedings of the 2021 International Conference on Management of Data, pp. 2614–2627 (2021)
3. Li, W., Zhang, Y., Sun, Y., Wang, W., Zhang, W., Lin, X.: Approximate nearest neighbor search on high dimensional data – experiments, analyses, and improvement. IEEE Trans. Knowl. Data Eng. **32**(8), 1475–1488 (2020). https://doi.org/10.1109/TKDE.2019.2909204
4. Aumüller, M., Bernhardsson, E., Faithfull, A.: ANN-Benchmarks: a benchmarking tool for approximate nearest neighbor algorithms. Inf. Syst. **87**, 101374 (2020)
5. Deng, J., Guo, J., Zhou, Y., Yu, J., Kotsia, I., Zafeiriou, S.: RetinaFace: single-stage dense face localisation in the wild. In: Proceedings of the IEEE/CVF International Conference on Computer Vision, pp. 5203–5212 (2019). https://doi.org/10.1109/ICCV.2019.00530
6. Deng, J., Guo, J., Yang, J., Xue, N., Kotsia, I., Zafeiriou, S.: ArcFace: additive angular margin loss for deep face recognition. In: Proceedings of the IEEE/CVF Conference on Computer Vision and Pattern Recognition, pp. 4690–4699 (2019)
7. Caron, M., et al.: Emerging properties in self-supervised vision transformers. In: Proceedings of the IEEE/CVF International Conference on Computer Vision, pp. 9638–9648 (2021)
8. Taipalus, T.: Vector database management systems: fundamental concepts, use-cases, and current challenges. Cogn. Syst. Res. **85**, 101216 (2024). https://doi.org/10.1016/j.cogsys.2024.101216
9. Li, Y., Manoharan, S.: A performance comparison of SQL and NoSQL databases. In: Proceedings of the IEEE Pacific Rim Conference on Communications, Computers, and Signal Processing, pp. 1–6. IEEE (2013). https://doi.org/10.1109/PACRIM.2013.6625441
10. De Mel, V.L.B.: Survey of evaluation metrics in facial recognition systems. Preprint (2023). https://doi.org/10.13140/RG.2.2.10974.20805

11. Nguyen, H. V., Bai, L.: Cosine similarity metric learning for face verification. In: Asian Conference on Computer Vision, pp. 709–720. Springer (2010)
12. Malkov, Y.A., Yashunin, D.A.: Efficient and robust approximate nearest neighbor search using hierarchical navigable small world graphs. IEEE Trans. Pattern Anal. Mach. Intell. **42**(4), 824–836 (2018). https://doi.org/10.1109/TPAMI.2018.2889473
13. Vadrevu, V.S.P.K., Xing, L., Aref, W.G.: Tutorial: the ubiquitous skiplist, its variants, and applications in modern big data systems. arXiv preprint: arXiv:2304.09983 (2023)
14. Milvus: an open-source vector database. https://milvus.io/. Accessed 13 Mar 2025
15. Weaviate: an open-source vector database. https://weaviate.io/. Accessed 13 Mar 2025
16. Chroma: an open-source vector database. https://www.trychroma.com/. Accessed 09 Apr 2025
17. PGVector: an open-source vector similarity search for Postgres. https://github.com/pgvector/pgvector. Accessed 09 Apr 2025
18. PostgreSQL: the world's most advanced open source relational database. https://www.postgresql.org/. Accessed 09 Apr 2025
19. Elasticsearch: an open-source vector database. https://www.elastic.co/elasticsearch. Accessed 09 Apr 2025
20. Qdrant: an open-source vector database. https://qdrant.tech/. Accessed 09 Apr 2025
21. ANN benchmarks: a benchmark suite for approximate nearest neighbor algorithms. https://ann-benchmarks.com/. Accessed 14 Mar 2025
22. Qdrant: Qdrant benchmarks. https://qdrant.tech/benchmarks/. Accessed 17 Mar 2025
23. Rao, P.: LanceDB study. https://github.com/prrao87/lancedb-study?tab=readme-ov-file. Accessed 17 Mar 2025
24. Weaviate: Weaviate benchmarks. https://weaviate.io/developers/weaviate/benchmarks. Accessed 17 Mar 2025
25. MyScale: Weaviate vs. chroma performance analysis. https://myscale.com/blog/weaviate-vs-chroma-performance-analysis-vector-databases/. Accessed 17 Mar 2025
26. Timescale: PGVector vs. Pinecone. https://www.timescale.com/blog/pgvector-vs-pinecone. Accessed 17 Mar 2025
27. Pinecone: Pinecone Algorithms Set New Records for BIGANN. https://www.pinecone.io/blog/pinecone-algorithms-set-new-records-for-bigann/. Accessed 17 Mar 2025
28. Picture taken from DataStax. Hierarchical Navigable Small Worlds (2023). https://www.datastax.com/guides/hierarchical-navigable-small-worlds. Accessed 25 Mar 2025

PUL: Pre-load in Software for Caches Wouldn't Always Play Along

Arthur Bernhardt[1], Sajjad Tamimi[2], Florian Stock[2], Andreas Koch[2], and Ilia Petrov[1(✉)]

[1] Data Management Lab, Reutlingen University, Reutlingen, Germany
ilia.petrov@reutlingen-university.de
[2] Embedded Systems and Applications Group, TU Darmstadt, Darmstadt, Germany

Abstract. Memory latencies and bandwidth are major factors, limiting system performance and scalability. Modern CPUs aim at *hiding* latencies by employing large caches, out-of-order execution, or complex hardware prefetchers. However, software-based prefetching exhibits higher efficiency, improving with newer CPU generations.

In this paper we investigate software-based, post-Moore systems that offload operations to intelligent memories. We show that software-based prefetching has even higher potential in near-data processing settings by maximizing compute utilization through compute/IO interleaving.

Keywords: Near Data Processing · Processing-in-Memory · pre-fetching · pre-loading

1 Introduction

Memory Latencies and Bandwidth are Major Factors, Limiting System Performance and Scalability. Memory is getting colder as over the last 20 years DRAM capacity improved 128×, bandwidth 20×, latencies only 1.3× [21] and it takes several hundred nanoseconds to access it [11,28]. To address the so called *memory wall* modern CPU hardware aims at *hiding* latencies by employing large caches, out-of-order execution, and complex hardware prefetchers. Unfortunately, they quickly reach their limits on non-predictable access patterns. Aggravating matters, data-intensive systems face low cache-efficiency due to low data locality, and extensive data transfers of growing datasets.

Software-Based Prefetching Presents a Promising Direction Delivering Steadily Improving Results with Newer Hardware Generations. Software prefetching is a well-studied technique [4,5,14,15,17,18,20,23] that allows software to provide hints to the CPU data prefetcher about future data accesses and hard to predict patterns using special instructions. Recent study by Kühn et al. [16] thoroughly examines the merits and caveats of software prefetching on CPUs, indicating that its efficiency improves with each of the

P. K. Chrysanthis et al. (Eds.): ADBIS 2025, LNCS 16043, pp. 44–59, 2026.
https://doi.org/10.1007/978-3-032-05281-0_4

newer hardware generations. Notably, by hinting at upcoming accesses software prefetching becomes most effective if coupled to data structures, their layouts, operations and cache-optimisations [4,5,14,15,20,23]. However, CPU-based software prefetching does not cover the complete problem space.

Novel Post-moore Architectures Introduce Additional Challenges. Modern post-Moore architectures take a different avenue to mitigate the impact of high latencies and bandwidth limitations, by attempting to offload data processing operations to some kind of processing element (PE) located close to or even within the physical memory. As a result Near-Data Processing (NDP) becomes possible *in-memory* (as so called Processing-in-Memory – PIM), *in-storage* (on smart/computational storage), or *in-network* (on SmartNICs). To remain *economical* and comply with the *commodity model, under which memory and storage are sold* these are simple in-order PEs that are *slow* (150 – 500 MHz) and with *cache and memory hierarchies* different than those on CPUs. Such intelligent memories/devices typically operate on various memories with a broad latency spectrum (e.g., FLASH, NVM, DRAM, HBM, SRAM). What's more, the in-situ *cache/memory organisation differs* from those on CPUs and involves scratchpad memories instead of caches, therefore well-known cache-optimisations do not work, even if host-CPUs and in-situ PEs operate on the same data and layouts. Furthermore, the weak PEs get slowed down further by high memory latencies which, depending on the underlying memory technologies, may have a wide range. On the upside, there is ample bandwidth and throughput in-situ.

Therefore software prefetching emerges as a technique for *maximizing the compute utilisation on the weak PEs by interleaving compute and I/O.*

PUL: Custom Software Pre- and Un-Loading to the Rescue. In this paper, we introduce novel preloading techniques that facilitate interleaving compute and data transfers and leverage PE's low compute power by better utilizing the in-situ scratchpads. Preloading enables byte-addressable asynchronous application-defined transfers from arbitrary addresses into small PE-attached scratchpad memories (SRAM), while the PE performs useful work. In the mean time, persisting temporary or final PE results suffers from the inverse problem, as the PE is slowed down by expensive flushes from scratchpads to persistent storage. To this end, we employ *unloading* in combination with scratchpad double-buffering strategies to asynchronously write out results.

We distinguish between software *prefetching/flushing* and *preloading/unloading*. The former is inherently tied to CPU-specific cache hierarchies as well as vendor-defined instructions to *hint* at fixed-size cache-line transfers in advance without explicit control over data cache-placement or cache-eviction. The latter (focus of the present paper) employs externally attached SRAM-based "cache-like" scratchpad memory with explicitly managed, variable-sized data transfers. Both mechanisms require clear differentiation as they are not mutually exclusive.

Our Initial Experiment. demonstrates the effectiveness of PUL for utilizing the compute/IO interleaving (Fig. 1). To this end we investigate the different algorithmic/operational intensities for DRAM and slow NVM with different

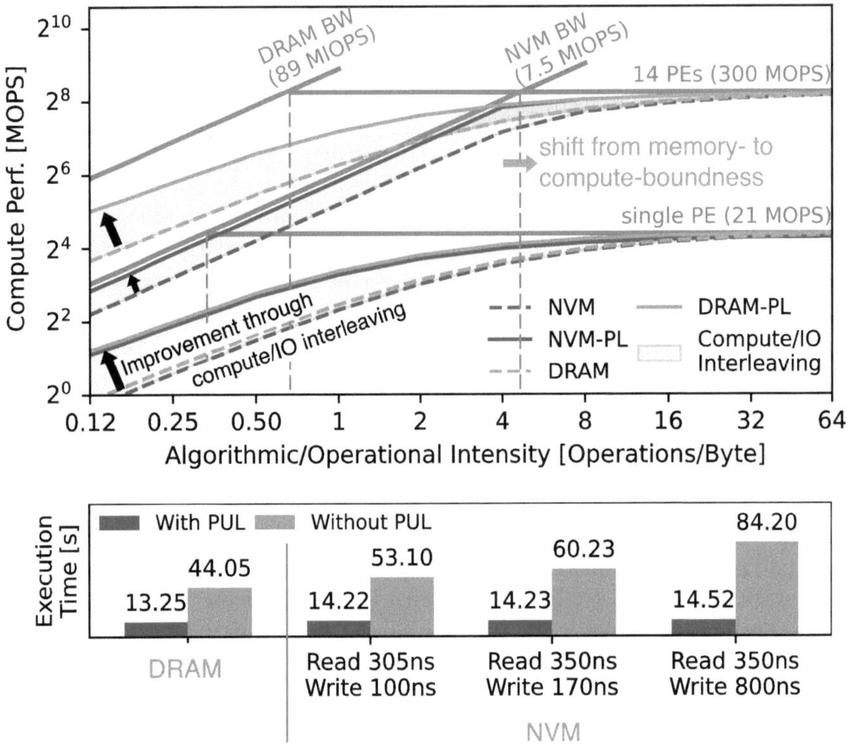

Fig. 1. (top) Roofline analysis showing the potential for pre-loading. PUL increases compute utilisation $\geq 2\times$ through compute/IO interleaving with low algorithmic intensities on DRAM and NVM in NDP settings. (bottom) Mitigation of the impact of higher latencies through PUL.

numbers of PEs. The compute/IO interleaving is especially prominent for typical DBMS operations that tend to have low operational intensity [8]. Furthermore, PUL-based software preloading and unloading enables hiding high NVM latencies and improving bandwidth utilisation (Fig. 1, bottom).

The **contributions** of this paper are:

- We investigate the effectiveness of PUL in novel, post-Moore architectures such as PIM and NDP, exploring the potential for compute and I/O interleaving on real hardware.
- We evaluate the impact of different memory latencies (e.g., DRAM, NVM) on performance and efficiency of compute/IO interleaving strategies and provide insights for further optimizations.
- We introduce *unloading* to complement preloading and prefetching techniques, providing a new opportunity for optimizing compute performance and enhanced system efficiency.

- We demonstrate a novel DMA-engine, specifically tailored for PUL and present insights in designing, optimizing, and employing such custom DMA-engine in NDP.

2 PUL Architectures and Interfaces

One of the most noticeable differences between traditional CPU-architectures and emerging paradigms like PIM and NDP lies in their raw compute capabilities per core. In contrast to recent CPU systems, which already reach clock speeds of 4–6 GHz, commercially available PIM systems like the PIM-enabled memory DIMMs from UPMEM only reach 350 MHz and are expected to increase to 466 MHz in later generations. Furthermore, recently published architectures for NDP on Field-Programmable Gate Arrays (FPGAs) [3,25,26,29] have demonstrated PE clock frequency rates as low as 150MHz. This disparity highlights the significance of maximizing the utilization of all available clock cycles in such systems.

Listing 1.1. Example of software pre- and un-loading in PUL.

```
PRELOAD_SET_SIZE(64);
...
PRELOAD_WAIT();                                // Synchronize
// Begin I/O and Compute Interleaving ============
// Async. I/O
PRELOAD(rand_ptr[4], bram_ptr[4]); // ---
PRELOAD(rand_ptr[5], bram_ptr[5]); //    | Preload-
PRELOAD(rand_ptr[6], bram_ptr[6]); //    | Distance
PRELOAD(rand_ptr[7], bram_ptr[7]); //    | of 4
// Interleaved Compute                   |
result += bram_ptr[0];  // <-------------
result += bram_ptr[1];
result += bram_ptr[2];
result += bram_ptr[3];
// Async. I/O
UNLOAD(bram_ptr[0], nvm_ptr, 64*4);
// End Interleaving ==============================
PRELOAD_WAIT();                                // Synchronize
```

PEs are designed to be resource-conscious, featuring in-order operation and lean memory hierarchies without multiple cache levels or complex HW prefetching mechanism found in traditional CPU architectures (Fig. 2). This simplicity enables the dense packaging of a large number of PE units within a single device. For instance, UPMEM supports up to 128 PEs (called DPU) per DIMM [6]. The high degree of parallelism is facilitated by the systems being inherently non-cache-coherent. This does present its own challenges in terms of programming model and synchronization, but is considered out of scope for this paper. However, the benefits of reduced cache contention, particularly in async.

preloading/unloading and managing cached data, make this design choice highly economical.

PUL leverages architectures that employ explicit memory management as in PIM or NDP, where software takes full control over memory access and caching. Data requests are directly dispatched to DMA engines, which asynchronously transfer data to fast, cache-like scratchpad memories, allowing PEs to access it in a single cycle. Notably, these systems do not rely on the traditional virtual memory abstraction, instead performing direct access to physical memory addresses. This allows for more efficient request processing, as the system can bypass intermediate layers of abstraction and operate at the hardware level. For example, TLB misses on CPU-based architectures and SW-prefetching pose an considerable overhead [16], requiring applications to utilize *huge pages*, physically contiguous memory allocated in advance, to mitigate the issue.

To fully leverage the benefits of PUL, it is crucial to interleave compute and I/O operations seamlessly. This mandates that the DMA engine concurrently accepts new requests, while simultaneously processing already dispatched ones, without blocking the PE. Furthermore, the PE must be able to continue processing operations without waiting for the DMA-transfer from the local DRAM to the PE-local scratchpad to complete. Prior works [2,9] utilize UPMEM's PEs to leverage PUL through a programming model that employs software kernels and multiple Tasklets. Tasklets are lightweight software threads in the UPMEM programming model that can be executed in parallel on a single PE (UPMEM DPU). By dispatching DMA requests on some Tasklets, while other Tasklets concurrently utilize clock cycles, the PE interleaves I/O and compute operations.

Devices built upon FPGAs allow a high degree of freedom on HW design, including custom memory hierarchies. We leverage this flexibility in our smart storage devices (Fig. 2-NDP), which employ soft-core PEs for in-situ processing, by designing and deploying a custom DMA engine specifically for PUL. This custom DMA-engine, bundled with fast scratchpad memory, e.g. FPGA BRAM, allows PEs to efficiently manage data transfer and compute. Through a simple

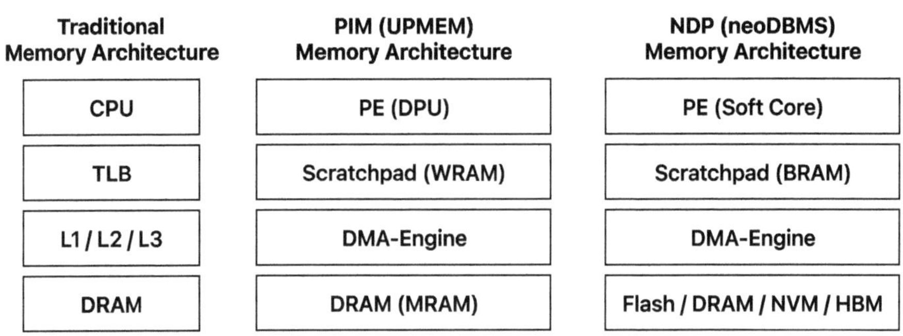

Fig. 2. Characteristics of Post-Moore architectures differ greatly from traditional CPU-based architectures.

programming interface (Listing 1.1), the PE can dispatch both DMA preload and unload requests to the custom DMA-engine by writing physical source and destination addresses, as well as transfer sizes, into exposed HW registers. The custom DMA-engine features two FIFO queues, one for preload and one for unload requests, each capable of storing up to 64 requests simultaneously. Synchronizing the asynchronous non-blocking DMA requests is achieved by reading from an exposed HW status-register. This register returns the status of the FIFO, indicating whether all requests have completed or if some remain pending in the queue. While this simple synchronization approach has proven sufficient for PUL, as preload and unload requests can already be synchronized independently, we are still investigating the potential benefits of a more fine-grained FIFO status handling, i.e. identifying specific pending requests, to further optimize PUL efficiency. Additional optimization techniques employed in our design include register value buffering. When DMA requests do not alter register values, for example, the transfer size remains the same, we can skip writing to the register altogether. As an external module, our PUL HW does not assume any specific characteristics of the underlying PEs or HW.

Factors Impacting the Quality of Preloading. Most importantly, data structures/layouts and algorithms are designed and optimized for CPU-architectures and caches, where prefetching operates at fix-sized cacheline transfers. Configurable transfer sizes in PUL present opportunities to explore new strategies in layout and access-based optimizations. Similarly, data structures have direct impact on the execution models and might pose constraints on gathering and firing sufficient requests in advance. Proposed solutions like grouping [4,15], coroutines [10,14,23,30], or fine-grained tasks [20], have yet to be reevaluated for such new memory architectures that do not run the risk of premature cache-evictions, or require deliberate prefetch-timing as discussed in [13].

3 Experiments and Insights

Experimental Setup. We evaluate PUL both in a PIM and an in-storage NDP architecture. The PIM experiments are conducted on an Intel Xeon Silver 4110 CPU equipped with a PIM-enabled UPMEM memory PIM-DIMM (Codename P21). NDP experiments are conducted on an ARM Neoverse N1 System Development Platform [1] equipped with 4x ARM N1-CPU cores. A Xilinx Alveo U280 FPGA (AU280) card serves as a smart/computational storage device and is connected via PCIe Gen3 x16 to the host. NDP processing is offloaded onto a PE-array, employing 14 MicroBlaze [19] soft cores (64bit) operating at 150 MHz. PUL HW-modules are attached to each PE, providing up to 64KiB scratchpad memory (BRAM). In addition, a configurable HW-module (NVMulator [27]) emulates a variety of read and write latencies over conventional DRAM is part of the HW-design. For NVM emulation, we employ latencies 350ns read, 170ns write based on [12,22,24] allowing us to quantify the impact of latencies on PUL. Although Intel Optane PMEM has been discontinued, our results enable projections for upcoming NVM technologies like DRAM+ [7].

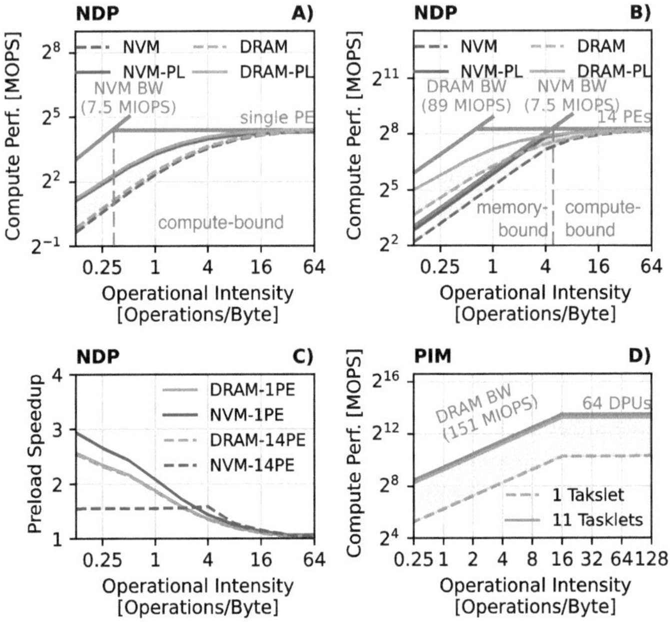

Fig. 3. Roofline analysis of pre-loading and different operational intensities. PUL achieves speedups through compute/IO interleaving in NDP and PIM on DRAM and NVM.

Workload and Dataset. The *workload* and the *dataset* are based on the prefetching micro-benchmark from Kühn et al. [16]. The dataset is stored in an array and randomly accessed by a pre-generated trace. A transfer size of 64B is used if not stated otherwise. In NDP the preload-distance is configured to 64 if not explicitly stated in the experiments.

Experiment 1: PUL consistently improves compute efficiency. The performance of systems with weak PEs and low clock-frequencies highly depends on their ability to efficiently utilize available compute cycles. In fact, even with fast memories I/O wait times waste cycles and reduce overall system performance. However, slower memories such as NVM or Flash exacerbate matters, highlighting the need for software-based preloading.

To quantify the benefits of preloading (PL) in both PIM and NDP architectures, we investigate random cacheline-sized data transfers, while varying algorithmic/operational intensities (Fig. 3) to cover a wide range of DB operations. We compare the impact of DRAM and NVM latencies and evaluate the performance under two scenarios: (1) separate I/O and compute phases (no PL or 1 Tasklet), and (2) interleaved I/O and compute (with PL or 11 Tasklets).

Preloading consistently improves performance in NDP for operations with low algorithmic intensities, effectively hiding the NVM and DRAM access latency. As a result, PUL achieves the *same* compute performance with NVM and

DRAM *despite* the access latency gap of 3.5× (Fig. 1). Noticeably, higher NVM-latencies offer a greater potential for compute/IO interleaving with preloading, yielding higher speedups of up to 2.9× compared to DRAM's 2.5× (Fig. 3-C). The results also reveal that preloading with 14 PEs saturates the available NVM bandwidth, but still delivers up to 1.5× performance gain. In contrast, preloading on DRAM maintains its improvements with 14 PEs across all intensities as the high available bandwidth and throughput are being underutilized due to the slow individual PE compute performance (Fig. 3-B). In PIM settings we observe a different behavior, yielding speedups of 7.8× with low and 11× with high operational intensity (Fig. 3-D). Notably, the maximum PIM compute performance strongly correlates with the utilization of at least 11 Tasklets, which is not achievable in lower intensities due to bandwidth saturation.

Insights: The preload speedups achieved by PUL align closely with those attained by software prefetching on common CPU architectures. However, newer architectures, like AMD Zen4 or Intel Sapphire Rapids, demonstrate that software prefetching is becoming more advantageous, with speedups of 4× to 5× [16]. Yet, PUL provides a more consistent interface decoupled from PE-specific hardware/-software features and ensures predictable behaviour by explicitly managing data cache-placement and the timing of cache-evictions.

Experiment 2: Interleaving DB Operations and I/O Allows for Higher Computational Intensity at Constant Execution Time. Slower memories like NVM or Flash significantly increase the I/O wait time, opening up the opportunity to interleave operation execution with I/O requests. As a result, higher computational intensity can be tolerated, while the execution time remains constant.

To understand the intricate trade-offs between optimal compute and I/O-latency in preloading, we investigate aggregations in PIM. We increase the operational complexity by increasing the number of attributes being aggregated, thereby increasing the amount of computation, but keeping I/O-latency constant as attributes are stored row-wise and accessed in a single transfer. We measure both the execution time (Fig. 4-A) and the instructions per cycle (Fig. 4-B) to determine the optimal balance. Furthermore, we assess the compute intensity of common DB operations in NDP and position them in interleaving space on NVM (Fig. 4-C).

Our experimental results demonstrate that simply aggregating a single attribute in PIM falls short of efficiently interleaving compute and I/O, resulting in a low IPC value of 0.58 (max. IPC 1.0). Furthermore, we observe that increasing the number of attributes aggregated has minimal impact on execution time, but rather yields improvement in IPC values (Fig. 4-B). The analysis of wait times in NDP reveals that they are significantly higher than the compute costs associated with common DB operations, such as aggregations and MVCC visibility checking. These can be interleaved multiple times during a single data request (Fig. 4-C), highlighting the need to improve PE utilization by interleaving more compute.

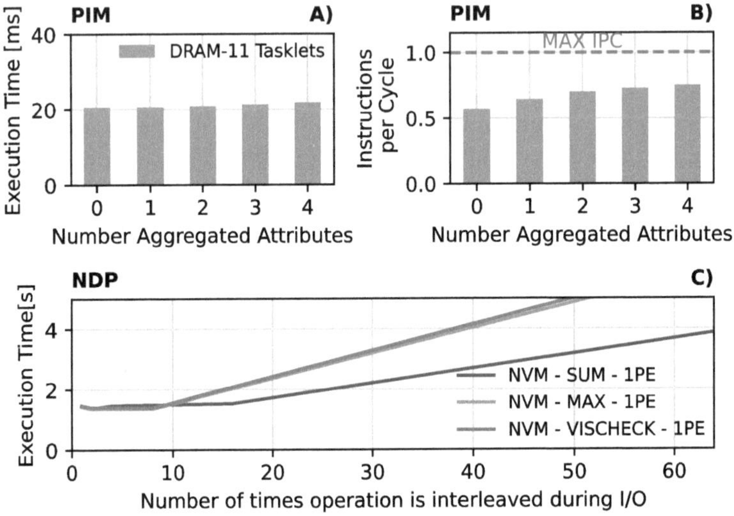

Fig. 4. PUL interleaves common DB-operations with data transfers without impacting execution times.

Insights: Attaining optimal I/O and PE utilization with PUL mandates a delicate balance between compute-latency and I/O-latency. Higher memory latency, while potentially beneficial for interleaving compute-intensive operations, is challenging for data-intensive operations. Their low operational intensity and inherent data dependencies lead to poor PE utilization. Increases in PE clock frequency aggravates matters. However, interleaving *multiple* I/O requests increases I/O throughput utilization, improving PE utilization during data-intensive operations as we demonstrate next.

Experiment 3: Preload Distance. The *temporal preload window* between a data request and actual data consumption from scratchpad has a profound impact on preload efficiency. As the preload distance increases, the preload window for requesting data and ensuring timely availability expands, allowing higher flexibility in request scheduling, latency hiding, and an overall improvement in I/O throughput. Listing 1.1 shows an example with a preload distance of 4, preloading the fourth element in advance of computing on the zeroth element.

To evaluate the effect of preload distance, we incrementally increase the preload distance, while keeping the operational intensity constant and low by employing a `SUM` operation. Our results (Fig. 5-A) demonstrate a continuous reduction in operation execution times as preload distance increases. Notably, it reaches a plateau at `d16`, Fig. 5-A, indicating that memory latency is fully mitigated and longer distances yield diminishing returns, which is confirmed by a separate examination of both the compute and I/O-latency (Fig. 5-B). Conversely, the reduction in overall latency through preloading improves I/O throughput and PE utilization (Fig. 5-C).

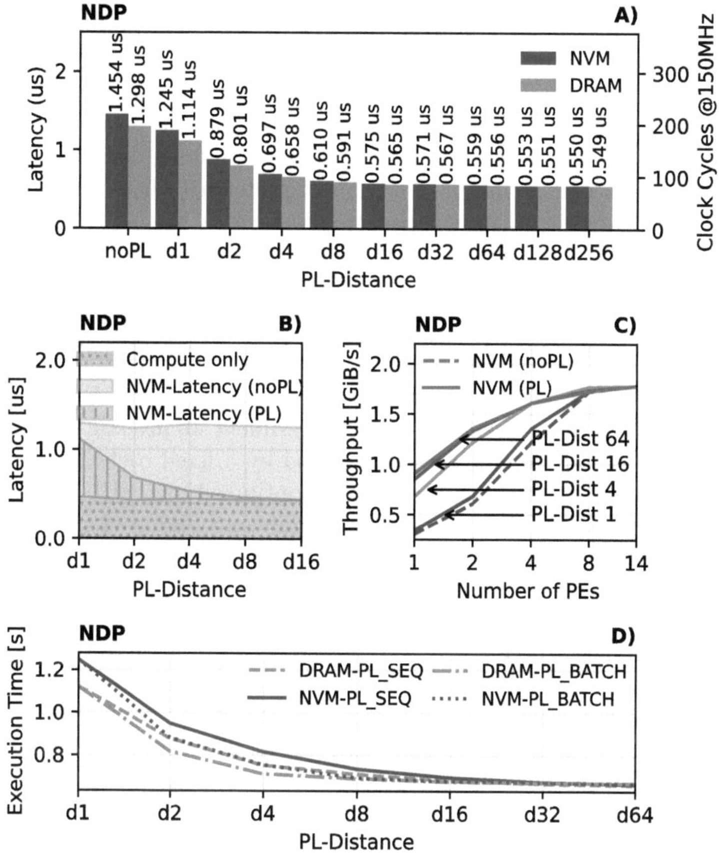

Fig. 5. Interleaving multiple I/O-requests with PUL can mitigate both NVM and DDR access latency, increase memory throughput, and maximize PE utilization.

Next, we investigate two distinct preload-strategies (Fig. 5-D). First, we explore sequential interleaving, where alternating preload requests and compute operations are overlapped (e.g., `PL[4]`→ `compute[0]`→`PL[5]`→`compute[1]`...). Secondly, we examine batch-wise execution, wherein multiple requests (number corresponds to the distance) are sent first, followed by processing the data of the previous preload-batch. Our results indicate that batch-wise execution outperforms the sequential interleaving strategy in terms of overall I/O throughput, eliminating the computational delays intrinsic to the sequential approach. Notably, as the preload distance begins to fully compensate the memory latency (>`d16`), both methods provide similar performance, indicating that a batch-wise execution is beneficial for I/O interleaving while it has no effect on compute interleaving.

Insights: By leveraging deep I/O request queues, our system can accommodate a high number of parallel pending data requests, outperforming the capabilities of traditional host-side prefetching (HW dependent) and PIM (limited by the number of Tasklets), overcoming the limitations reported in [16]. Moreover, preloading far into the future allows for optimal PE utilization, albeit constrained by available bandwidth and compute resources. Noticeably, in contrast to CPU-side cache-based prefetching methods, where excessive prefetch distances and cache contention can lead to premature eviction of prefetched data [16], PUL's non-shared, explicitly managed scratchpads minimize this risk and do not require deliberate timing as discussed in [13]. In conclusion, practical preload distances are short (<`d16`), require small scratchpad space and ensure high utilization of the weak PEs, while hiding latencies. Furthermore, by supporting longer preload distances PUL become robust against varying operational intensities.

Experiment 4: PUL Offers Configurable Transfer Sizes Enabling Layout and Access Optimisations. In this experiment we investigate two claims. First, while CPU-based prefetchers are limited by a fixed cacheline-sized granularity, which drives many layout and access optimizations, PUL overcomes the limitation with configurable transfer sizes, significantly improving performance. Secondly, leveraging larger transfers, PUL achieves high memory bandwidth, exceeding that of CPU-based software prefetchers.

While a single large data transfer results in a higher latency compared to a smaller one, multiple sequential transfers have worse total latency compared to the larger one, due to lower request management overhead on the PE. This effect is particularly pronounced in memory architectures lacking HW prefetchers and complex caching hierarchies. The higher latencies of larger transfers can, however, be effectively hidden through PUL I/O-interleaving. PUL increases bandwidth and reduces the amount of PE resources required to fully utilize the memory bandwidth. In addition, the higher access latency associated with larger transfers offer the potential for more compute interleaving.

To investigate the effects of transfer sizes on execution time, memory bandwidth, and I/O-interleaving, we conduct multiple experiments, in which we execute the same sequence of requests per PE with varying transfer sizes, while maintaining minimal compute-overhead through the use of a simple `SUM` operation. Furthermore, we increase the algorithmic intensity by varying the number of values for the `SUM` until further interleaving of computations becomes impossible and report this compute interleaving potential for increasing transfer sizes. In addition, varying the number of PEs allocated during execution allows us to determine how much resources are required to fully utilize the bandwidth with and without PUL. Increasing the number of PEs even further naturally increases bandwidth-boundness, allowing us to evaluate its impact on PUL.

In the case of NVM (Fig. 6-A), we observe that a single PE can interleave larger transfer sizes, up to 512B, without affecting execution time. Furthermore, we find that increasing the number of allocated PEs leads to an increase in execution time, as the system quickly reaches bandwidth saturation. This also effects I/O interleaving efficiency, as larger transfers increase pressure on memory bandwidth, resulting in worse execution times. In terms of memory throughput (Fig. 6-C), we observe that PUL is able to saturate bandwidth by leveraging 2–3

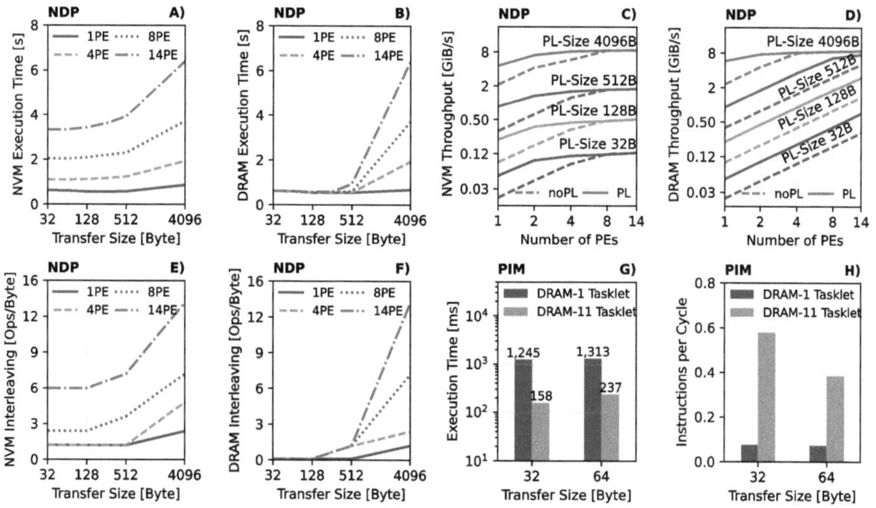

Fig. 6. PUL greatly reduces the number of PEs required to fully utilize bandwidth while efficiently interleaving larger transfers to increase throughput.

PEs, whereas without PUL, an equivalent level of throughput can only be reached using at least 8 PEs. As expected, the transfer size greatly improves memory bandwidth utilization. However, the system throughput is saturated at transfer sizes of 4096B, preventing us from reaching speeds exceeding 8 GiB/s. In comparison, DRAM's much higher throughput allows PUL to perfectly interleave requests up to 512B, while utilizing the maximum number of PEs (Fig. 6-B). Yet, DRAM is also constrained by our current system, also limiting the bandwidth to 8 GiB/s (Fig. 6-D). While we do not observe significant impact on compute interleaving at larger transfer sizes without saturating the NVM bandwidth (Fig. 6-E), with DRAM we do (Fig. 6-F), beyond 128B. These experiments also reveal a notable correlation between the degree of bandwidth-boundness and its direct impact on the overall prefetch latency, as well as a related effect on reduced PE-utilization with higher potential for compute interleaving.

We also evaluate the impact of transfer sizes on PIM execution times (Fig. 6-G) and PE utilization (Fig. 6-H). PIM PEs immediately become bandwidth-bound executing a simple SUM, resulting in an IPC value of 0.58 utilizing 32B transfers. Increasing the transfer size further has a detrimental effect on performance, leading to higher access latencies, worse execution times, and lower IPC values as the additional data is not sufficiently amortized by compute.

Insights: Compared to fixed cacheline-sized CPU-prefetchers, configurable transfer sizes in PUL allow data requests to be more easily optimized and aligned to a variety of data layouts/structures and algorithms, while efficiently supporting compute and I/O interleaving with less PE resources. These benefits cannot be achieved on traditional CPUs due to the limitations imposed by unpredictable cache evictions and cache contention.

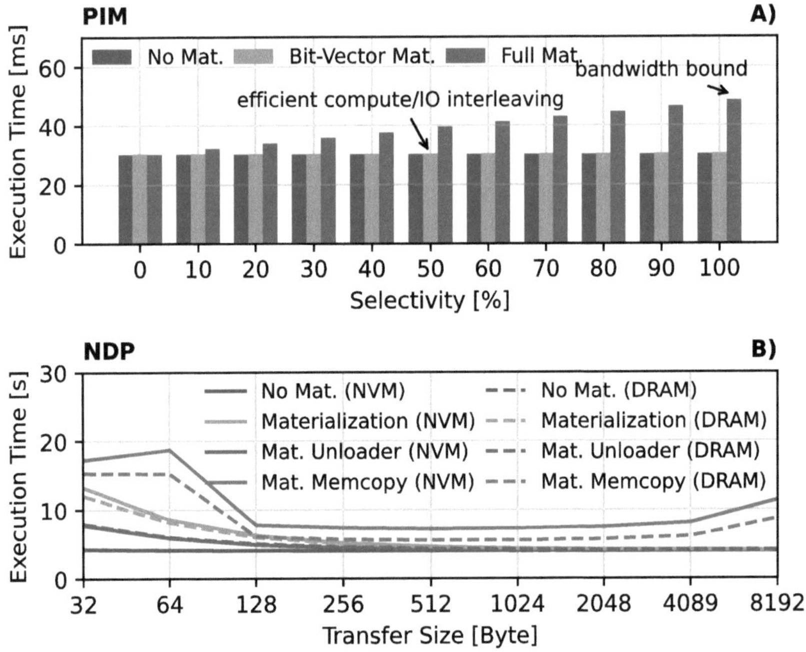

Fig. 7. Leveraging PUL's unloading capabilities during result materialization provide equally important improvements to preloading.

Experiment 5: Unloading. Materializing final or temporary PE results incurs a problem inverse to preloading, i.e., PEs are slowed down by expensive flushes from scratchpad memory to persistent storage. Unloading enables efficient interleaving of those flushes, which can occur either in several small random locations, or sequential blocks.

To demonstrate the effectiveness of I/O interleaving in *unloading* settings, we offload a simple filter operation onto PIM and measure different selectivities. When records are not filtered out, we materialize them in sequential blocks, representing the result-set. Our results show that fully materializing the whole record is not compute intensive, but has a significant impact on the bandwidth and, due to the already bandwidth-bound filter operation, leads to an increasingly worse performance (Fig. 7-A). By introducing a positional bit-vector to encode addresses of filtered records instead, we reduce the required bandwidth while also optimizing PE utilization by interleaving the addition compute for the bit-vector creation. This allows us to fully mitigate the materialization overhead.

We also investigate the impact of unloading in NDP and analyze the effects of varying transfer sizes. We iteratively update values stored in the scratchpad and periodically trigger flushing, when a threshold transfer size is exceeded. Our results (Fig. 7-B) compare the execution time of materialization leveraging PUL with a baseline excluding PUL. We also include numbers for standard `memcpy`, which however falls back to cache-line sized transfers.

Flushing larger data sizes benefits most materialization strategies due to the reduction on flush management overheads and better PE utilization, with the exception of memcpy, which at 2048b incurs too many requests, saturating bandwidth. PUL provides the best improvement with frequent random flushes that are also accompanied by low operational intensity. This is consistent with our previous experimental findings.

Insights: The exclusive PE managed scratchpad enables the unloading of large blocks of data at once, overcoming the limitations of cache-line write back (CLWB) instructions on the CPU. Compared to preloading, where timely availability is crucial due to data dependencies, persisting results often have relaxed timing constraints, as they only require periodic flushing until the end of execution. However, when synchronization mechanisms like locking and updating crucial data structures like indices are involved, ensuring that all necessary data is persisted before releasing locks becomes critical. Leveraging PUL's explicit software synchronization mechanisms makes sure that data is persisted in such cases.

4 Conclusions

In this paper, we presented PUL and a custom DMA-engine to address the challenges posed by post-Moore architectures such as PIM and NDP. Notably, the cache/memory organization in these systems differ from traditional CPUs as they rely on scratchpad memory and explicitly controlled DMA-engines, making widely adopted cache optimizations techniques ineffective, while still operating on the same data and layouts. PUL's flexibility in interleaving compute and I/O via software significantly improves compute performance and resource utilization, aligned with software prefetching techniques on common CPUs, while providing developer-friendly interfaces decoupled from vendor specific knowledge.

Acknowledgments. This work has been partially supported by *DFG neoDBMS.2 – 419942270*.

References

1. ARM: Neoverse N1 system development platform (SDP) - documentation and support. https://developer.arm.com/ToolsandSoftware/NeoverseN1SDP. Accessed Oct 2023
2. Bernhardt, A., Koch, A., Petrov, I.: pimDB: from main-memory DBMS to processing-in-memory DBMS-engines on intelligent memories. In: DaMoN '23 (2023). https://doi.org/10.1145/3592980.3595312
3. Bernhardt, A., et al.: neoDBMS: In-situ snapshots for multi-version DBMS on native computational storage. Proc. ICDE (2022). https://doi.org/10.1109/ICDE53745.2022.00290

4. Chen, S., Ailamaki, A., Gibbons, P., Mowry, T.: Improving hash join performance through prefetching. In: Proc. ICDE, pp. 116–127 (2004).https://doi.org/10.1109/ICDE.2004.1319989
5. Chen, S., Gibbons, P.B., Mowry, T.C.: Improving index performance through prefetching. In: Proc. SIGMOD, SIGMOD '01, pp. 235–246. Association for Computing Machinery, New York (2001). https://doi.org/10.1145/375663.375688
6. Devaux, F.: The true processing in memory accelerator. In: IEEE Hot Chips (2019). https://doi.org/10.1109/HOTCHIPS.2019.8875680
7. Ferroelectric Memory GmbH: Feroeelectric memory company. https://ferroelectric-memory.com. Accessed May 2025
8. G mez-Luna, J., Hajj, I.E., Fernandez, I., Giannoula, C., Oliveira, G.F., Mutlu, O.: Benchmarking a new paradigm: experimental analysis and characterization of a real processing-in-memory system. IEEE Access **10**, 52565–52608 (2022). https://doi.org/10.1109/ACCESS.2022.3174101
9. G mez-Luna, J., Hajj, I.E., Fernandez, I., Giannoula, C., Oliveira, G.F., Mutlu, O.: Benchmarking a new paradigm: experimental analysis and characterization of a real processing-in-memory system. IEEE Access **10** (2022). https://doi.org/10.1109/ACCESS.2022.3174101
10. He, Y., Lu, J., Wang, T.: CoroBase: coroutine-oriented main-memory database engine. Proc. VLDB Endow. **14**(3) (2020). https://doi.org/10.14778/3430915.3430932
11. Huang, H., et al.: Shuhai: a tool for benchmarking high bandwidth memory on FPGAS. IEEE Trans. Comput. **71**(5), 1133–1144 (2022). https://doi.org/10.1109/TC.2021.3075765
12. Izraelevitz, J., et al.: Basic performance measurements of the intel optane DC persistent memory module (2019). https://doi.org/10.48550/ARXIV.1903.05714
13. Jamilan, S., Khan, T.A., Ayers, G., Kasikci, B., Litz, H.: APT-GET: profile-guided timely software prefetching. In: Proc. EuroSys (2022). https://doi.org/10.1145/3492321.3519583
14. Jonathan, C., Minhas, U.F., Hunter, J., Levandoski, J., Nishanov, G.: Exploiting coroutines to attack the "killer nanoseconds". Proc. VLDB Endow. **11**(11), 1702–1714 (2018). https://doi.org/10.14778/3236187.3236216
15. Kocberber, O., Falsafi, B., Grot, B.: Asynchronous memory access chaining. Proc. VLDB Endow. **9**(4), 252–263 (2015). https://doi.org/10.14778/2856318.2856321
16. Kühn, R., Mühlig, J., Teubner, J.: How to be fast and not furious: Looking under the hood of CPU cache prefetching. In: Proc. DAMON, DaMoN '24. Association for Computing Machinery, New York (2024). https://doi.org/10.1145/3662010.3663451
17. Lee, J., Kim, H., Vuduc, R.: When prefetching works, when it doesn t, and why. ACM Trans. Archit. Code Optim. **9**(1) (2012). https://doi.org/10.1145/2133382.2133384
18. Mao, Y., Kohler, E., Morris, R.T.: Cache craftiness for fast multicore key-value storage, EuroSys '12, pp. 183–196. Association for Computing Machinery, New York (2012). https://doi.org/10.1145/2168836.2168855
19. MicroBlaze, Xilinx Inc.: Microblaze soft processor core. https://www.xilinx.com/products/design-tools/microblaze.html. Accessed Jan 2025
20. Mühlig, J., Teubner, J.: MxTasks: how to make efficient synchronization and prefetching easy. In: Proceedings of the 2021 International Conference on Management of Data, SIGMOD '21, pp. 1331–1344. Association for Computing Machinery, New York (2021). https://doi.org/10.1145/3448016.3457268

21. Mutlu, O.: Processing data where it makes sense in modern computing systems. In: MECO (2018)
22. Peng, I.B., Gokhale, M.B., Green, E.W.: System evaluation of the intel optane byte-addressable nvm. In: Proc. MEMSYS (2019)
23. Psaropoulos, G., Legler, T., May, N., Ailamaki, A.: Interleaving with coroutines: a practical approach for robust index joins. Proc. VLDB Endow. **11**(2), 230–242 (2017). https://doi.org/10.14778/3149193.3149202
24. van Renen, A., Vogel, L., Leis, V., Neumann, T., Kemper, A.: Persistent memory i/o primitives. In: Proc. DaMoN (2019)
25. Tamimi, S., Bernhardt, A., Stock, F., Petrov, I., Koch, A.: CINDA: using cache-coherent interconnects for accelerating databases by enabling near-data processing of update transactions. IEEE Trans. Comput. **74**(7) (2025). https://doi.org/10.1109/TC.2025.3558028
26. Tamimi, S., et al.: DANSEN: database acceleration on native computational storage by exploiting NDP. ACM Trans. Reconfigurable Technol. Syst. (2024). https://doi.org/10.1145/3655625
27. Tamimi, S., Stock, F., Bernhardt, A., Petrov, I., Koch, A.: NVMulator: a configurable open-source non-volatile memory emulator for FPGAS. In: Applied Reconfigurable Computing. Architectures, Tools, and Applications. Springer International Publishing (2023)
28. Teubner, J., Woods, L.: Data Processing on FPGAs. Morgan & Claypool Publishers (2013)
29. Vinçon, T., et al.: Near-data processing in database systems on native computational storage under HTAP workloads. Proc. VLDB Endow. **15**(10), 1991–2004 (2022). https://doi.org/10.14778/3547305.3547307
30. Zhi, X., Yan, X., Tang, B., Yin, Z., Zhu, Y., Zhou, M.: CoroGraph: bridging cache efficiency and work efficiency for graph algorithm execution. Proc. VLDB Endow. **17**(4) (2023). https://doi.org/10.14778/3636218.3636240

Explainable AI

Responsible AI: Training Deep Learning Model Efficiently

Shwetha Salimath[1,2](✉), Francesca Bugiotti[1], and Sylvain Wlodarczyk[2]

[1] LISN, Universite-Paris-Saclay, Gif-sur-Yvette, France
shwetha.nagaraj.salimath@gmail.com
[2] SLB, Montpellier, France

Abstract. There has been a significant surge in the adoption of deep learning methods across various industrial and research domains. This increase requires efficient training strategies to improve convergence speed, robustness, and overall model optimization. The paper presents a benchmark analysis that highlights the benefits of learning rate (LR) schedulers. The benchmark has been conducted using four training methods that integrate the LR finder, the LR scheduler, and the LR optimizers like Adam. The first three training methods use the LR finder to compute the correct range and then use, respectively, the AdaptiveCycle LR scheduler, the OneCycle LR scheduler, and the static LR for training; the last training method uses only a static LR. The OneCycle LR scheduler has been used as it balances exploration and exploitation during training by adjusting the LR cyclically. Additionally, we introduce a novel scheduling method, AdaptiveCycle, that modifies the LR based on validation loss plateaus over a cyclic LR curve, further optimizing the training process. We have performed the benchmark using diverse time-series data sets from the UCR and UCE repositories. The benchmark is performed across eleven deep learning models, simple such as LSTM, FCN, and complex models such as Inception-time, XCM, and LSTM-FCN. The accuracy, F1 score, and computational time allow us to highlight the adaptability of training methods across various datasets and models. The findings emphasize the advantages of employing LR schedulers for efficient training. The AdaptiveCycle scheduler achieves optimal performance across different models and datasets. The benchmark results will encourage the adoption of the LR scheduler in research and industry.

Keywords: Deep Learning · Responsible AI · Training framework · Benchmark · LR schedulers

1 Introduction

The use of deep learning has increased significantly in the past decade across various applications. It has revolutionized the fields of computer vision, NLP, and autonomous systems. The two main reasons for this are the advancement in

P. K. Chrysanthis et al. (Eds.): ADBIS 2025, LNCS 16043, pp. 63–78, 2026.
https://doi.org/10.1007/978-3-032-05281-0_5

GPUs [16] and the enormous amount of data available [21]. Current technology allows us to collect, store, and process vast amounts of data to provide insights and uncover complex data patterns using deep learning. With GPUs being accessible and affordable, they have played a pivotal role in enabling researchers and developers to experiment with deep learning technologies.

With the increasing usage of neural networks, it is important to apply optimal training dynamics for their efficient training. Training dynamics refers to the process by which the neural networks update the weights associated with neurons in different layers of the model during training. Optimizing this process would lead to faster model convergence, saving computational time and resources while not compromising model robustness [1]. The LR is the most important hyperparameter in the neural network training process [11]. The LR is used to control the step size for weight updates during model training. An LR that is too high may lead to failed convergence due to unstable training, while a too low LR requires many iterations to converge, leading to slow training [15].

Three main strategies are available to set the LR: LR finder, which helps to select the best LR range for training; adaptive learning rate methods for optimization, such as Adam [13], RMSprop, and LR schedulers to dynamically change it during training. In this paper, we have benchmarked the results of four different combinations of the above methods. The main focus of this paper is to understand that a static LR for training is disadvantageous, as it fails to adapt the training process according to the optimization process. In optimal training, two subprocesses occur: exploration and exploitation. We need to explore the best model weights for convergence; for this, a high LR is required at the start of the training. Then, we need to fine-tune the model weight for an optimal solution, which is referred to as exploitation, and it requires a low LR. Thus, a static LR cannot satisfy both conditions, and a compromise needs to be made.

As stated before, we can use an adaptive optimizer like Adam, which will dynamically adjust the weight for each parameter according to the gradient information. This is the most widely practised solution. The second is to use LR schedulers; although they have gained popularity, they are still not widely used. Popular LR schedulers include exponential decay, step decay, cosine annealing, and cyclical learning rates [10]. In contrast to most of these methods, this paper focuses mainly on the OneCycle policy [18], which does not rely on a monotonically decreasing LR schedule.

In OneCycle, the LR first starts small, gradually increases for exploration during high LR, and then finally starts decreasing back to the lower LR for fine-tuning the model during exploitation. The LR change process is almost identical to that required for optimal training. OneCycle also adjusts the momentum for the adaptive optimizers to counter the effect of the varying LR, which might make the model unstable. More details on these schedulers are provided in Sect. 2, explaining the existing related work.

In this, we introduce a new scheduler called AdaptiveCycle that could adopt the LR depending on validation loss during training. AdaptiveCycle uses the OneCycle policy on the backhand, which allows us to adjust the LR if the validation loss has been increasing or on a plateau. Section 4, on methodology, explains in detail the algorithm of AdaptiveCycle to dynamically update the LR. The dependence on validation loss allows the model to overfit. While doing a literature review, we found the limited benchmark studies conducted in this domain across diverse tasks, especially across functions with varying data characteristics, and imbalanced and noisy datasets. The datasets used are time series datasets from the UCR/UCE, having different features, as stated in Sect. 3, benchmark setup. Time-series data were chosen to perform the benchmark experiments as it is used in many industries, such as finance, economics, healthcare, energy, telecommunications, retail, logistics, manufacturing, and environment. The adaptability is tested across different deep learning models with various sizes and complexities. We have evaluated the performance across four main categories: accuracy, convergence speed, stability, and model size. Brief explanations of the models used are provided in Sect. 3, the benchmark setup.

Section 5 consists of the results and analysis. The conclusion and recommendations of best practices are stated in Sect. 6. The main goal of the paper is to provide the advantages of LR schedulers and to encourage their usage among researchers and industry professionals, allowing more efficient training and a robust, well-generalized model.

2 Related Work

This section gives a brief overview of the different adaptive optimizers and LR schedulers available. Stochastic gradient descent (SGD) is a foundation optimizer that uses gradients (first moments) to update parameters using a randomly selected data subset. Unlike traditional optimizers such as SGD, adaptive optimizers modify the LR for each parameter using squared gradients (second moment). Adgrad [5] updates the LR based on the historic sum, while RMSprop uses the moving average of the squared gradients. Adam uses an exponential moving average of both moments. Adam is the most widely used as it is computationally efficient, requires less memory, and works well with large models and datasets.

The StepLR, Cosine annealing, and OneCycle are the most widely used LR schedulers [10]. StepLR reduces the LR by a fixed factor after a specified number of epochs, called the step size. It is simple with a gradual LR decay and is effective for model training, especially for fine-tuning. Cosine annealing reduces the LR following a cosine curve. It initially starts from a high LR and gradually approaches the minimum LR, and a warm restart is performed at the end of training. It usually provides a smooth loss decay, but still needs a larger number of epochs. The warm restart helps escape local minima.

OneCycle adjusts the LR cyclically over one smooth cycle [19]. Initially, the LR starts at an assigned minimum value to warm up. Then it gradually increases to the maximum value; having a high LR accelerates training. Finally, the LR decreases back to the initial minimum LR or even to a slightly lower LR for fine-tuning. The momentum is also adjusted inversely to the LR rate curve to balance the impact of high and low LRs. The high LR regions result in the exploration of global minima and help avoid local minima. This allows the model to avoid overfitting with better generalization and faster convergence. OneCycle is used prominently for computer vision and NLP tasks, especially for large datasets and models.

Studies have highlighted that schedulers like cosine annealing and OneCycle improve the convergence rate, especially with OneCycle, which creates balanced training for large datasets and models. However, most of these studies were conducted on image datasets such as MNIST and CIFAR, over just a few models. There is a lack of understanding of how it performs on smaller datasets and models [22].

3 Benchmark Setup

This section elaborates on the dataset employed, emphasizing its various features. Additionally, an outline of the models applied in the research is presented, along with five training scenarios. Details on the evaluation metrics and the experimental setup are also provided.

3.1 Datasets

We primarily used the existing UCR benchmark dataset, as our results can be compared with other benchmarks. We have selected four univariate and four multivariate datasets from both UCR and UCE [3]. Table 1 illustrates these datasets, provides a detailed summary of various datasets, and highlights key attributes for each. In particular, we organized these signals into distinct categories based on the origin and nature of the time series.

Most of the parameters of deep learning models depend on time series attributes, such as input length, input features, and the number of output classes and labels. Hence, we also group the datasets according to characteristics that may impact the formation and performance of the model. Each dataset is characterized by several attributes: Length, Strength, Dimension, Multiclass, Split-ratio, Balance, and Data-type as shown in Table. The length of the dataset refers to the sequence length of the time series. The shortest dataset is Electric Devices, which has a length of 96, while the longest is Ethanol Concentration, which has a length of 1751. Strength refers to the number of total samples in a dataset. Electric devices have the highest strength, followed by Pen Digits. Self-regulation SCP1 has the lowest strength with 561 samples. Dimensionality is the number of input features; the higher the dimension, the more complex. We similarly perform categorization depending on the number of output classes, referred to as multiclass. Electric Devices and Pen Digits are multiclass datasets, while Star Light Curves and Ethanol Concentration are not. This attribute is crucial for classification tasks. All of our datasets are uni-labelled; the output is a single dimension.

For the benchmark to be consistent with the provided dataset to understand the impacts of the model, we used the original train-test split provided. We calculate the split ratio, which is the proportion between the training and testing datasets. A ratio around one suggests a balanced division, whereas a ratio greater than one means there are too few testing samples compared to training, and less than one indicates an excess of testing samples relative to training. For the balance of the class distribution, we again have two categories, balanced or unbalanced, which are assigned depending on the class ratio.

Table 1. Dataset characteristics

Dataset	Length	Strength	Dimension	Multiclass	Split-ratio	Balance	Data-type
Star Light Curves	1024	9236	1	3	0.12	Unbalanced	Sensor
Electric Devices	96	16637	1	7	1.15	Unbalanced	Device
ECG 5000	140	5000	1	4	0.12	Unbalanced	ECG
Ethanol Concentration	1751	1004	1	5	1	Balanced	Spectro
Pen Digits	8	10992	2	10	2.14	Balanced	Motion
Self Regulation SCP1	896	561	6	2	1	Balanced	EEG
Racket Sports	30	303	6	4	1	Balanced	HAR
NATOPS	51	360	24	6	1	Balanced	HAR

3.2 Deep Learning Models

In this section, we examine a range of deep learning models for time series classification utilized in the study. The models are generally categorized into three types: basic simple models, extensive deep models, and hybrid models integrating both CNN and RNN structures in their design.

LSTM network [8] - Effective for tasks like time forecasting and anomaly detection, it can learn long-term dependencies through its memory structure. The Bi-LSTM processes sequences in both forward and backward directions.

FCN network [14] - Utilizes 1D CNN blocks for time series, featuring layers that upsample and downsample data. Average pooling reduces output dimensions before feeding into a linear layer to obtain class probabilities as output. We have also tested the same architecture using 2dCNN, which is known as **Convent2d**.

Resnet Network [7] - Resnet blocks are used to build the network. The main concept introduced is residual connections in the block, which help in the convergence and stabilization of the network. These blocks are used to upsample the input data and then apply average adaptive pooling, followed by a linear layer.

InceptionTime [9] - It is an ensemble of six Inception modules [20] with different randomly initialized weight values. The inception module is composed of CNN layers with varying filter lengths. This allows the neural network to extract relevant features from long and short-time series (Fig. 1).

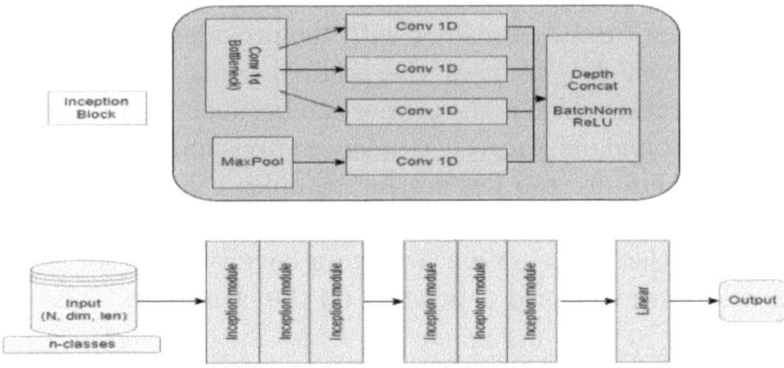

Fig. 1. InceptionTime architecture.

XceptionTime [17] - Its architecture integrates depthwise separable CNN with different filter sizes, adaptive average pooling, and a non-linear normalization technique. They are concatenated in series to capture temporal and spatial information without manually extracting features.

dCNN [2], short for dimension-wise Convolutional Neural Network, was introduced recently to overcome the limitations of traditional CNNs in handling multivariate time series data. The architecture of dCNN is similar to that of a traditional CNN model, the difference lies in the input to the model, which is transformed into a cube, in which each row contains a combination of all dimensions. This helps the model learn temporal and dimensional discriminant information. We have also used dResnet similar to dCNN as mentioned in the original paper [2].

XCM [6] - Designed for multivariate time series, it merges 2D and 1D CNNs in a parallel architecture with upsampling and downsampling layers. Their output is concatenated in the first dimension and passed through an upsampling CNN layer. A linear layer at the end follows a pooling layer to get the output.

LSTM-FCN [12] - Combines LSTM and FCN in parallel, allowing for concatenation of results and the potential addition of an attention mechanism to weigh the importance of sequence components (Fig. 2).

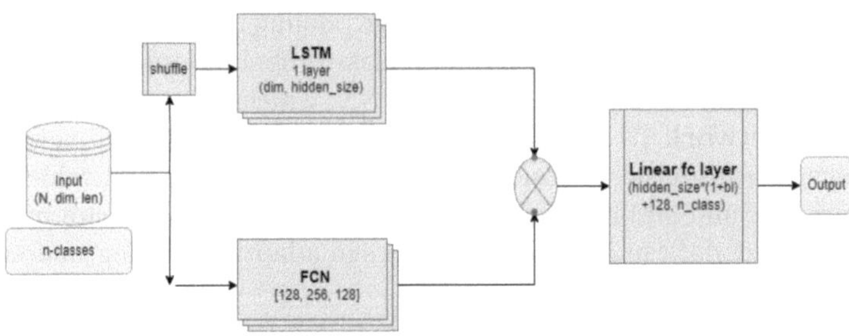

Fig. 2. LSTM-FCN model architecture.

We have introduced a smaller hybrid architecture LSTM-2dCNN similar to the LSTM-FCN. **LSTM-2dCNN** consists of a parallel network of LSTM and 2dCNN layers. The output of the 2dCNN is reshaped to allow for the concatenation. In this case, after the concatenation of the outputs of the parallel network, the result is passed through the dropout and linear layer (Fig. 3).

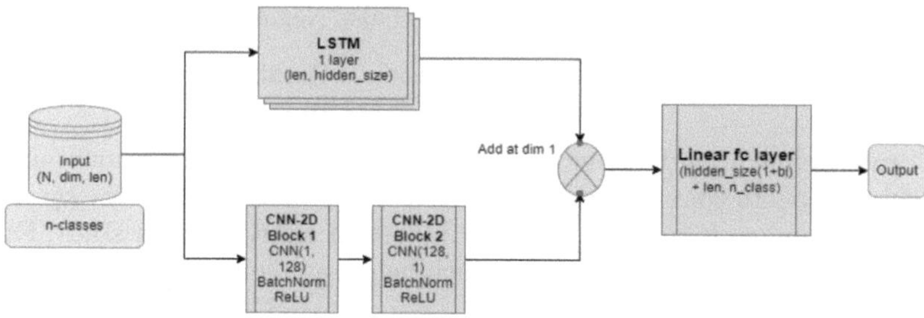

Fig. 3. LSTM-2DCNN model architecture.

3.3 Training Scenarios

In this study, we have experimented with four different training algorithms. This section describes the different training scenarios.

1. The LR finder is first used to find the correct range, and then use the AdaptiveCycle LR scheduler, which we have introduced in the paper.
2. Similarly, the LR finder is used to find the correct range and then the OneCycle LR scheduler.
3. Using LR finder to find the optimal LR (the LR resulting in minimum loss in the loss curve) to train the model and then use static LR for training.
4. Using a static LR to train the neural network, which has been set to 1e-5.

3.4 Evaluation Metrics

The different evaluation metrics used for the benchmarking model and the five training scenarios performances are explained in this subsection. These will help us to understand the effect of various characteristics on performance.

Accuracy is a percentage of correctly classified inputs over the total number of inputs in the testing dataset. The higher the accuracy, the better is modal performance. However, in the case of an imbalanced dataset, it might be misleading to use accuracy, as it cannot detect the model bias due to the imbalanced dataset.

F1score is the harmonic mean of precision and recall. It is useful, especially in the case of imbalanced data, as we also take into account the false positives and false negatives to calculate the precision and recall. Similar to accuracy, the higher the F1 score, the better the model performance.

$$\text{Precision} = \frac{\text{True Positives}}{\text{True Positives} + \text{False Positives}} \tag{1}$$

$$\text{Recall} = \frac{\text{True Positives}}{\text{True Positives} + \text{False Negatives}} \tag{2}$$

$$F1\ \text{Score} = 2 \times \frac{\text{Precision} \times \text{Recall}}{\text{Precision} + \text{Recall}} \tag{3}$$

Convergence speed, in our case, has been quantified as the amount of time a model requires to converge, that is, to have an optimal performance by stable loss during training. It is one of the most important metrics for our study. Faster convergence means efficient training, which helps save computational resources.

3.5 Experimentation Setup

We have used cross-entropy loss for backpropagation during training and Adam as an adaptive optimizer for all four scenarios mentioned previously. We have used Pytorch to build all our functions and the customer schedulers instead of fastai, as it provides more transparency, greater clarity, and control over the training process. An L4 GPU of 16 GB from Nvidia has been used for training and testing these models. We have used PyTorch dataloaders and normalized the data. Normalizing the data provides scaled uniformity and helps avoid inconsistencies caused by batch-level statistics during training.

4 AdaptiveCycle Methodology

In this section, we go through the main process of training the deep learning model with a detailed explanation of the LR finder, OneCycle LR, and AdaptiveCycle LR schedulers. Parameters from the dataset, such as sequence length, number of classes, and dimension, are extracted and used for model creation. The loss criteria and optimization are initialized.

LR Finder is used to calculate the optimal range of LR. This range is calculated by increasing the LR exponential after each batch iteration. Usually, the LR finder fits for just one epoch, but this might lead to unstable results when the dataset is small, as it might not provide enough iterations to form a stable curve. We have modified the existing LR finder to be used over more epochs. The user can select this by setting the multi-epoch value as True or False. Figure 4 shows the difference between with and without multiple epochs for the ECG5000 dataset. In this case, it is more beneficial to use multi-epochs. In the Results section, we discuss the scenarios depending on the dataset in which setting the multi-epochs as True is more beneficial.

Another important hyperparameter affecting the training and model optimization is the weight decay (Wd) we set for the optimizers. Figure 5 shows the

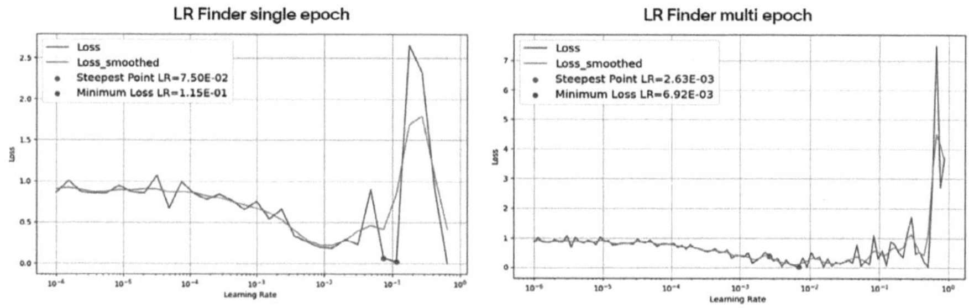

Fig. 4. LR finder results for ECG using LSTM-FCN model.

LR finder loss curve initialized with different Wd, and we can see that it does have an effect. Usually, lower Wd works well for simple models, as there is less chance of overfitting. At the same time, for complex architectures, a higher Wd can help regularise the model by preventing heavy reliance on specific features. We suggest performing a grid search of Wd using an LR finder. It will efficiently capture the effects of the Wd and help pick the best one according to the loss curves. The ideal Wd should have a smooth loss curve with less divergence. We provide analysis on Wd in the Results section.

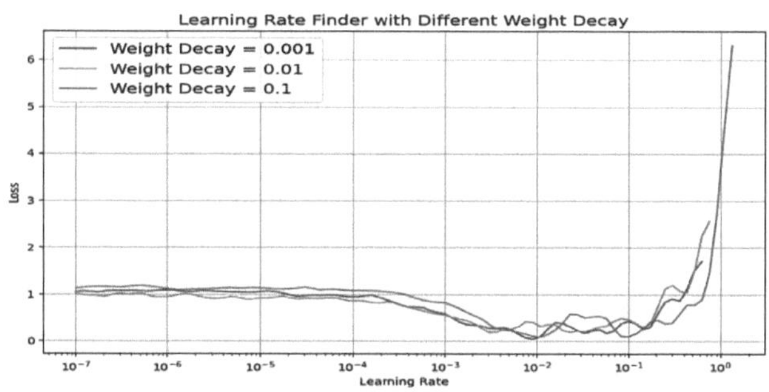

Fig. 5. Effect of weight decay on loss for ECG using LSTM-FCN model.

Before going into detail about the AdaptiveCycle LR, it is important to understand the OneCycle LR and why there is a need to upgrade it to a more customizable and controllable LR scheduler. OneCycle can be explained by dividing the training into four phases depending on the LR curve during training. We start with a warm-up phase where the LR starts low at the set lr_min and gradually increases. This helps the model stabilize and avoid large weight updates at the start of training. Then, the high LR phase, where it is around the set lr_max value. It helps train faster and explore the lost landscape to avoid local minima. In the decay phase, the LR decreases and returns to lr_min, where the model is fine-tuned. Then, in the final phase, the LR is reduced further for just a few iterations, allowing the model to settle well into its optimal values.

The momentum follows a cycle inverse to the LR if set to true. It starts from high, goes to a minimum, and then back to the original high value. However, during the final phase of LR, the momentum also decreased slightly to avoid a prominent change in model weight in the final phase. We can set the percentage of the cycle used for the increasing LR, which is usually set to 30 or 50.

AdaptiveCycle LR is a custom scheduler that is a combination of OneCycleLR with the concept of reducing the LR value when there is no improvement in the validation and training accuracy. This allows the validation performance to have an impact on the LR adjustments. This increases our control over the LR scheduler by offering to flexibly set the parameters controlling conditions to reduce LR and reinitializing the OneCycle LR with updated parameter values. The reduction of LR depending on validation loss helps stabilize the learning, especially during the warm-up and high LR phases, preventing overfitting and improving generalization. It helps in better fine-tuning the model. The algorithm of the AdaptiveCycle LR helps us understand the concept better Algorithm 1.

Algorithm 1. AdaptiveCycle Learning Rate Scheduler

Require: initial minimum learning rate lr_{min}, maximum learning rate lr_{max}, final minimum learning rate lr_{fmin}, cycle length T, epochs E, patience p, factor f
  Initialize OneCycleLR($lr_{min}, lr_{max}, lr_{fmin}, T, E$)
  least_validation_loss $= \infty$
  for epoch $= 1$ to N **do**
    Start OneCycleLR: $lr \leftarrow lr_{min}$
    $epochs\_patient \leftarrow 0$
    if epochs_patient $> p$ **then**
      When epoch $< 1, current\_loss \leftarrow$ Validation Loss()
      if $current\_loss < least\_validation\_loss$ **then**
        $least\_validation\_loss \leftarrow current\_loss$
      else
        $epochs\_patient \leftarrow epochs\_patient + 1$
      end if
      for Iteration per batch **do**
        $lr_{new} \leftarrow$ OneCycleLR($lr_{min}, lr_{max}, lr_{fmin}, T, E$)
      end for
    else if epochs_patient $> p$ **then**
      $lr\_old = lr\_current$
      Apply Reduction policy $lr\_new = lr\_current * f$
      Reinitialise the OneCycleLR
      OneCycleLR($lr_{min} = (lr\_new)0.5, lr_{max} = lr\_new, lr_{fmin}, T, E$)
    end if
  end for

Figure 6 shows the AdaptiveCycle LR applied in the dataset of Electric devices and Racket sports when trained using the LSTM-FCN model. Observing the LR curves helps us better understand the algorithm. For the electric device dataset, we have set the cycle length as default, which means that after

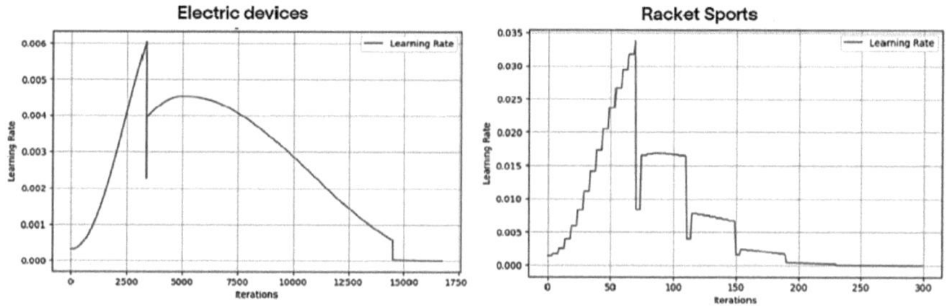

Fig. 6. AdaptiveCycle LR curve when training LSTM-FCN model.

each batch iteration, the algorithm changes the LR value, resulting in a smooth curve. For the Racket Sports dataset, we have set the cycle length to $T_{original}/5$. Thus, the LR update now occurs every 5 iterations, and can be observed in the image as steps. When a plateau in validation loss is reached, the LR reduction occurs, and the OneCycle is reset.

5 Benchmark Results

In this section, we provide a discussion on the results of the experiments conducted with different training methods and on the evaluation metrics mentioned in the section on benchmark setup. We also make inferences with respect to the dataset features and try to generalize the best model in each case.

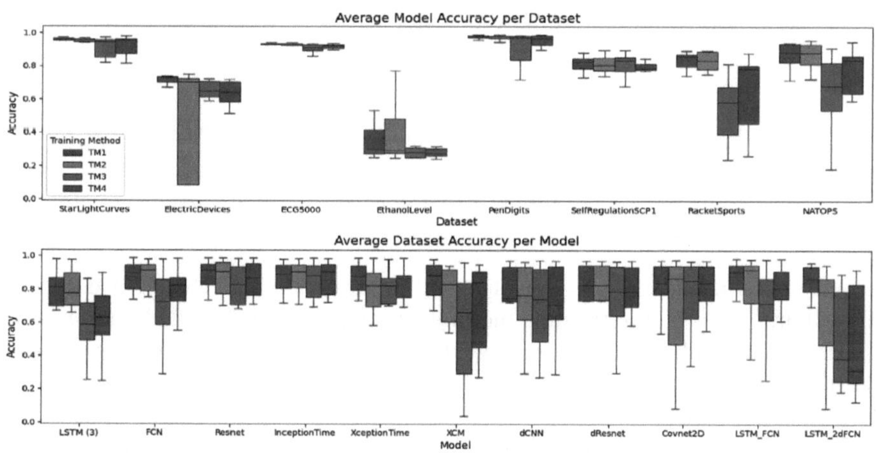

Fig. 7. Accuracy score plots.

The accuracy plot in Fig. 7 consists of two subplots, the first representing a box plot of average accuracy across the eleven models used per dataset and the

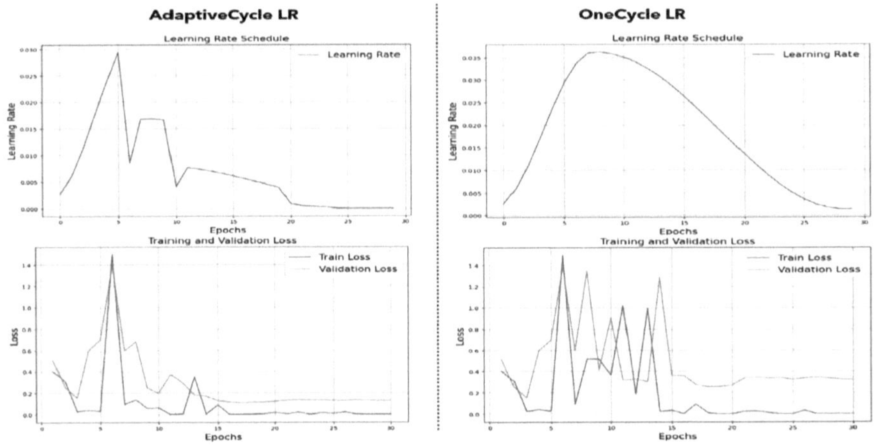

Fig. 8. Loss curves for Racket sports with LSTM-FCN model.

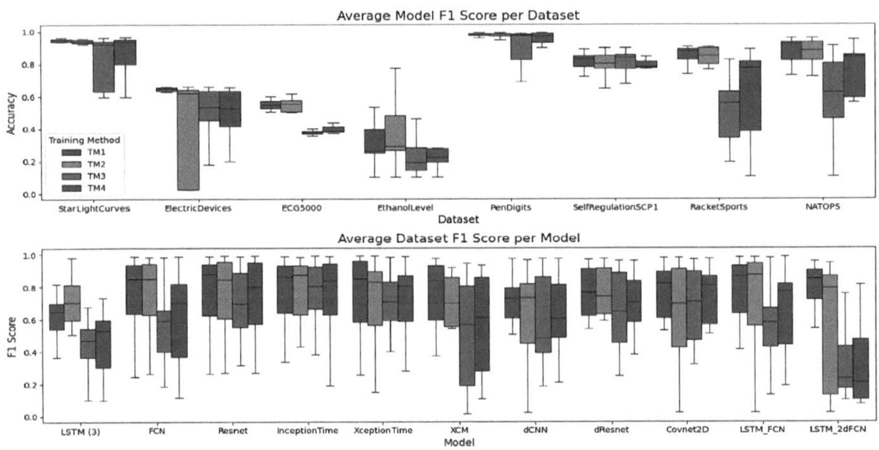

Fig. 9. F1 score plots.

second representing the box plot of model accuracies across the eight datasets. We will first focus on the training methods; both AdaptiveCycle LR and OneCycle LR have provided better accuracy when compared to training without schedulers. We can see that the variance for AdaptiveCycle LR is the least across models and datasets.

Figure 8 represents the training and validation loss curve when using the two schedulers; it can be seen that the AdaptiveCycle LR proposed provides greater stability over the training compared to OneCycle LR. In the second half of the training cycle, due to the decreasing LR, both methods lead to stability, but the gap between the train and validation loss is greater at the end in OneCycle. We also found that the variation across different initial seeds is more when using OneCycle LR vs the AdaptiveCycle.

Similar to the accuracy plot, we also have the F1 score plot displayed in Fig. 9. The F1 score plot is important to observe when the training dataset has an imbalanced class distribution. We have a skewed distribution of classes in the Starlight Curves, Electric devices, and ECG datasets. Among them, the F1 score does not decrease compared with accuracy for Starlight Curves and Electric devices; this might be due to the larger dataset, indicating that the model can still learn the class attributes when enough samples are available. This is not the case for ECG5000, as we have less than 20 samples for three of the five classes, leading to a performance decrease. It can also be observed that the performance of the LSTM model is lower when compared to the accuracy, indicating that the model has not been generalized well across the different classes.

The model scores per dataset plot allows us to compare the different training methods, while the dataset scores across models will enable us to categorize the model's performance. After comparing the accuracy and the F1 score plots, we can conclude that the Resnet and InceptionTime are stable models converging well across the different training methods. Both models comprise numerous CNNs and feature deep convolutional architectures. The LSTM-2dCNN has is the best performing model when using AdaptiveCycle training method, but is highly volatile with its performance when using OneCycle. LSTM-2dCNN, with LSTM and XCM are the models have a huge variations in score when compared across different training methods.

We have observed that the OneCycle needs more fine-tuning when selling the lr-max and lr-min. Incorrect settings may result in the model being unable to reach the optimal solution, as observed for the Electric devices dataset. Especially when the lr-max is a bit higher, it leads to an unstable model. As OneCycle follows a predefined pattern of a fixed increase schedule followed by a decreasing LR, it is inflexible. The AdaptiveCycle can address this issue as it changes dynamically and is controlled by the validation loss; it helps prevent scenarios of large oscillation during training due to high LR.

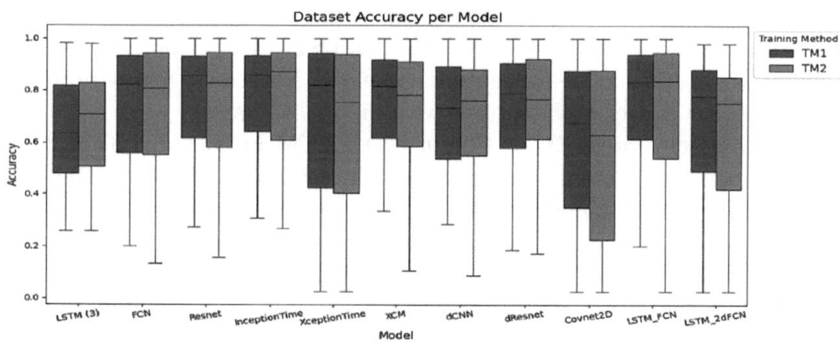

Fig. 10. Average Model accuracy over 40 datasets.

Additional experiments were conducted using only the training methods that employed learning rate schedulers, utilizing 40 datasets from the UCE and UAE collections. This was done to assess whether the previously observed pattern holds consistently across a broader range of datasets. Figure 10 represents the average accuracy, and it can be seen that the AtaptiveCycle provides better optimisation.

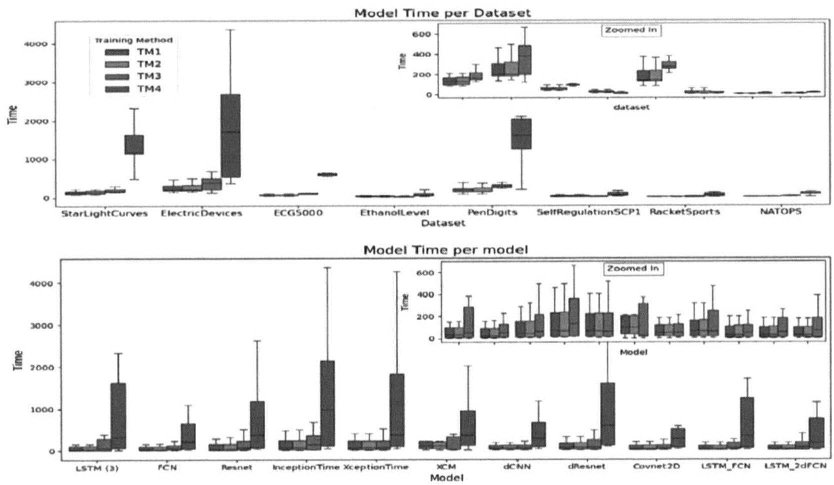

Fig. 11. Training time plots.

For evaluating the performance against computational time, we will be referring to Fig. 11. Schedulers would be less time-consuming than traditional learning with a low static learning rate. Each figure contains a zoomed plot; they have the same consumption time since we run the schedulers for about 50 epochs. Among the datasets, Electric devices have taken the most time, as they have the largest data samples for training. Resulting in more batches which leads to more iterations to be performed per epoch. When comparing across models, it can be seen that the LSTM, FCN, Convnet2D, LSTM-FCN, and LSTM-2dFCN are among the faster models. Inception Time and Xception Time require much more as they are deeper architectures with more CNN layers.

The parameter size of XCM, Convnet2D, LSTM-FCN, and LSTM-2dFCN are heavily dependent on the sequence length of the dataset, while the others are more dependent on the dimension of the input data. However, the parameter size has a negligible effect on the model's computational time and accuracy scores. This is due to the use of GPUs, designed to handle large-scale parallel computations efficiently. GPUs' high memory bandwidth and parallel processing capabilities allow them to manage large parameter sizes without significant performance degradation.

6 Conclusion

Using an LR scheduler is the way to train deep learning models efficiently. We can obtain better and faster convergence without affecting the model performance. Both the OneCycle and AdaptiveCycle use the same initial strategy of warmup of LR for exploration, followed by fine-tuning the model during the LR descent. The OneCycle is easy to use, with its implementation being available in many libraries such as Pytorch and fastai. Its only disadvantage is that during the high LR, the model oscillates a lot, which may, in some cases, lead to a non-optimal modal solution, as observed in the paper.

AdaptiveCycle, proposed by us, helps resolve most of the disadvantages of using just OneCycle. It gives more fine-grained control over the learning rate adjustments based on validation performance while still keeping the advantages of OneCycle with faster convergence. It provides improved stability and performance by decreasing the LR in cases when the oscillation of loss occurs, especially in handling large, complex datasets. We hope the results of our benchmark study will lead to more people using schedulers, as it will allow the research time to be efficiently spent on data and model architecture rather than training models, while also saving resources. We would also release open-source code for our project, allowing people to use the customer scheduler.

Acknowledgments. We would like to acknowledge the UCR and UCE Time Series Classification Archive [4] for providing the datasets used in this study.

References

1. Akl, A., El-Henawy, I., Salah, A., Li, K.: Optimizing deep neural networks hyperparameter positions and values. J. Intell. Fuzzy Syst. **37**(5), 6665–6681 (2019)
2. Boniol, P., Meftah, M., Remy, E., Palpanas, T.: DCAM: dimension-wise class activation map for explaining multivariate data series classification. In: Proceedings of the 2022 International Conference on Management of Data, SIGMOD/PODS 2022, pp. 1175–1189. ACM (2022). https://doi.org/10.1145/3514221.3526183
3. Darke, P., Missier, P., Bacardit, J.: Benchmark time series data sets for PyTorch - the torchtime package (2022). https://doi.org/10.48550/arXiv.2207.12503
4. Dau, H.A., et al.: The UCR time series archive. CoRR abs/1810.07758 (2018). http://arxiv.org/abs/1810.07758
5. Duchi, J., Hazan, E., Singer, Y.: Adaptive subgradient methods for online learning and stochastic optimization. J. Mach. Learn. Res. **12**, 2121–2159 (2011)
6. Fauvel, K., Lin, T., Masson, V., Fromont, É., Termier, A.: XCM: an explainable convolutional neural network for multivariate time series classification. Mathematics **9**(23), 3137 (2021). https://doi.org/10.3390/math9233137
7. He, K., Zhang, X., Ren, S., Sun, J.: Deep residual learning for image recognition. CoRR abs/1512.03385 (2015). http://arxiv.org/abs/1512.03385
8. Hochreiter, S., Schmidhuber, J.: Long short-term memory. Neural Comput. **9**(8), 1735–1780 (1997). https://doi.org/10.1162/neco.1997.9.8.1735
9. Ismail Fawaz, H., et al.: Inceptiontime: finding alexnet for time series classification. Data Min. Knowl. Disc. **34**(6), 1936–1962 (2020)

10. Johnson, O.V., XinYing, C., Johnson, O.E., Khaw, K.W., Lee, M.H.: Learning rate schedules and optimizers, a game changer for deep neural networks. In: Saeed, F., Mohammed, F., Fazea, Y. (eds.) Advances in Intelligent Computing Techniques and Applications, pp. 327–340. Springer, Cham (2024). https://doi.org/10.1007/978-3-031-59711-4_28
11. Kadhim, Z.S., Abdullah, H.S., Ghathwan, K.I.: Artificial neural network hyperparameters optimization: a survey. Int. J. Online Biomed. Eng. **18**(15), 59–87 (2022)
12. Karim, F., Majumdar, S., Darabi, H., Chen, S.: LSTM fully convolutional networks for time series classification. IEEE Access **6**, 1662–1669 (2018). https://doi.org/10.1109/access.2017.2779939
13. Kingma, D.P., Ba, J.: Adam: a method for stochastic optimization (2017). https://arxiv.org/abs/1412.6980
14. Long, J., Shelhamer, E., Darrell, T.: Fully convolutional networks for semantic segmentation. In: 2015 IEEE Conference on Computer Vision and Pattern Recognition (CVPR), pp. 3431–3440 (2015). https://doi.org/10.1109/CVPR.2015.7298965
15. Napoli Spatafora, M.A., Ortis, A., Battiato, S.: GLR: gradient-based learning rate scheduler. In: Foresti, G.L., Fusiello, A., Hancock, E. (eds.) Image Analysis and Processing - ICIAP 2023, pp. 269–281. Springer, Cham (2023). https://doi.org/10.1007/978-3-031-43148-7_23
16. Pandey, M., et al.: The transformational role of GPU computing and deep learning in drug discovery. Nat. Mach. Intell. **4**(3), 211–221 (2022)
17. Rahimian, E., Zabihi, S., Atashzar, S.F., Asif, A., Mohammadi, A.: Xceptiontime: independent time-window xceptiontime architecture for hand gesture classification. In: ICASSP 2020 - 2020 IEEE International Conference on Acoustics, Speech and Signal Processing (ICASSP), pp. 1304–1308 (2020). https://doi.org/10.1109/ICASSP40776.2020.9054586
18. Smith, L.N.: Cyclical learning rates for training neural networks (2017). https://arxiv.org/abs/1506.01186
19. Smith, L.N., Topin, N.: Super-convergence: very fast training of neural networks using large learning rates (2018)
20. Szegedy, C., et al.: Going deeper with convolutions. In: 2015 IEEE Conference on Computer Vision and Pattern Recognition (CVPR), pp. 1–9 (2015). https://doi.org/10.1109/CVPR.2015.7298594
21. Wang, W., Zhang, M., Chen, G., Jagadish, H., Ooi, B.C., Tan, K.L.: Database meets deep learning: challenges and opportunities. ACM SIGMOD Rec. **45**(2), 17–22 (2016)
22. Xu, Z., Dai, A.M., Kemp, J., Metz, L.: Learning an adaptive learning rate schedule (2019). https://arxiv.org/abs/1909.09712

Exploring Local Feature Influences with Hierarchical Explanation Trees

Emmanuel Doumard[1,2](✉), Julien Aligon[2], Paul Monsarrat[3], Nicolas Labroche[1], and Alexandre Chanson[1]

[1] University of Tours, Tours, France
{emmanuel.doumard,nicolas.labroche,alexandre.chanson}@univ-tours.fr
[2] IRIT, Toulouse, France
{emmanuel.doumard,julien.aligon}@irit.fr
[3] University of Toulouse 3 Paul Sabatier, Toulouse, France
paul.monsarrat@univ-tlse3.fr

Abstract. When applying clustering techniques to data exploration, ensuring the practical usefulness of clusters by aligning them with expert knowledge is highly desirable. A recent approach, known as supervised clustering, addresses this by selecting a target feature and constructing a Target Explanation Space (TES) using a supervised model combined with local feature attribution methods such as LIME or SHAP. While TES enhances clustering performance, its lack of interpretability remains a significant limitation. To address this, we introduce the first hierarchical supervised clustering pipeline that generates interpretable, non-overlapping rules directly in the original data space—while still leveraging the improved clustering achieved in TES. Experimental results demonstrate that the rules produced are not only more concise but also more comprehensive than those from existing methods, enabling experts to effectively balance interpretability and predictive accuracy.

Keywords: Supervised clustering · XAI · Explainable clustering

1 Introduction

Clustering is a widely studied approach in data exploration which aims at "organizing data into sensible groupings", "according to measured or perceived intrinsic characteristics or similarity" [1,18]. It is crucial for specialists targeting a deeper understanding of the phenomena under analysis, even in scenarios where labeled datasets exist. In healthcare, a single diagnostic category can encompass a wide variety of patient profiles, as individuals with the same condition may present different causes or levels of severity—highlighting the need for personalized treatment approaches [29].

In high-stakes settings, it is crucial to make clustering results actionable by aligning them closely with domain expert knowledge. The challenge of integrating expert knowledge into clustering has long been addressed within the

P. K. Chrysanthis et al. (Eds.): ADBIS 2025, LNCS 16043, pp. 79–94, 2026.
https://doi.org/10.1007/978-3-032-05281-0_6

community through the development of semi-supervised methods [4,10]. These approaches typically represent expert knowledge as constraints that are defined either at the instance level [2,5], cluster level [17], partition level [12] or feature level [26] and then fed to adapted clustering approaches to generate clustering that more closely matches expert expectations. However, as indicated in [26], these methods rely solely on data and are "agnostic of the expert" and may not be easy to understand during the course of an exploration, not to mention the difficulty of defining the input constraints [31].

In this paper, we propose a simpler scenario where domain experts explore data with respect to a specific chosen feature as a target and formulate hypotheses about its role in the phenomenon under investigation, similar to the basic principles of Exploratory Data Analysis (EDA) [13] and the recent Evaluative AI paradigm [24]. More precisely, we consider the problem of **supervised clustering** [8,22] that builds upon the pairing of (i) supervised machine learning (ML) models trained on the target feature (ii) with local feature attribution methods from XAI [21,27]. The output is a space similar to the data space, but where each value corresponds to the influence of a feature on the ML model's prediction. This space, which we call the Target Explanation Space (TES), can be explored with clustering algorithms by grouping instances based on their values in the TES, benefiting from insights into how the ML model considers features to be important with respect to the target, and thus producing a partition optimized with expert knowledge.

Recent works put this framework into practice on specific domains such as diseases symptomatology [8,15] or multi-principal element alloys chemistry [19], yielding promising results. However, by shifting to the TES representation, experts lose the ability to use their knowledge of both the domain and the data to align clusters with their expectations.

To solve this issue, this paper presents the first explainable-by-design supervised clustering approach named **Hierarchical Explanation Tree** (HET). The main idea is to perform hierarchical clustering in the TES and simultaneously explain the output of this clustering by building a tree that mimics the hierarchical dendrogram at each split point. The duality between the original data and the TES allows to build an explanation tree with data-oriented interpretable rules, while preserving the relevance induced by the local influence space. A method for pruning unnecessary nodes is also proposed. Contrary to existing approaches in the explainable clustering such as explainable k-means [25], our approach benefits from the expert guidance via the selection of a target feature, and also from the pathways of prediction generated by the ML model.

Our distinct contribution lies in addressing the largely unexplored domain of explainable clustering [11] while integrating principles from supervised clustering [22], resulting in a clustering approach inherently designed to be explainable and tailored to a specific target. Contributions are as follows:

1. A new explainable-by-design hierarchical clustering combining benefits from supervised and explainable clustering to produce data-focused interpretable

and non-overlapping rules while maintaining the relevance induced by the local influences space. A method to prune unnecessary nodes is also proposed.
2. A baseline based on an alternative use of SkopeRules inspired by RuleFit [16] to frame influence-based TES clusters with such data-based rules.
3. Experiments are conducted on three distinct datasets, comparing methods approximating the same hierarchical clustering. The results show a tradeoff between rules approximation quality and their number, overlap and coverage. We also provide insights regarding the use of local influences for explainable supervised clustering.

This paper is organized as follows, we first introduce the HET approach, problem statement, and algorithm along with a motivating example in Sect. 2. Then in Sect. 3, we evaluate HET on three well-known datasets. Section 4 then present a state-of-the-art regarding supervised clustering. Finally, we conclude in Sect. 5 and share some perspectives.

2 Hierarchical Explanation Tree

2.1 Overview

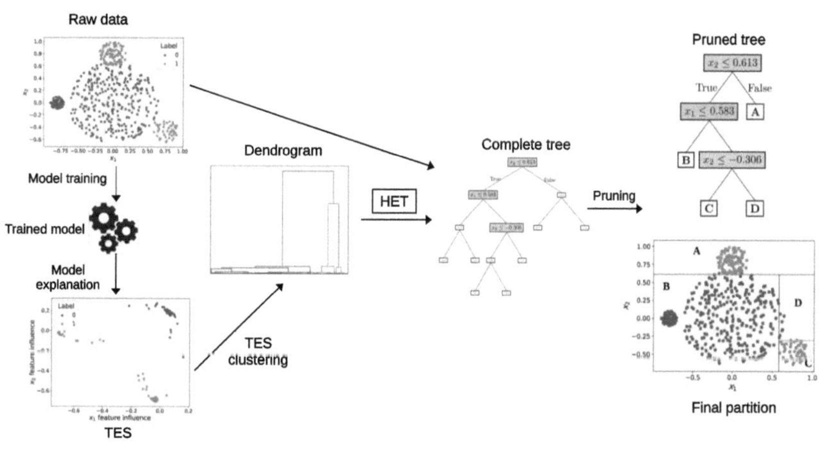

Fig. 1. General overview of the proposed pipeline for explainable supervised clustering.

This section introduces our new explainable-by-design supervised clustering pipeline as depicted in Fig. 1. Similar to other supervised clustering approaches, this pipeline relies on the identification of a target feature to train a supervised model. The use of a local influence post-hoc explainable method such as LIME or SHAP [21,27] on the model output defines a new space that we call Target Explanation Space hereafter. More formally, we present preliminary definitions:

Dataset. Let $X \in \mathbb{R}^{n \times d}$ be an initial dataset and $Y \in \mathbb{R}^n$ a target feature (regression or classification).

Model. Without losing generality, let f be a trained model that takes an input $x \in \mathbb{R}^d$ and returns $\hat{y} = f(x)$ the predicted value for this input.

Target Explanation Space (TES). Finally, let g be a local feature influence model, that takes a model f and an input x and computes the vector $\xi = g(f, x) \in \mathbb{R}^d$. This vector represents the influence of each input feature for the model prediction of a particular instance x. A high value (positive or negative) indicates a strong influence, while a low value indicates a weak influence. This way, we can compute $\Xi = \{g(f, x) | x \in X\} \in \mathbb{R}^{n \times d}$ the Target Explanation Space. It represents a new input space that has potential for exploratory data analysis for two primary reasons. Firstly, it shares the same cardinality (lines and columns) as the original input data and is naturally linked to it, allowing the analysis to have a reference point in the data space where features bear intrinsic meaning. Secondly, it transforms the data so that their values can be made comparable across features by standardizing them to the same "unit". Consequently, this can be regarded as a multidimensional normalization process that serves to reduce the noise induced by mostly irrelevant features and amplify the influence of discriminant features. Furthermore, it facilitates the concurrent consideration and comparison of both continuous and categorical data, since both types are transformed into numerical continuous influences.

Figure 2 shows a toy 2D dataset on the left and the influence data space on the right, where each data point is positioned relative to the influence of its data values on the prediction of the model. In this case, it is obvious that the influence space (TES) offers a much simpler separation between the two labels. This illustrative example will be used throughout the remainder of this paper as a straightforward, visual example to demonstrate the method.

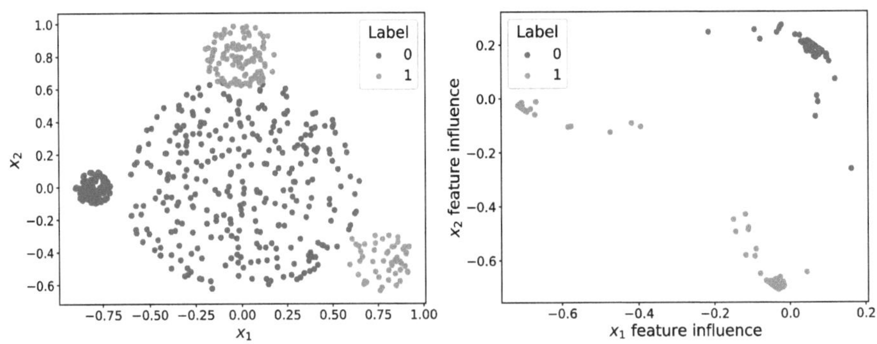

Fig. 2. Illustration of the potential of influence space for clustering: raw values (left) and influences values (right).

As shown in the middle of Fig. 1, a hierarchical clustering is subsequently carried out in TES, leveraging the ability to discriminate feature influence to generate a dendrogram.

Finally, the right part of Fig. 1 illustrates the core contribution of this paper: constructing an explanation tree using simple literals from the raw data space, derived from the supervised clustering in TES. The resulting rules remain easily interpretable while leveraging the enhanced clustering provided by TES.

2.2 Problem Formulation

A dendrogram can be seen as a binary decision tree, each node corresponding to either a split in divisive Hierarchical Clustering (HC), or a merge in agglomerative HC. However, this tree is not interpretable: these operations are often based on a distance metric (e.g. Euclidean distance) that considers all features simultaneously, making it difficult to express the split as a simple rule. Moreover, in our case, the clustering is built based on influence values but we want the explanations to be based on data values because they are more intuitive for users. For example, the rule "age $\leq$ 40" is easier to understand and provides more direct information to the user than the rule "age influence ≤ 0.32".

Therefore, the goal of the present work is to build a tree that approximates the actual tree that would be built from the HC dendrogram, with a single feature/threshold pair per node split. The objective is to produce a set of explainable rules that are simpler, with fewer literals, still providing a good accuracy on the original clusters, and thus more interpretable.

Hierarchical Explanation Tree. We formulate the discovery of rules in the original raw data space as problem of learning a tree from the clusters discovered in TES. The key distinction from a traditional decision tree [6] lies in our objective: instead of focusing solely on clusters or class labels from a flat partition, we aim to maximize a coverage criterion at each split of the dendrogram.

2.3 Building of the Explanation Tree

Iterative Explanation Tree Construction. The main algorithm for constructing the tree is outlined in Algorithm 1. Here, we denote $X_\mathcal{C}$ as the subset of data corresponding to cluster $\mathcal{C}$, and X_{node} as the subset of data corresponding to the points that adhere to the rules from the tree root to the considered node. In addition, the term $\mathcal{T}_\mathcal{C}$ denotes the tree node that is associated with the cluster $\mathcal{C}$.

Beginning at the root node of the tree, representing the entire dataset X, which constitutes the sole cluster in $\mathcal{D}_1$, the algorithm proceeds iteratively: at step i, the partition $\mathcal{D}_i$ is examined to identify the cluster C that is split into two subclusters, C_L and C_R, in $\mathcal{D}_{i+1}$ (line 4).

The CART criterion on entropy [20] is used to efficiently find the best split on data values, thus providing a "feature $\leq$ threshold" rule to explain the split of the node C into its children C_L and C_R (line 9). Each child node is subsequently assigned the label of its corresponding cluster. Lines 7 and 8 serve to correct errors from earlier steps by restricting consideration to only those points that actually fall within the node being split. Line 6 allows for the early termination

Algorithm 1. Hierarchical Explanation Tree

Inputs: $X \in \mathbb{R}^{n \times d}$ the original input dataset and $\Xi \in \mathbb{R}^{n \times d}$, the influence matrix.
Output: $\mathcal{T}$ ▷ Explanation tree

$\mathcal{D} \leftarrow HC(\Xi)$ ▷ Perform hierarchical clustering on influences
$\mathcal{T} \leftarrow node(X, \mathcal{D}_1)$ ▷ Initialize tree with root node
for $i = 1$ to $n-1$ **do**
  $C, C_L, C_R \leftarrow HCSplit(\mathcal{D}_i, \mathcal{D}_{i+1})$ ▷ Split cluster C into left and right
  $node \leftarrow \mathcal{T}_C$ ▷ Retrieve corresponding node in tree
  if $X_{C_L} \cap X_{node} \neq \emptyset$ **and** $X_{C_R} \cap X_{node} \neq \emptyset$ **then**
    $X_{rule} \leftarrow X_C \cap X_{node}$
    $y_{rule} \leftarrow \{\mathbf{1}_j\}_{j \in X_{C_L} \cap X_{node}} \cup \{\mathbf{0}_j\}_{j \in X_{C_R} \cap X_{node}}$ ▷ Error correction
    $node$.rule $\leftarrow find\_best\_split(X_{rule}, y_{rule})$
    $node$.left_child $\leftarrow \mathcal{T}_{C_L}$
    $node$.right_child $\leftarrow \mathcal{T}_{C_R}$
  end if
end for

of a split if the node contains no elements from either of the corresponding subclusters. As the algorithm follows a hierarchical clustering approach, the procedure may be halted once each data instance occupies a distinct leaf node.

However, since misclassified instances are discarded at each step, it is possible that a node selected for the next split contains no members from one of the two subclusters identified by the hierarchical clustering algorithm. As a result, one of the corresponding child nodes in the tree would be empty. This condition acts as a stopping criterion once all nodes have been expanded. For large-scale datasets and to ensure computational efficiency, constraints may be imposed on the maximum number of terminal clusters (leaves) or on the depth of the tree.

Rules Simplification. The complete rule corresponding to a given cluster can be constructed by traversing the path from the root node to the cluster's node and aggregating the decision rules along the way. The maximum number of rule elements associated with a node is k, where k denotes the depth of the node. However, this number may be smaller if a particular feature is used multiple times along the path. The following cases outline potential rule simplifications, assuming without loss of generality that $\theta_0 \leq \theta_1$.

(i) $\{x_i \leq \theta_0\} \cap \{x_i \leq \theta_1\} \rightarrow \{x_i \leq \theta_0\}$
(ii) $\{x_i \geq \theta_0\} \cap \{x_i \geq \theta_1\} \rightarrow \{x_i \geq \theta_1\}$
(iii) $\{x_i \geq \theta_0\} \cap \{x_i \leq \theta_1\} \rightarrow \{\theta_0 \leq x_i \leq \theta_1\}$

(i) and (ii) are used iteratively to get at most 2 rules per feature used in the path, subsequently merged with (iii) if 2 rules are remaining.

Pruning. We implement a pruning strategy to minimize the number of nodes and leaves. This step serves two primary purposes. Firstly, reducing the depth of the tree leads to more concise decision rules, making the explanations clearer and easier to understand. Secondly, errors such as misassigned instances can occur due to the approximation of the original cluster hierarchy. Although some errors are inevitable and tolerable, it is crucial to measure their frequency and avoid further node splitting if it results in a significant number of errors.

Algorithm 2. Recursive_Pruning

Inputs: *node*, the current tree node and α_b, the corrected threshold for p-value.

```
if node is terminal then
    return True
else
    left_pruned ← Recursive_Pruning(node.left_child, α_b)
    right_pruned ← Recursive_Pruning(node.right_child, α_b)
    if left_pruned and right_pruned and fisherp ≥ α_b then
        prune(node)
        return True
    end if
end if
return False
```

To ensure that each node and leaf represents a significant grouping of instances, unnecessary splits are systematically removed. Our **prune** procedure transforms a non-terminal node into a terminal node by removing its children, thereby simplifying the structure of the tree. The pruning process starts at the root of the tree, and then employs a recursive approach detailed in Algorithm 2 to eliminate nodes whose splits fail to meet the statistical threshold determined by Fisher's exact test, using the confusion matrix of the split as its basis. The same value of threshold is used for Fisher's exact test throughout the process.

To address the problem of multiple comparisons, the threshold for Fisher's exact test is adjusted using a Bonferroni correction, calibrated to account for the number of splits, which corresponds to the number of non-terminal nodes within the tree. Here, $\tilde{n}$ denotes the count of terminal nodes, while $\tilde{n} - 1$ represents the number of non-terminal nodes. In our proposal, the threshold for Fisher's exact test is corrected as $\alpha_b = \alpha/(\tilde{n} - 1)$, with $\alpha \in [0, 1]$ with a default value set to 0.05.

This methodology ensures the statistical allocation of instances into appropriate subclusters at each division. Pruning is conducted exclusively for nodes whose children are either terminal leaves or nodes that similarly fail the pruning criteria. This selective approach prevents the premature removal of entire branches due to the presence of a single underperforming node, preserving the potential for meaningful splits further along the hierarchy. Consequently, the method operates as a bottom-up pruning mechanism, ceasing its action upon encountering a statistically significant split.

2.4 Motivating Example

Figure 3 illustrates the HET algorithm applied to our introductory example. On the left, Fig. 3 shows the tree constructed from the TES hierarchical clustering dendrogram along with the corresponding raw data rules built by HET. Initially, the tree contains 257 nodes. After pruning, only the most straightforward and effective nodes for label separation are preserved, reducing the structure to 7 clusters and 4 terminal nodes. For clarity, the right side of the figure displays these tree-based splits projected onto a 2D representation of the raw data.

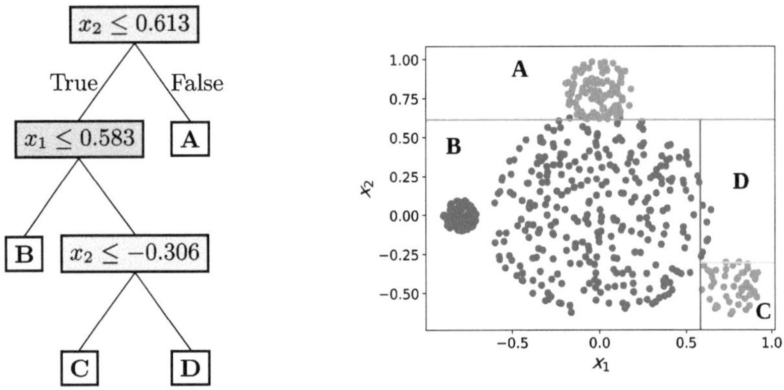

(a) HET built from the illustrative example, after pruning

(b) Data with HET splits

Fig. 3. Application on illustrative example.

It can be seen in Fig. 3 that the splits that best differentiate between the positive and negative labels are readily identified. It is worth noting that neither the dendrogram nor the tree contains a split that separates the small, blue cluster on the left (characterized by low values for x_1) from the rest of the data. This is expected and intended, as this data cluster is not distinctly explained compared to other regions of the data; the influence of its features is similar to at least part of the larger cluster.

To further exemplify the advantages of this new method in realistic scenarios, larger-scale experiments are presented in the next section.

3 Experiments

This section details the experiments conducted to validate our approach. We present hereafter in a first subsection our protocol containing the method, the evaluation criteria and datasets. Then, in a second subsection the comparative study to assess the qualities of our pipeline for explainable clustering.

3.1 Experimental Protocol

Hierarchical Clustering. To compute hierarchical clustering, agglomerative clustering was chosen with Ward's criterion that uses euclidean distance, implemented in python package `scipy (v1.10.1)`.

Model and Explanation Method. The focus of our study does not lie on the ML model or the XAI method. Therefore, we employ a single type of ML model, specifically the XGBoost algorithm, which is a variant of gradient-boosted trees implemented in Python library `xgboost (v3.0.0)`. This choice is motivated by XGBoost's superior performance on tabular data and compatibility with efficient local attributive explanation methods. We use TreeSHAP [22], a specialized and efficient version of SHAP tailored for tree-based models implemented in the `shap (v0.40.0)` library, to provide local, attributive explanations.

Comparison Method. A key challenge lies in the fact that, to the best of our knowledge, our work is among the first in the field of explainable supervised clustering, as both explainable clustering and supervised clustering are relatively new domains. To assess HET's performance, we propose a comparison with SkopeRules, a well-established method derived from RuleFit [16], which constructs decision rules based on a dataset and class labels and that is targeted towards the performance of rules with heuristics to remove low-performing rules.

To explain k clusters in the hierarchy, we label data according to their cluster and apply SkopeRules cluster by cluster. We use default implementation values but set minimum precision and recall to zero to force it to return one rule per cluster. For each cluster, we configure SkopeRules to use the same number of literals found in the corresponding HET node to frame said cluster. This way, as k increases and cluster nodes are found deeper in HET, SkopeRules is allowed to make longer rules for the cluster, resulting in rules of the same size that we can compare. By iterating over k between 2 and 20, we will frame the behavior of each method. For SkopeRules, as there is some stochasticity involved, each experiment is reproduced 10 times for each number of clusters for robustness, this is not needed for HET as it is deterministic.

Evaluation Metrics. The evaluation relies on several complementary metrics, as summarized in Table 1. These metrics were chosen according to explanations desiderata found in the XAI community [3], adapted to our pipeline. Precision and recall assess the predictive accuracy of the explainable rules generated by HET and SkopeRules, while sparsity, conciseness, and overlap quantify their interpretability. Non-coverage reflects the reliability of an explanation, indicating the extent to which an expert can trust the provided rules. Lastly, computation time reflects the usability of each method.

Table 1. Summary of Evaluation Metrics

Metric	Description
Precision	Proportion of data points correctly assigned to their corresponding cluster
Recall	Proportion of data points correctly assigned to the true cluster
Sparsity	Measures feature distribution in rules; high sparsity indicates diverse feature usage
Conciseness	Number of unique rules; HET reuses rules due to tree structure
Overlap	Instances framed by multiple rules; zero for HET due to exclusive rule assignment
Non-coverage	Instances not covered by any rule; occurs due to pruning or independent rule construction
Computation time	Time required to generate explanations; excludes ML model training and clustering computation

Datasets. The evaluation is conducted on three datasets presented in Table 2, covering both classification and regression tasks, with varying feature complexity and instance counts.

Table 2. Summary of the datasets used. C stands for Classification and R for Regression

Name	Task	# Inst.	# Feat.	Description
SA-Heart	C	462	10	Medical dataset indicating the presence or absence of cardiovascular disease based on high-level clinical features
Ionosphere	C	351	34	Radar data measuring electromagnetic signal responses from the ionosphere, based on raw complex-valued attributes
House Prices	R	1460	81	Dataset containing housing attributes (e.g., number of rooms, property characteristics) for predicting house sale prices

3.2 Results

This section outlines the results obtained from the three previously mentioned datasets.

SA-Heart. Figure 4 presents the results for the SA-Heart dataset. In terms of precision and recall, the rules generated by SkopeRules outperform those derived from the explanation tree, with both methods significantly surpassing random assignment. Regarding sparsity, HET produces explanations that utilize fewer features, enhancing interpretability. The conciseness analysis confirms our initial hypothesis: as the number of clusters increases, HET employs fewer unique rules on average compared to SkopeRules. While the latter scales linearly with the number of clusters, the former quickly stabilizes at a plateau.

Since HET rules are inherently non-overlapping, they do not appear in the overlap analysis. In contrast, SkopeRules exhibits a substantial overlap rate, with 20–30% of instances in the SA-Heart dataset covered by multiple rules simultaneously. Additionally, SkopeRules has a higher non-coverage rate than HET. Even with a lower number of clusters, up to 10% of instances remain uncovered, a figure that rises to 15–25% as the number of clusters increases. In comparison, HET maintains low non-coverage levels up to 22 clusters, beyond which pruning eliminates lower nodes, leading to non-discriminated clusters.

Finally, the computation time results align with our expectations: for a small number of clusters, SkopeRules is more efficient than HET. However, beyond approximately 10 clusters, HET becomes computationally advantageous. Notably, HET requires only a single computation to structure the tree, allowing efficient retrieval of cluster information thereafter, justifying its higher initial cost.

Ionosphere. For the Ionosphere dataset, the results presented in Fig. 5 exhibit similar trends. Notably, an interesting distinction emerges in the precision and recall curves: for a small number of clusters, HET achieves lower precision and recall than SkopeRules, yet they remain higher than the recall of SkopeRules. This can partly be explained by the better coverage of HET, giving every instance an associated cluster. This also suggests a tendency for SkopeRules to prioritize high precision at the expense of recall.

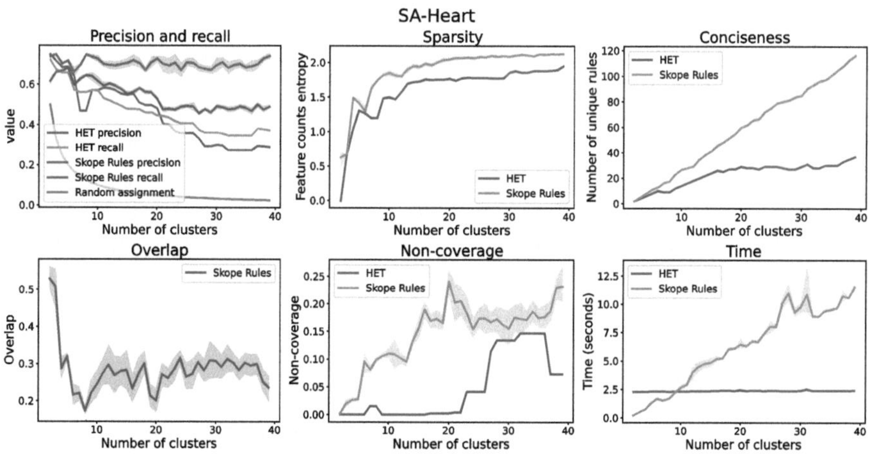

Fig. 4. Results on SA-Heart dataset.

For a large number of clusters, HET exhibits approximately 10% instance non-coverage, contributing to a decline in performance. In terms of conciseness, the number of rules in HET appears to scale linearly with the number of clusters, similar to SkopeRules. However, HET requires approximately half the number of rules to achieve the same level of clustering explainability.

Finally, although the overlap rate in SkopeRules is significantly lower than in SA-Heart, it tends to increase as the number of rules grows, potentially indicating insufficient specificity in the generated rules for higher number of clusters.

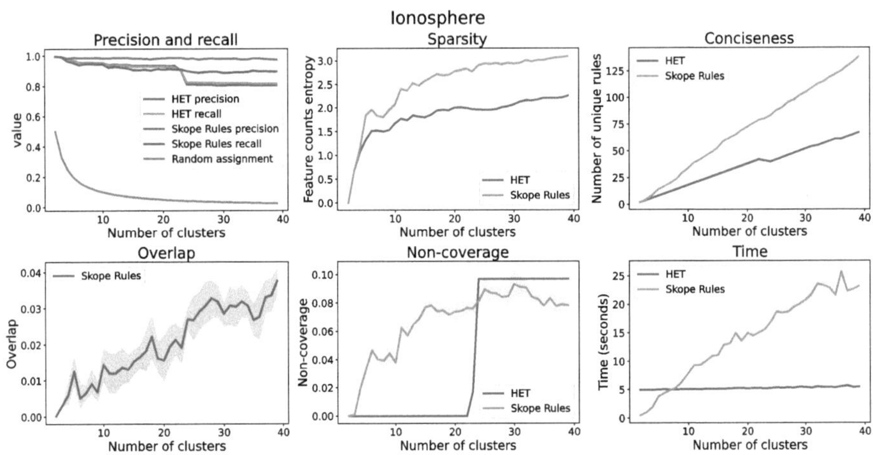

Fig. 5. Results on Ionosphere dataset.

House Prices. Again, results for House Prices presented in Fig. 6 confirm previously seen trends. Interestingly, the fact that it is a regression task does not seem to impact results. We also note the increased computation time needed by SkopeRules for this bigger dataset, indicating that this method may scale poorly with the number of instances and features in datasets.

Overall, these 3 experiments show the repeatability of our results by their consistencies. Despite having slightly lower precision and recall that SkopeRules, our proposal shines by its ability to produce organized sets of rules, negating any overlapping and improving on coverage. Its one-time computation guarantees a faster exploration of clusters, and a much faster computation for high number of clusters, even when considering larger datasets.

4 Related Works

Finding relevant and intelligible groups of data and influence points has been a rising topic divided into two main axes: influence clustering and explainable clustering. The former usually applies clustering techniques on the influence space, while the latter focuses on making user-interpretable clusters.

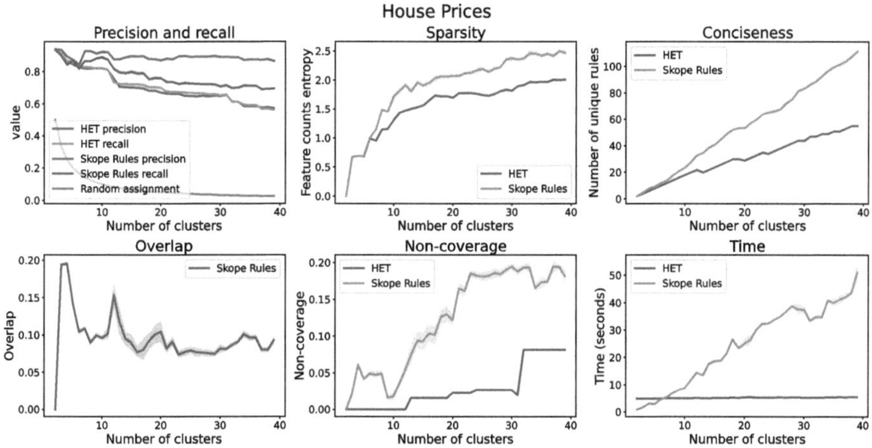

Fig. 6. Results on House Prices dataset.

4.1 Influence Clustering

The use of local feature influence methods usually constitutes the end result of the XAI pipeline. However, these local feature influences can serve as foundational data for deeper analysis, leading to the concept of "supervised clustering" where clustering is performed on the influence values [22], using the feature influence matrix generated by the XAI method. SHAP values adhere to properties such as consistency and accuracy, making them more reliable for use as input in clustering algorithms [22]. Other works [8,15,19] have applied this strategy to specific use cases, such as urinary disease and Covid-19 symptomatology, and multi-principal element alloys chemistry, yielding promising results. It has been shown experimentally [14] that influence-based clusters can be more homogeneous and of better quality than their data-based counterparts, even with low-performing ML models.

4.2 Explainable Clustering

Many clustering techniques encounter a common challenge: they fail to elucidate the underlying rationale behind the clusters they generate. Dimensional reduction techniques such as PCA, UMAP [23] or t-SNE [30] are often used to visualize clusters computed on high dimensional data, or to perform clustering on reduced dimensions. However, this approach does not facilitate the user's ability to discern the characteristics or signatures of the clusters in relation to the original data, which can be exceedingly complex depending on the nature and dimensionality of the data.

A popular approach to this problem is to explain each cluster through a set of rules. Thus, an instance is deemed to belong to a cluster if and only if it adheres to the given set of rules. While Anchors [28], a method based on LIME, doesn't aim to explain clusters directly, it endeavors to refine the local boundaries of a local explanation by creating a minimal decision rule sufficient to explain the predicted influence. While using decision rules to locally explain prediction makes the explanation easier to understand for humans, Anchors identifies a distinct set of rules for every single instance, making the process of explaining an entire dataset for a model tedious and unsuitable for clustering.

In [25], the authors propose Explainable k-Means, an algorithm that aims to approximate a k-means clustering by building a threshold tree efficiently. Such an algorithm starts from a set of cluster centers and builds a binary threshold tree in a top-down iterative fashion by minimizing the number of mistakes (instances affected to a leaf not corresponding to their original cluster) at each split. The algorithm stops when the tree achieves k leaves, each leaf corresponding to a cluster. The path from the root to a leaf describes the cluster membership of instances. An enhancement of this algorithm has been proposed [9], offering a trade-off between the size of the tree and the degree of approximation.

Regarding hierarchical clustering, DIVCLUS-T [7] enables the interpretation of a dendrogram as a binary tree, explaining each cut with a single rule. This is accomplished by minimizing the inertia criterion, akin to k-means, but only for single-rule splits. DIVCLUS-T is applicable to both numerical and categorical data, although not to mixed data types. Comparisons with Ward and k-means clustering demonstrate similar proportions of inertia explained.

In [11], the authors deviate from previous approaches that mainly relied on post hoc explanation tools from the classification world. They propose an intrinsically explainable clustering framework that relies on representative points to explain the whole partition. While we follow a hierarchical approach, Davidson et al. [11] propose solving a global optimization problem simultaneously by building clusters (optimizing for compactness) and selecting representative points.

While the literature suggests the promise of using binary decision trees to explain clusterings, to our knowledge, no prior work has attempted to define such rules based on feature influences, particularly within a hierarchical context. Moreover, existing studies employ the same data for both performing and explaining clusterings. That means that influence clustering cannot be explained using these methods, as the influence space yields no intrinsic meaning for domain experts.

5 Conclusion

This work introduces a new explainable-by-design supervised clustering pipeline that relies on hierarchical clustering and a novel explanation tree model to produce human-readable and interpretable, i.e. short, sparse and non-overlapping feature-based rules.

We suggest three main leads to improve HET. (i) Incorporating user interaction and feedback mechanisms into the hierarchical explanation tree methodology would enhance its usability and effectiveness. An interactive interface could allow users to explore and navigate the tree structure, visualize clusters, and examine feature influences in a more intuitive and user-friendly manner. (ii) As the number of clusters quickly grow while exploring the tree, implementing cluster-specific metrics to identify highly relevant ones and potentially ranking them based on user-defined preferences could streamline the exploration process and help users discover data-embedded patterns. (iii) Developing heuristic methods to estimate values in the influence space, indeed as HET only relies on distances between points in this space we ascertain heuristics may yield important performance gain by degrading the partitions constructed by HET.

References

1. Aggarwal, C.C., Reddy, C.K.: Data Clustering: Algorithms and Applications. CRC Press (2013)
2. Antoine, V., Labroche, N., Vu, V.V.: Evidential seed-based semi-supervised clustering. In: Proceedings of SCIS-ISIS (2014)
3. Barredo Arrieta, A., et al.: Explainable artificial intelligence (XAI): concepts, taxonomies, opportunities and challenges toward responsible AI. Inf. Fusion **58**, 82–115 (2020)
4. Basu, S., Banerjee, A., Mooney, R.J.: Semi-supervised clustering by seeding. In: Proceedings of the 19th ICML, ICML 2002, pp. 27–34. Morgan Kaufmann Publishers Inc. (2002)
5. Bilenko, M., Basu, S., Mooney, R.J.: Integrating constraints and metric learning in semi-supervised clustering. In: Proceedings of the 21st ICML, p. 11. ACM (2004)
6. Breiman, L., Friedman, J.H., Olshen, R.A., Stone, C.J.: Classification and Regression Trees. Wadsworth (1984)
7. Chavent, M., Lechevallier, Y., Briant, O.: DIVCLUS-T: a monothetic divisive hierarchical clustering method. Comput. Stat. Data Anal. **52**(2), 687–701 (2007)
8. Cooper, A., Bourke, A., Doyle, O.: Supervised clustering for subgroup discovery: an application to COVID-19 symptomatology. In: Proceedings of ECML/PKDD Conference, vol. 1525, pp. 408–422. Springer (2021)
9. Dasgupta, S., Frost, N., Moshkovitz, M., Rashtchian, C.: Explainable k-means clustering: theory and practice. In: XXAI Workshop. ICML (2020)
10. Davidson, I., Basu, S.: A survey of clustering with instance level constraints. ACM Trans. Knowl. Discov. Data 1–41 (2007)
11. Davidson, I., Livanos, M.J., Gourru, A., Walker, P.B., Velcin, J., Ravi, S.S.: An exemplars-based approach for explainable clustering: complexity and efficient approximation algorithms. In: Proceedings of the 2024 SIAM International Conference on Data Mining, SDM, Houston, USA, pp. 46–54. SIAM (2024)
12. Davidson, I., Qi, Z.: Finding alternative clusterings using constraints. In: Proceedings of ICDM Conference, pp. 773–778 (2008)
13. Mast, J., Kemper, B.: Principles of exploratory data analysis in problem solving: what can we learn from a well-known case? Qual. Eng. **21**, 366–375 (2009)
14. Escriva, E., Aligon, J., Excoffier, J.B., Monsarrat, P., Soulé-Dupuy, C.: How to make the most of local explanations: effective clustering based on influences. In: Advances in Databases and Information Systems, vol. 13985, pp. 146–160. Springer (2023)
15. Escriva, E., Doumard, E., Excoffier, J.B., Aligon, J., Monsarrat, P., Soulé-Dupuy, C.: Data exploration based on local attribution explanation: a medical use case. In: New Trends in Database and Information Systems, vol. 1850, pp. 315–323. Springer (2023)

16. Friedman, J.H., Popescu, B.E.: Predictive learning via rule ensembles. Ann. Appl. Stat. **2**(3) (2008)
17. Ganganath, N., Cheng, C.T., Tse, C.K.: Data clustering with cluster size constraints using a modified k-means algorithm. In: 2014 International Conference on Cyber-Enabled Distributed Computing and Knowledge Discovery, pp. 158–161
18. Jain, A.K.: Data clustering: 50 years beyond k-means. Pattern Recognit. Lett. **31**(8), 651–666 (2010)
19. Lee, K., Ayyasamy, M.V., Ji, Y., Balachandran, P.V.: A comparison of explainable artificial intelligence methods in the phase classification of multi-principal element alloys. Sci. Rep. **12**(1), 11591 (2022)
20. Loh, W.Y.: Classification and regression trees. Wiley Interdisc. Rev. Data Mining Knowl. Discov. **1**(1), 14–23 (2011)
21. Lundberg, S., Lee, S.I.: A Unified Approach to Interpreting Model Predictions (2017)
22. Lundberg, S.M., Erion, G.G., Lee, S.I.: Consistent Individualized Feature Attribution for Tree Ensembles (2019)
23. McInnes, L., Healy, J., Melville, J.: UMAP: Uniform Manifold Approximation and Projection for Dimension Reduction (2020)
24. Miller, T.: Explainable AI is dead, long live explainable AI!: Hypothesis-driven decision support using evaluative AI. In: FAccT 2023, pp. 333–342 (2023)
25. Moshkovitz, M., Dasgupta, S., Rashtchian, C., Frost, N.: Explainable k-means and k-medians clustering. In: International Conference on Machine Learning, pp. 7055–7065. PMLR (2020)
26. Moussawi, A.E., Giacometti, A., Labroche, N., Soulet, A.: MAPK-means: a clustering algorithm with quantitative preferences on attributes. Intell. Data Anal. **24**(2), 459–489 (2020)
27. Ribeiro, M.T., Singh, S., Guestrin, C.: "Why should i trust you?": explaining the predictions of any classifier. In: Proceedings of the 22nd ACM SIGKDD International Conference on Knowledge Discovery and Data Mining, pp. 1135–1144. ACM (2016)
28. Ribeiro, M.T., Singh, S., Guestrin, C.: Anchors: high-precision model-agnostic explanations. In: Proceedings of the AAAI Conference on Artificial Intelligence, vol. 32, no. 1 (2018)
29. Tavazzi, E., Longato, E., Vettoretti, M., Aidos, H., et al.: Artificial intelligence and statistical methods for stratification and prediction of progression in amyotrophic lateral sclerosis: a systematic review. Artif. Intell. Med. (2023)
30. Van der Maaten, L., Hinton, G.: Visualizing data using t-SNE. J. Mach. Learn. Res. **9**(11) (2008)
31. Vu, V., Labroche, N., Bouchon-Meunier, B.: Improving constrained clustering with active query selection. Pattern Recogn. **45**(4), 1749–1758 (2012)

Towards Trace Variant Explainability

Iuliana Malina Grigore[1(✉)], Gabriel Marques Tavares[2,3], Vincenzo Pasquadibisceglie[4], Thomas Seidl[2,3], and Sylvio Barbon Junior[1]

[1] Università degli Studi di Trieste, Trieste, Italy
iulianamalina.grigore@phd.units.it, sylvio.barbonjunior@units.it
[2] LMU Munich, Munich, Germany
{tavares,seidl}@dbs.ifi.lmu.de
[3] Munich Center for Machine Learning (MCML), Munich, Germany
[4] University of Bari Aldo Moro, Bari, Italy
vincenzo.pasquadibisceglie@uniba.it

Abstract. Understanding process variability remains a key challenge in process mining, especially when analyzing trace variants that capture diverse behavioral patterns in modern systems characterized by high data volume and complexity. In this paper, we propose a multi-view explanation framework that enables interpretable analysis of trace variants through both global and local explanations. Our approach first encodes traces using multiple views—including activity, transition, case, and event-level information—applying nominal and time-based strategies suited to structural and performance aspects. Then, it introduces a dual-layer explanation mechanism: at the global level, variant representatives are compared to a global behavioral profile to highlight key differentiating features; at the local level, individual traces are contrasted with their assigned representatives to reveal the specific factors driving their inclusion. The framework is compatible with various variant analysis techniques, as long as a representative (e.g., medoid or central node) can be defined. We demonstrate the method's effectiveness using benchmark event logs and interpretable visualizations, showing how it supports transparency, trust, and diagnostic insight in trace variant analysis.

Keywords: Process mining · Conformance Checking · Clustering · Explainability · Variant analysis · Multi-perspective

1 Introduction

Process mining is a discipline that leverages advanced techniques to extract valuable insights from business processes [1]. The main goal is to improve the execution quality of such processes, i.e., reducing cost, throughput time, and resource consumption. Nevertheless, there are still many open challenges for visualizing the complex processes for analysis, which often lead to unexplained models that are difficult to understand without a given context and background

P. K. Chrysanthis et al. (Eds.): ADBIS 2025, LNCS 16043, pp. 95–110, 2026.
https://doi.org/10.1007/978-3-032-05281-0_7

information. For instance, inherent process complexity comes from (i) unstructured processes with high behavioral variability (e.g., decisions are dependent on human actors), or (ii) high data volume from large organizations [15]. In this context, process variants become essential for understanding how business processes are executed. The complexity of the process leads to considerable variability, meaning that while a process may be designed to follow a standardized sequence of activities, the reality is often far more complex. Each unique sequence of activities recorded in an event log represents a different variant—a different path that a case can take through the process. Variant analysis techniques support organizations uncover unusual behaviors by comparing different variants, offering valuable insights into process variability. Understanding this variability is crucial for reducing complexity and enhancing the effectiveness of automated process discovery methods, ultimately enabling higher stakeholder awareness and leading to an optimized workflow. A wide range of techniques have been defined for analyzing process variants, with most focusing exclusively on variants as unique sequences of process activities [22]. However, more recent extensions have proposed to include multi-perspective approaches [5,16] or object-centric event logs [2] that provide a more comprehensive view of process behavior. In parallel, unsupervised learning methods, such as clustering and community detection, have become prevalent in process mining to uncover trace variants and further improve the ability to recognize patterns and structuresent logs [6,11,21]. Further studies have adopted cluster-based techniques to identify promising process variants to overcome spaghetti-like process structures. For example, in the field of process discovery, several studies [4,7] use clustering-based approaches to derive simpler process models and improve conformance metrics.

However, the majority of trace variant analysis methods, based on unsupervised learning, often lack transparency, making it difficult to understand why traces are grouped into specific variants [11]. Explainable Artificial Intelligence (XAI) aims to address the transparency issue by making the inner workings and decisions of complex models interpretable to human users [13]. In the context of process mining, XAI has gained increasing attention for its potential to clarify why certain process behaviors emerge, particularly when data-driven methods are used for variant analysis [19]. Trace variant analysis—which focuses on identifying and characterizing behavioral deviations across process executions—can benefit significantly from XAI techniques that provide both global explanations, highlighting structural and performance-level differences between groups, and local explanations, which clarify why individual traces belong to specific variants. Existing methods can be categorized as agnostic, when they provide explanations independently of the internal mechanics of the detection algorithm, or non-agnostic, when they are tightly coupled with specific model structures [14]. Despite recent advances, XAI for trace variant analysis remains an open challenge, particularly in unsupervised settings where no ground truth or explicit decision boundaries exist. This gap motivates our work, which proposes a flexible, interpretable framework to bridge this limitation by enabling transparent, multiview explanations of trace variants derived from any unsupervised method.

We propose a dual-layer explanation framework that combines multiview encoding with both global and local interpretability mechanisms to offer a comprehensive, human-understandable characterization of process variability. At the global level, the framework distinguishes behavioral patterns across variant representatives, while at the local level, it elucidates the reasons behind the assignment of individual traces to specific variants. Importantly, the approach is agnostic to the underlying variant analysis strategy—be it clustering- or graph-based—as long as a representative trace can be identified for each variant. This flexibility enables broad applicability and seamless integration with existing trace variant analysis techniques, supporting a wide spectrum of process mining scenarios with interpretable, multiview-driven explanations. The main contributions of this work can be summarized as follows:

- We introduce a multiview trace variant explanation framework that integrates multiple behavioral perspectives (activity, transition, case, and event views) to support comprehensive and interpretable analysis.
- We propose a dual-layer explanation strategy, combining *global variant-level interpretation*—based on representative traces and average feature profiling—with *local trace-level interpretation*, highlighting the alignment and deviation of individual traces within their assigned variants.
- Our approach is agnostic to the variant analysis method, allowing the use of clustering- or graph-based techniques, as long as variant representatives (e.g., medoids or central nodes) can be identified.
- We provide a set of simple interpretable visual artifacts, such as heatmaps, that allow practitioners to understand key differentiating features between variants, including activity patterns, control-flow deviations, and time-based behaviors.

The remainder of this work is organized as follows: Sect. 2 presents details of the proposed approach. Section 3 details an interpretable visualization for the global and local explanation of variants. Section 4 draws the conclusions and paves the way for future research.

2 Proposed Approach

Our approach suits trace variant analysis methods to enable human-interpretable explanations at both the trace and variant levels. Figure 1 illustrates the proposed framework for trace variant explanation based on multiview profiling. Our approach is divided into Multiview Trace Profiling and Multiview Variant Explanation, as discussed in the next subsection.

2.1 Interpretable Multiview Trace Profiling

The profiling process begins with the ingestion of an event log, from which traces are extracted and analyzed across multiple perspectives: `Activity View`,

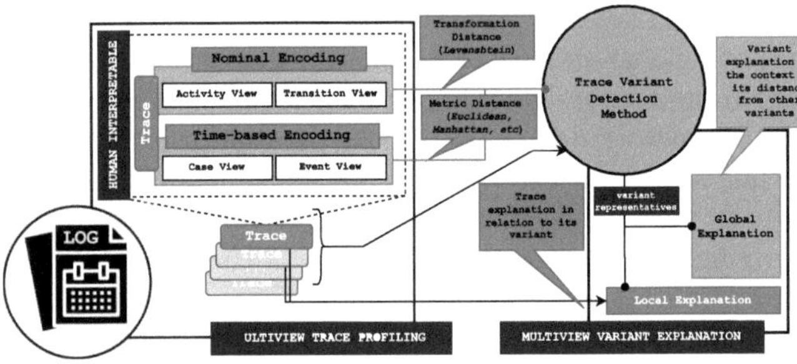

Fig. 1. Multiview trace variant explanation framework, integrating multi-view and distance measures to interpretable global and local explanations of variant.

`Transition View`, `Case View`, and `Event View`, with a particular emphasis on the control-flow dimension. Each view captures distinct structural or behavioral aspects of a trace and is subsequently encoded using either nominal or time-based strategies, depending on the nature of the data. Given our focus on human-interpretable explanations, both encoding types are employed to enrich the representation and support meaningful comparisons.

Nominal encoding focuses on categorical attributes, particularly related to control-flow, such as activity occurrences and their transitions. We apply one-hot encoding [20] to transform categorical features into binary vectors, which facilitates the use of transformation-based distance measures (e.g., Levenshtein distance) that preserve the structural differences. Let $\mathcal{T} = \{\tau_1, \tau_2, \ldots, \tau_n\}$ be a set of traces extracted from an event log $\mathcal{L}$. Each trace τ_i is a finite sequence of activities $\tau_i = \langle a_1, a_2, \ldots, a_k \rangle, a_j \in \mathcal{A}$ where $\mathcal{A} = \{a_1, a_2, \ldots, a_m\}$ is the set of all distinct activity labels in the log.

Define an encoding function $\phi : \mathcal{T} \to \{0,1\}^m$ that maps each trace τ_i to a binary vector $v_i \in \{0,1\}^m$, such that:

$$v_i[j] = \begin{cases} 1, & \text{if } a_j \in \tau_i \\ 0, & \text{otherwise} \end{cases} \quad \text{for } j = 1, \ldots, m$$

This representation captures the presence or absence of each activity in τ_i, ignoring order and frequency. To include transitions, we extend the encoding with the set of directly-follows pairs $\mathcal{T}_A = \{(a_i, a_j) \mid a_i, a_j \in \mathcal{A},\ a_i \to a_j \text{ in some } \tau \in \mathcal{T}\}$. Thus, each trace is encoded as a binary vector in $\{0,1\}^{|\mathcal{A}|+|\mathcal{T}_A|}$, representing both activity occurrences and their transitions.

Time-based encoding captures continuous temporal features extracted from event logs. Let $\mathcal{E} = \langle e_1, e_2, \ldots, e_k \rangle$ be a trace consisting of events, where each event $e_i = (a_i, t_i)$ pairs an activity $a_i \in \mathcal{A}$ with a timestamp $t_i \in R^+$. All timestamps t_i are relative to the same case instance and share a common temporal context. From these timestamps, we derive a set of real-valued features: *Execution time*: $\delta_i = t_i^{\text{end}} - t_i^{\text{start}}$, when both timestamps are available. *Accumulated time*: $\alpha_i = t_i - t_1$, the time since the start of the trace. Additional temporal features are computed using calendar decomposition (e.g.,(hour_of_day),day_of_week, month_of_year, etc.). All time-based features are standardized $\hat{x}_{ij} = \frac{x_{ij}-\mu_j}{\sigma_j}$ where x_{ij} is the j-th temporal feature of event e_i, and μ_j, σ_j are the dataset-wide mean and standard deviation of that feature.

Each standardized feature vector $\hat{x}_i \in R^d$ can then be clustered using a metric distance function such as Euclidean distance. We adopt the time feature extraction methodology proposed by Oyamada et al. [17]. The output of this step is an encoded representation that can be used in trace variant analysis across a wide range of methods.

Numerical Example: Consider an event log $\mathcal{L}$ with activities A={Start, Approve, Reject, End} and observed transitions T_A = {(Start, Approve), (Approve, End), (Start, Reject), (Reject, End)}. Given a trace $\tau = \langle$Start (t_1), Approve (t_2), End (t_3) $\rangle$, the nominal encoding results in a vector $v_{\text{nominal}} \in \{0, 1\}$, where:

- Activities present: `Start`, `Approve`, `End` $\rightarrow$ 1
- Activity absent: `Reject` $\rightarrow$ 0
- Transitions present: (`Start`, `Approve`), (`Approve`, `End`) $\rightarrow$ 1
- Transitions absent: (`Start`, `Reject`), (`Reject`, `End`) $\rightarrow$ 0

Thus, $v_{nominal} = (1, 1, 0, 1, 1, 1, 0, 0)$. Moreover, assuming extracted and standardized time-based features such as execution times and accumulated times yield v_{time} = (0.2, 0.4, 0.1, 0.5, 0.7, 1.0), the final trace vector is v_{final} = (1, 1, 0, 1, 1, 1, 0, 0, 0.2, 0.4, 0.1, 0.5, 0.7, 1.0).

2.2 Trace Variant Analysis

Trace variants are identified using an analysis method, such as those proposed in [3,8,12]. The only requirement imposed by our framework is that the chosen analysis method must enable the computation of trace similarity or dissimilarity. These comparisons must be based on one of two distance paradigms: transformation distances (e.g., Levenshtein), which capture structural differences between traces [8], and metric distances (e.g., Euclidean), which operate on vectorized representations of continuous features [3,12]. These distance measures support the analysis by providing a quantitative basis for grouping similar traces, identifying representative variants, and detecting deviations or anomalies within the event log.

Once the distances are calculated, the framework assigns each trace to a variant class and extracts a representative trace for each class. The definition of the representative depends on the analysis method: for clustering-based approaches, it could be the medoid or the centroid; for graph-based methods, it could be the node with the highest degree or centrality. These representatives serve as surrogates for the overall behavior of the corresponding variants.

2.3 Interpretable Multiview Variant Explanation

Following the variant analysis step, our framework advances to an interpretable multiview explanation phase, which supports both global and local interpretability of the variant structure. This component relies on the representative traces extracted for each variant, which are assumed to capture its central behavior. Their primary role is to serve as interpretable surrogates for summarizing the behavior of the variant they represent.

Representative Trace. Let $V = \{\sigma_1, \sigma_2, \ldots, \sigma_n\}$ be the set of traces associated with a given variant, and let $d : V \times V \rightarrow \mathbb{R}_{\geq 0}$ denote a distance function quantifying dissimilarity between traces. A *representative trace* $\sigma^* \in V$ is defined as the trace that minimizes the average distance to all other traces in V, that is:

$$\sigma^* = \arg\min_{\sigma \in V} \frac{1}{n} \sum_{i=1}^{n} d(\sigma, \sigma_i)$$

The representative trace serves as a concrete, observed example that encapsulates the central behavioral characteristics of the variant, thereby enhancing interpretability by offering a real, easily comprehensible reference for human analysis.

Global Explanations. Let $\mathcal{V} = \{v_1, v_2, \ldots, v_K\}$ denote the set of variants identified by a trace variant analysis method, where K is the number of variants. For each variant $v_k \in \mathcal{V}$, let $r_k \in R^d$ represent the *feature vector of the variant representative*, extracted according to the strategy defined by the underlying detection method (e.g., medoid, centroid, or highest-degree node). These feature vectors lie in a common multiview-encoded space of dimension d, capturing the structural and temporal characteristics of the traces. We define the *global average representative vector* as $\bar{r} = \frac{1}{K} \sum_{k=1}^{K} r_k$. Next, we compute the distance between each representative r_k and the global average $\bar{r}$ using a distance function $D : R^d \times R^d \rightarrow R_{\geq 0}$. For metric-based encodings (e.g., time-based or continuous case/event features), we use the Euclidean distance: $D(r_k, \bar{r}) = \|r_k - \bar{r}\|_2$. For nominal encodings (e.g., activity or transition), we employ the Levenshtein distance $D(\tau_k, \bar{\tau}) = \text{Lev}(\tau_k, \bar{\tau})$, where τ_k is the activity sequence of the representative trace for variant v_k, and $\bar{\tau}$ is a representative sequence derived from the overall process behavior (e.g., a medoid of all representatives). $\text{Lev}(\cdot, \cdot)$ denotes the Levenshtein distance, measuring the minimum number of changes required to transform one sequence into another.

To understand *why* a representative r_k deviates from the global behavior, we compute the *difference vector*, i.e., $\Delta_k = r_k - \bar{r} \in R^d$. Each component $\Delta_k^{(j)}$ of this vector quantifies the contribution of feature j to the overall deviation of variant v_k. The magnitude and sign of $\Delta_k^{(j)}$ indicate how much and in what direction feature j differs from the average behavior: (i) when $\Delta_k^{(j)} > 0$, feature j is overrepresented in variant v_k; (ii) when $\Delta_k^{(j)} < 0$, feature j is underrepresented or absent in v_k; (iii) when $\Delta_k^{(j)} \approx 0$, feature j aligns with the global behavior.

These vectors Δ_k can be visualized through heatmaps, where the color intensity reflects the relative deviation per feature. Such visual artifacts enable an intuitive and human-interpretable understanding of variant-specific behaviors about the global process landscape.

Local Explanations. The local explanations focus on characterizing the relationship between an individual trace and its assigned variant to understand why a given trace was grouped into a particular behavioral class. Let τ_i be a trace assigned to variant v_k, and let $x_i \in R^d$ denote its multiview feature vector, obtained via the encoding described in Sect. 2.1. Let $r_k \in R^d$ be the feature vector corresponding to the representative of variant v_k. To evaluate how well τ_i aligns with the behavioral pattern captured by its representative, we compute a *deviation vector*: $\delta_i = x_i - r_k \in R^d$.

The component $\delta_i^{(j)}$ measures the deviation of feature j in trace τ_i from the expected value in the representative. This difference can be interpreted as follows: (i) when $\delta_i^{(j)} \approx 0$, feature j of τ_i is consistent with the representative behavior; (ii) when $\delta_i^{(j)} > 0$, feature j is overexpressed in τ_i relative to the variant representative; (iii) when $\delta_i^{(j)} < 0$, feature j is underexpressed or missing in τ_i compared to the representative.

To quantify the overall alignment, we may compute a similarity or dissimilarity score between τ_i and r_k using an appropriate distance metric $D(x_i, r_k)$ (e.g., Euclidean or cosine distance). Higher values of D indicate that τ_i deviates more significantly from the prototypical behavior of variant v_k. This analysis enables localized interpretability: by examining the deviation vector δ_i, domain experts can identify the specific features, such as atypical activity transitions, outlier execution times, or rare case attributes, that justify the trace's inclusion within or potential exception from the variant.

2.4 Visual Explanations

Aiming to quantify and interpret the variability of execution trace variants and to provide a simple visualization of the obtained results, we propose a heatmap-based approach, as suggested in the visual explainability literature [18,23]. Our approach decomposes trace information into five analytical dimensions: activity frequency, self-loops, transition patterns, case-level time metrics, and average event timing. This decomposition enables fine-grained variance attribution and enables the visual inspection of structural and temporal deviations across traces.

By aligning each case as a row and each feature type as a column, the proposed schema enables immediate identification of high-variance regions, such as abnormal loop behavior or temporal drifts. The use of color-encoded difference-from-mean values allows for intuitive detection of outliers and patterns without requiring complex statistical interpretation. Moreover, this approach supports hypothesis generation regarding the sources of variance—e.g., performance bottlenecks, compliance violations, or role-based behavior differences—thus bridging the gap between numerical variance scores and causally relevant insights. Figure 2 presents a schema of the proposed usage of insights as a dashboard.

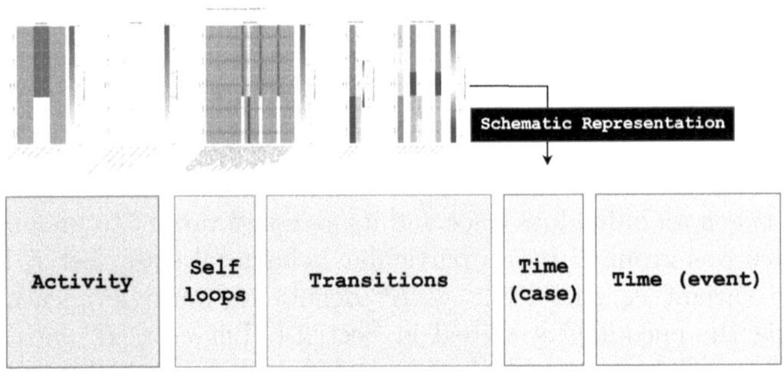

Fig. 2. Schematic overview of proposed explainability dashboard of trace variants.

3 Experiments and Discussion

To support the evaluation of our proposed method, we used a set of benchmark event logs that allow the analysis of trace variants from two perspectives: global and local explainability. For the global analysis, we employed a curated collection of event logs that exhibit distinct variant behaviors across dimensions such as activity sequences, self-loops, transitions, and timing. These logs were chosen to illustrate the interpretability capabilities of our approach in explaining structural and temporal variance. All benchmark event logs and implementation artifacts are openly available[1]. The benchmarks are divided into two categories: model examples and real datasets. The model examples are simplified event logs adapted to the approach proposed in [10], which builds on the requirements defined in [9]. They are used for illustration to support visual explanations. In contrast, the real-world datasets BPI Challenge 2017[2] and WABO[3] are used to validate the effectiveness of our approach in real-world scenarios. The BPI Challenge 2017 dataset is based on the loan application process of a Dutch financial institution. In contrast, the WABO dataset captures the execution records of the entry phase of a building permit process in an anonymized municipality. An overview of the characteristics of all benchmark event logs is provided in Table 1.

Table 1. Description of the benchmark event logs.

Event log	#Events	#Traces	#Variants	#Activities
BPI Challenge 2017	193849	42995	16	8
WABO	8577	1434	116	27
fullchange	6928	1500	2	6
tratimefrequent	8556	1500	2	5

1 https://github.com/sbarbonjr/xai4tracevariant.
2 https://data.4tu.nl/articles/dataset/BPI_Challenge_2017_-_Offer_log/12705737.
3 https://data.4tu.nl/datasets/2db2e3c1-9499-4699-9098-1a28c15a5913/2.

To analyze the trace variants in these benchmark event logs, we applied the k-means clustering algorithm using a maximum of 200 iterations as the termination criterion. To investigate the influence of different levels of granularity, we tested the method with several values for the number of clusters, particularly values ranging from 2 to 11. This approach used the medoid trace, the most central trace within each cluster, to represent the corresponding cluster.

3.1 Model Example Explanation

We have created some processes with only two variants to demonstrate the explanation of our visual approach. This choice was made to provide simple and intuitive examples to help users understand how visualizations can be used to explain the differences between variants.

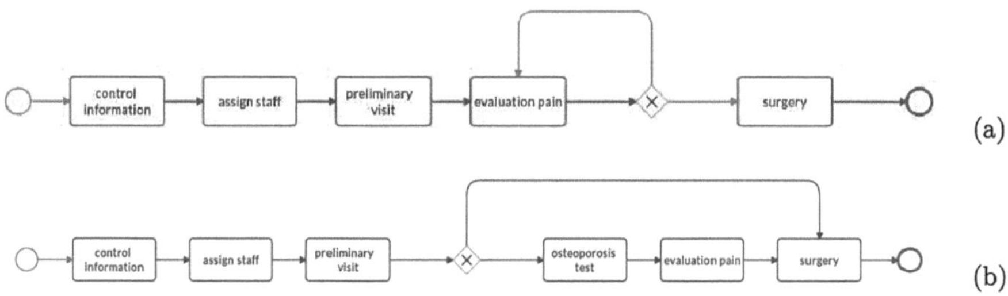

Fig. 3. Processes used for the experiments.

The activities and control flow of these processes are shown in Fig. 3, where the structural differences are immediately visible. For example, both processes include largely the same activities, with the exception that the *osteoporosis test* is present in process **b** only. In addition, the process **b** goes directly to the activity *surgery*, while the process **a** contains a self-loop with the activity *evaluation pain*. To illustrate the proposed approach, we decided to apply the k-means clustering method with a value of $k = 3$. As a result, the representative traces for each variant correspond to the medoids identified by the clustering algorithm.

Figure 4 shows the Activity heatmap, where the y-axis lists all activities in the event logs and the x-axis shows the representative traces for each variant. This example refers specifically to the process shown in Fig. 3b. In particular, the representative trace S_1056 shows negative values (e.g., -0.7) for activities such as *osteoporosis test* and *evaluation pain*, indicating that they do not occur in this variant. Conversely, positive values for the same activities in other nodes indicate their presence in these variants. This visualization immediately highlights the differences in the activity sequences between the variants. The difference is not limited to the presence or absence of certain activities, but also reflects the order in which they occur. To explore this further, we examine two additional heatmaps: Self-loops and Transitions.

Figure 5 shows the heatmap of the self-loop, which displays repeated activity sequences. Activities are shown on the y-axis and representative traces, on the x-axis, that relate specifically to the process in Fig. 3a. Zero values mean that there

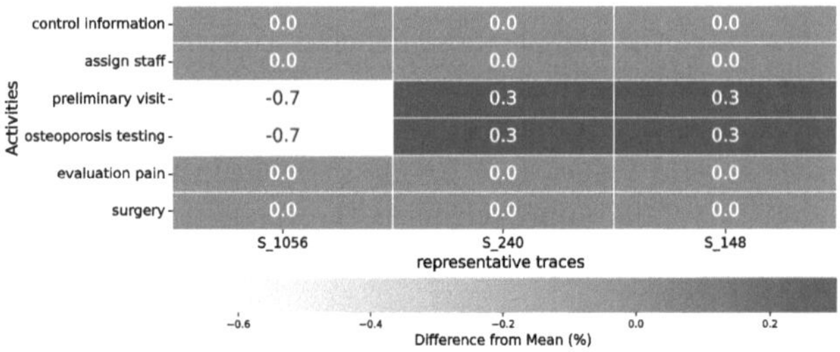

Fig. 4. Example of the Activity heatmap, where the y-axis represents the activities in the log, and the x-axis shows the representative traces for each variant.

are no loops, while negative values (e.g., −0.7) show that a loop is missing in the representative trace but is common in other variants. This provides information about behavioral nuances such as iterative diagnoses or repeated treatments.

Figure 6 shows the heatmap of transitions for all pairs of activities (y-axis), based on the process in Fig. 3b. Values close to zero indicate consistent transitions between the variants. Positive values highlight transitions that are typical for a particular variant, while negative values indicate their absence. Transitions such as *preliminary visit → osteoporosis test*, *preliminary visit → surgery*, *osteoporosis test → evaluation pain*, and *evaluation pain → surgery* help to distinguish the unique workflow patterns of each variant.

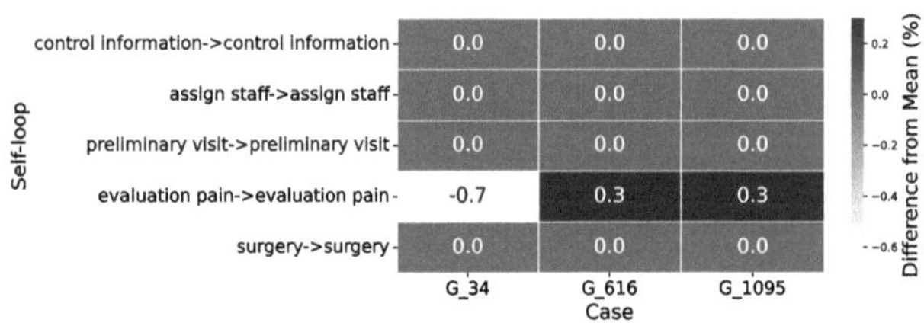

Fig. 5. Example of the Transition heatmap: shows all possible transitions between activities (y-axis), with the x-axis showing the representative traces for each variant.

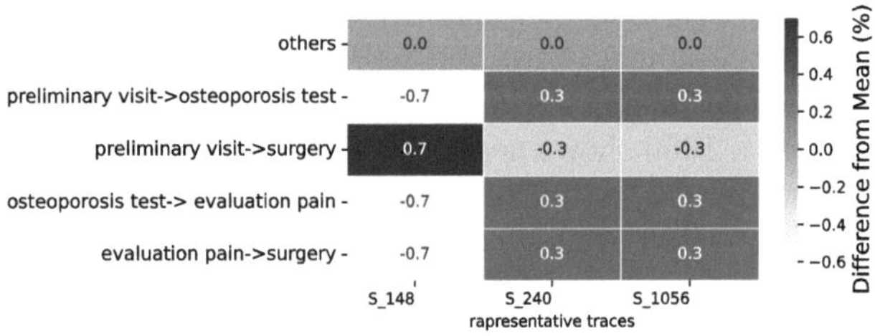

Fig. 6. Example of the Transition heatmap: illustrates all possible transitions between the activities found in the dataset, shown on the y-axis. And the x-axis shows the representative traces for each variant.

To examine the temporal differences between the variants, we use two heatmaps: Time (Traces), which shows execution and accumulated time, and Time (Avg. Event), which captures calendar-related features such as hour, minute, and day of the week. Figure 7, based on the process in Fig. 3b, supports this analysis.

As the dataset spans a short period, features such as `day_of_year` and `week_of_year` remain close to zero, reflecting limited variation. In contrast, features such as `day_of_week`, `hour_of_day`, and *min_of_hour* show clearer differences between the traces. Temporal features such as `day_of_year` and `week_of_year` show little variation, while `day_of_week` and `hour_of_day` highlight more significant differences between representative traces. Negative `day_of_week` values indicate days that are further away from a reference, while positive values indicate closer proximity. Even traces with similar `day_of_week` values (e.g., −0.3) may occur on the same day and hour but differ slightly in minutes, capturing subtle execution shifts.

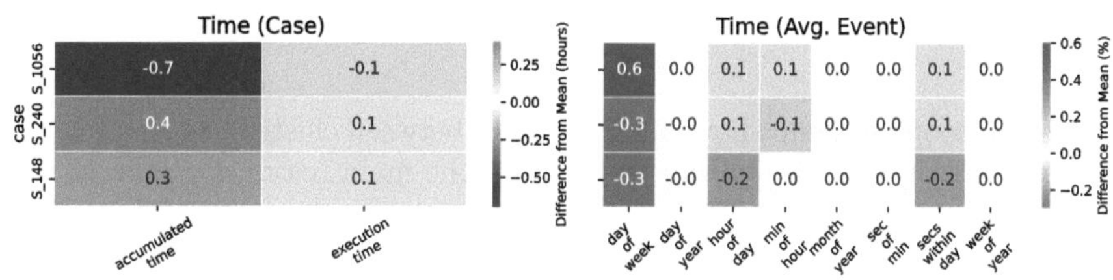

Fig. 7. Time-based encoding of Time (Traces) and Time (Avg. Event), with time features on the x-axis and representative traces for each variant on the y-axis.

3.2 Explaining a Real Event Log

This section illustrates how visual explainability is applied to real event logs. Figure 8 shows the visual explanation for the BPI 2017 event log. We analyzed this log with $k = 11$, a value that captures a more complicated interaction between the activities involved in the process and their relationship to each other.

The Activity heatmap highlights consistent behavior patterns in representative traces. Traces such as *Offer_1879441167*, *Offer_1776585564*, *Offer_1639775243*, *Offer_727616715*, and *Offer_595637015* show strong positive variations for the activities *O_Accepted*, and *O_Returned* indicating frequent engagement in this process. Conversely, traces such as *Offer_1490878929* and *Offer_625990241* show positive variations for *O_Cancelled* and negative variations for *O_Returned* and *O_Accepted*, indicating alternative, faster routes to process completion. Other cases, such as *Offer_1277835325* and *Offer_111109106*, are characterized by a positive occurrence of *O_Returned* and *O_Refused*, and a negative value of *O_Accepted*, which indicates a rejection of offers after the first contact. The presence of positive or negative values indicates the presence or absence of activity within a variant and provides a clear behavioral signature for each representative trace.

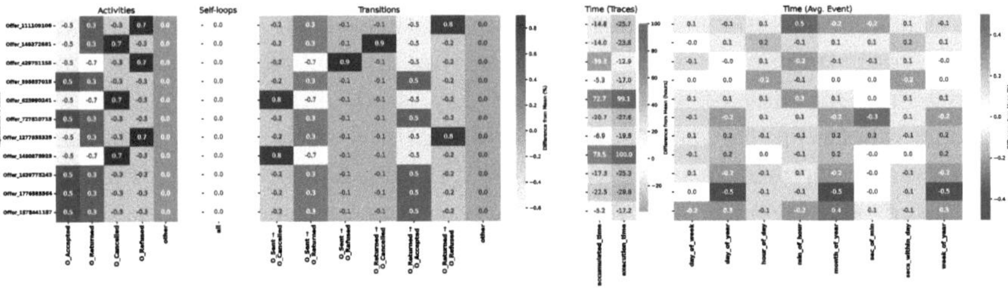

Fig. 8. Application of the global explanation method to the BPI event log, highlighting variations in behaviour between process variants. The visualisation highlights differences in the Activity, Self-loops, Transitions, Time (Traces), and Time (Avg. Event) heatmaps, capturing key deviations from the overall process behaviour.

The Self-loop heatmap shows little variance between clusters, suggesting that the repetition of activities within cases is not the main reason for process variability.

In contrast, the Transition heatmap shows clearer patterns. Traces with positive transitions from *O_Sent* to *O_Returned* and from *O_Returned* to *O_Accepted* are indicative of successful process completions. In particular, representative traces such as *Offer_1277835325* and *Offer_111109106* after *O_Returned* show transitions to *O_Refused*, which indicates a rejection behavior. Representative traces such as *Offer_1490878929* and *Offer_625990241* show increased transitions to *O_Cancelled*, with corresponding negative deviations in transitions involving *Returned* activities, reflecting an early withdrawal from the process. Similarly, *Offer_429751155* shows a positive transition from *O_Sent* to *O_Refused*, reinforcing the different process outcomes compared to other variants.

The temporal analysis enables further differentiation. The Time (Traces) shows considerable variability in the execution and accumulation times between the traces, which indicates that the process durations are not uniform. In particular, traces such as *Offer_1490878929* and *Offer_625990241* have a longer duration compared to others, indicating more complex or delayed decision-making processes. The Time (Avg. Event) shows larger fluctuations in features such as `day of the year`, `month of the year` and `week of the year`, while the changes in `day of the week` are minimal (about 0.1, 0.0 and −0.1 across all traces). This suggests that although the activities often fall on similar days of the week, they are distributed differently across the calendar. Each representative trace thus has a unique temporal footprint that shows how process behavior evolves.

Figure 9 shows a human-interpretable explanation of the WABO event log. In this analysis, we found that the result with $k = 9$ provides a more comprehensive insight into the activities involved and their relationships, highlighting greater variability in the number of variants used to describe the process. The Activity heatmap shows that most representative traces have a common core of activities, with a few notable exceptions. *Case-7935* shows positive deviations due to the inclusion of *T03 Adjust confirmation of receipt*, which does not occur in the other traces. Similarly, *case-4346* shows positive deviations, especially concerning *T07-1 Draft internal advice aspect 1.* In contrast, *case-7183* exhibits negative deviations, reflecting the absence of several activities typically performed in the other traces, including *T02 Check confirmation of receipt*, *T04 Determine confirmation of receipt*, *T05 Print and send confirmation of receipt*, *T06 Determine necessity of stop advice* and *T10 Determine necessity to stop indication.*

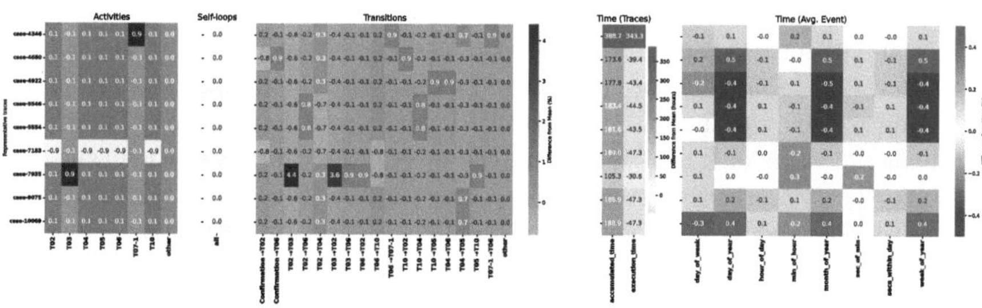

Fig. 9. Application of the global explanation method to the Wabo event log, highlighting behavioral variations across process variants.

The Self-loop heatmap reveals minimal variation between traces, indicating that all variants have similar activity frequencies and there is no evidence of repeated actions contributing to the behavioral differences.

The Transition heatmap highlights specific patterns of relational variations between activities. Negative values are observed in paths such as from *Confirmation* to *T02 Check confirmation of receipt*, from *T02 Check confirmation of receipt* to *T04 Determine confirmation of receipt*, and from *T06 Determine necessity of stop advice* to *T10 Determine necessity to stop indication.* Conversely, there are positive variations in representative traces such as *case-5554* and *case-5546*, especially in the transitions from *T02 Check confirmation of receipt* to *T06 Determine necessity of stop advice* and from *T10 Determine necessity to stop indication* to *T04 Determine confirmation of receipt.* A distinct pattern can be observed in *case-7935*, which shows strong positive values for the transitions between *T02 Check confirmation of receipt* and *T03 Adjust confirmation of receipt*, indicating the existence of a loop between these activities. Finally, *case-7183* shows consistently negative variations that match the missing activities in the Activity heatmap.

The Time (Traces) shows significant variations in execution and accumulation times in the representative traces, indicating that the process duration is not uniform. The Time (Avg. Event) heatmap shows that *case-4922*, *case-5546*, and *case-5554* have similar values for the feature `week of the year` (about −0.4), but differ for the feature `month of the year`, which indicates that *case-4922* occurs about one month earlier than the others. In contrast, the remaining representative traces show very similar values for both the `day of the year` and the `hour of the day`, indicating that these representative traces were performed within a very narrow time window. Interestingly, these temporally close traces belong to different variants, illustrating the coexistence and temporal overlap of different process behaviors within the same time.

3.3 Local Explanation

Understanding the explanation of a variant is crucial not only for interpreting its structure but also for clarifying why a particular trace is associated with it. In this analysis, we use a local-level comparison to observe how the variant evolves and select two representative traces associated with it: *Offer_1878441187* and *Offer_1776585564*. Figure 10 shows the local visual explanations for a trace belonging to this variant based on the BPI 2017 event log. For consistency, we use the same value of k that is used in the overall description of the event log.

Since the two traces represent the same variant, the heatmaps for Activities, Self-loops, and Transitions do not highlight significant deviations. However, there are notable differences in the temporal dimension. In particular, the Time (Traces) shows that the accumulated time and execution time have a positive deviation, which indicates a longer total duration in the trace.

At a finer level of granularity, additional time-based features such as `day of the year`, `minute of the hour`, `month of the year`, `seconds within the day`, and `week of the year` also show positive deviations. Among these, the `day of the year`, `month of the year`, and `week of the year` show the strongest deviations, indicating that the temporal differences in these dimensions are particularly pronounced. This positive deviation might indicate that the representative trace occurs in different months, on different days, and within different weeks compared to the other trace being compared.

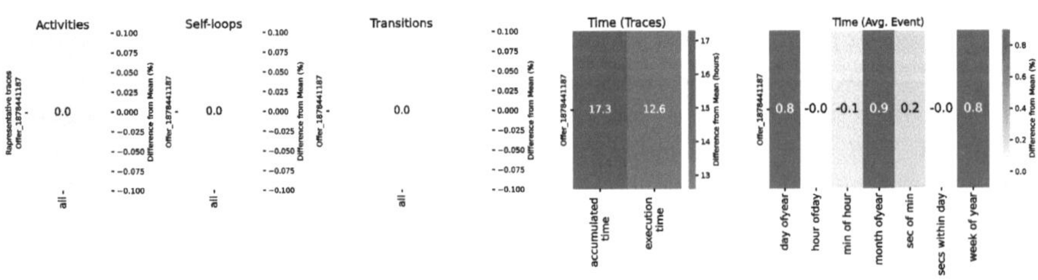

Fig. 10. Local explanation applied to a trace representative and *case 1015*.

4 Conclusion

We presented a two-layer explanatory framework that combines multiview coding with global and local interpretability for the analysis of process variants. This approach enables both high-level pattern recognition and fine-grained trace-level analysis, enhancing usability and insight generation. The framework uses variant analysis, which is supported by k-means clustering. Heatmaps are used to highlight key differentiators such as activity sequences and control flow divergences, providing a clear visual insight into process variability. At the local level, the system explains trace assignments by identifying behavioral and temporal deviations, offering interpretable justifications for variant membership. Overall, the framework provides a robust and customizable solution for exploring and understanding process variability. Future work will focus on developing a user-friendly interface to support the intuitive identification of behavioral differences across variants.

Acknowledgments. We acknowledge financial support under the National Recovery and Resilience Plan (NRRP), M4C2I1.1, funded by the European Union – NextGenerationEU– Project Title aRtificial intElligence for Process Analytics (REPA) - Grant Assignment Decree No. 2022CJWPNA by the Italian Ministry of Ministry of University and Research (MUR). Also, this research was carried out in the framework of the PORTRAIT project (Port to Rail Digital Twin in the Adriatic Region), funded under the PR FESR 2021–2027, Action A1.1.2, the funding was awarded through the DGR 784/2023 of the FVG Region (Italy). The work of Vincenzo Pasquadibisceglie was supported by PNRR project FAIR - Future AI Research (PE00000013), Spoke 6 - Symbiotic AI (CUP H97G22000210007) under the NRRP MUR program funded by the NextGenerationEU.

References

1. van der Aalst, W.M.P.: Process Mining - Data Science in Action. Springer (2016)
2. Adams, J.N., Schuster, D., Schmitz, S., Schuh, G., van der Aalst, W.M.: Defining cases and variants for object-centric event data. In: 2022 4th International Conference on Process Mining (ICPM), pp. 128–135. IEEE (2022)
3. Appice, A., Malerba, D.: A co-training strategy for multiple view clustering in process mining. IEEE Trans. Serv. Comput. **9**(6), 832–845 (2016)
4. Cirne, R., Melquiades, C., Leite, R., Leijden, E., Maciel, A., de Lima Neto, F.B.: Data mining for process modeling: a clustered process discovery approach. In: Ganzha, M., Maciaszek, L.A., Paprzycki, M. (eds.) Proceedings of the 2020 Federated Conference on Computer Science and Information Systems, FedCSIS 2020, Sofia, Bulgaria, 6–9 September 2020. Annals of Computer Science and Information Systems, vol. 21, pp. 587–590 (2020)
5. Cremerius, J., Patzlaff, H., Rahn, V.X., Leopold, H.: Data-based process variant analysis. In: HICSS, pp. 3255–3264 (2023)
6. De Koninck, P., Nelissen, K., vanden Broucke, S., Baesens, B., Snoeck, M., De Weerdt, J.: Expert-driven trace clustering with instance-level constraints. Knowl. Inf. Syst. **63**(5), 1197–1220 (2021)

7. Fani Sani, M., Boltenhagen, M., Aalst, W.: Prototype selection using clustering and conformance metrics for process discovery. In: Del Río Ortega, A., Leopold, H., Santoro, F.M. (eds.) Business Process Management Workshops, pp. 281–294. Springer, Cham (2020)
8. Grigore, I.M., Tavares, G.M., Junior, S.B.: Beyond flattening: detecting concurrency anomalies using K-NN graph-based modeling in object-centric event logs. In: Czekster, R.M., Milazzo, P. (eds.) From Data to Models and Back, pp. 116–136. Springer, Cham (2025)
9. Intrigila, B., Della Penna, G., D'Ambrogio, A.: A lightweight BPMN extension for business process-oriented requirements engineering. Computers **10**(12) (2021)
10. Intrigila, B., Della Penna, G., D'Ambrogio, A., Campagna, D., Grigore, M.: Process-oriented requirements definition and analysis of software components in critical systems. Computers **12**(9) (2023)
11. Koninck, P.D., Weerdt, J.D., vanden Broucke, S.K.L.M.: Explaining clusterings of process instances. Data Min. Knowl. Discov. **31**(3), 774–808 (2017)
12. Lee, D., Park, J., Pulshashi, I.R., Bae, H.: Clustering and operation analysis for assembly blocks using process mining in shipbuilding industry. In: Song, M., Wynn, M.T., Liu, J. (eds.) Asia Pacific Business Process Management, pp. 67–80. Springer, Cham (2013)
13. Longo, L., et al.: Explainable artificial intelligence (XAI) 2.0: a manifesto of open challenges and interdisciplinary research directions. Inf. Fusion **106**, 102301 (2024)
14. Nauta, M., et al.: From anecdotal evidence to quantitative evaluation methods: a systematic review on evaluating explainable AI. ACM Comput. Surv. **55**(13s) (2023)
15. Neubauer, T.R., Pamponet Sobrinho, G., Fantinato, M., Peres, S.M.: Visualization for enabling human-in-the-loop in trace clustering-based process mining tasks. In: 2021 IEEE International Conference on Big Data (Big Data), pp. 3548–3556 (2021)
16. Nguyen, H., Dumas, M., Rosa, M., Hofstede, A.H.M.: Multi-perspective comparison of business process variants based on event logs. In: Trujillo, J.C., et al. (eds.) Conceptual Modeling, pp. 449–459. Springer, Cham (2018)
17. Oyamada, R.S., Tavares, G.M., Barbon Jr., S., Ceravolo, P.: A scikit-learn extension dedicated to process mining purposes. In: CEUR Workshop Proceedings, vol. 3552, pp. 11–15. CEUR-WS (2023)
18. Rožanec, J.M., Šircelj, B., Fortuna, B., Mladenić, D.: Adaptive explainable artificial intelligence for visual defect inspection. Procedia Comput. Sci. **232**, 3034–3043 (2024)
19. Rubensson, C., Mendling, J., Weidlich, M.: Variants of variants: context-based variant analysis for process mining. In: Guizzardi, G., Santoro, F., Mouratidis, H., Soffer, P. (eds.) Advanced Information Systems Engineering, pp. 387–402. Springer, Cham (2024)
20. Tavares, G.M., Oyamada, R.S., Barbon, S., Ceravolo, P.: Trace encoding in process mining: a survey and benchmarking. Eng. Appl. Artif. Intell. **126**, 107028 (2023)
21. Tavares, G.M., Barbon Junior, S., Damiani, E., Ceravolo, P.: Selecting optimal trace clustering pipelines with meta-learning. In: Xavier-Junior, J.C., Rios, R.A. (eds.) Intelligent Systems, pp. 150–164. Springer, Cham (2022)
22. Taymouri, F., Rosa, M.L., Dumas, M., Maggi, F.M.: Business process variant analysis: survey and classification. Knowl.-Based Syst. **211**, 106557 (2021)
23. Viton, F., Elbattah, M., Guérin, J.L., Dequen, G.: Heatmaps for visual explainability of CNN-based predictions for multivariate time series with application to healthcare. In: 2020 IEEE International Conference on Healthcare Informatics (ICHI), pp. 1–8. IEEE (2020)

Evaluating Knowledge Generation and Self-refinement Strategies for LLM-Based Column Type Annotation

Keti Korini(✉) and Christian Bizer

Data and Web Science Group, University of Mannheim, Mannheim, Germany
{kkorini,christian.bizer}@uni-mannheim.de

Abstract. Understanding the semantics of columns in relational tables is an important pre-processing step for indexing data lakes in order to provide rich data search. An approach to establishing such understanding is column type annotation (CTA) where the goal is to annotate table columns with terms from a given vocabulary. This paper experimentally compares knowledge generation and self-refinement strategies for LLM-based CTA. The strategies include using LLMs to generate term definitions, error-based refinement of term definitions, and fine-tuning using examples and term definitions. We evaluate these strategies along two dimensions: effectiveness measured as F1 performance and efficiency measured in terms of token usage and cost. Our experiments show that using training data to generate label definitions outperforms using the same data as demonstrations for in-context learning for two out of three datasets using OpenAI models. The experiments further show that using the LLMs to refine label definitions brings an average increase of 3.9% F1 in most setups compared to the performance of the non-refined definitions. Combining fine-tuned models with self-refined term definitions results in the overall highest performance for gpt-4o, outperforming zero-shot prompting fine-tuned models by at least 3% in F1. The cost analysis shows that self-refinement via prompting is more cost-efficient than fine-tuning for use cases requiring smaller amounts of tables to be annotated.

Keywords: Table Annotation · Column Type Annotation · Large Language Models · Self-refinement · Fine-tuning

1 Introduction

With the growing number of datasets that are collected within corporate data lakes, data search and discovery methods are increasingly becoming critical to enable organizations to take advantage of the collected data [2,4]. However, identifying relevant data faces challenges such as schema heterogeneity [16,21] and lack of table metadata [2]. Annotating tabular data with terms from a single vocabulary can help alleviate these challenges [25]. Tasks that target the understanding of table semantics are grouped under the concept of *table interpretation* [18]. Column Type Annotation (CTA) is a sub-task of table interpretation

P. K. Chrysanthis et al. (Eds.): ADBIS 2025, LNCS 16043, pp. 111–127, 2026.
https://doi.org/10.1007/978-3-032-05281-0_8

where the goal is to link columns in a table to semantic types from a pre-defined vocabulary. For example, in the table in Fig. 1, the first column can be linked to the label "RecipeName" as it contains values that describe the names of different recipes.

RecipeName	RestrictedDiet	Duration
???	???	???
Asparagus and Arugula Salad	GlutenFreeDiet	PT30M
Dark/White Chocolate Ice Cream	VegetarianDiet	PT0H4M
Cheesy Baked Zucchini Fries	GlutenFreeDiet	PT30M

Fig. 1. An annotated table with column types. The CTA labels are shown above.

To tackle the task of CTA, different works build deep learning based methods [3,9], leverage knowledge graphs (KGs) [11,19,22], or fine-tune pre-trained language models (PLMs) [6,23,24]. With the recent advancements in natural language processing, large language models (LLMs) have been used in different setups on the task of CTA and have shown promise in reducing the need for large amounts of labeled training data which the previous methods required. In these works, prompting techniques are explored [12,13], as well as fine-tuning LLMs for the task of CTA or for several table tasks [8,15,29].

While previous papers [15,28] focused on training generic models to enhance performance across multiple table tasks, we focus on testing two methods to adapt LLMs to the domain of a specific dataset: knowledge generation prompting [17] and self-refinement [20]. As vocabulary terms can have different meanings and usages in different datasets, we experiment with knowledge generation by prompting an LLM to generate definitions of the vocabulary terms with the idea to guide the generated annotation with these definitions. To further adapt the generated definitions, we test a self-refinement method where the LLM refines and adapts the definitions using errors made during the annotation of a validation set. As in real-world use cases labeled data is expensive to obtain and table metadata can be missing, we focus on a scenario in which only a small amount of training data and no table metadata is available. For this scenario, we choose three datasets which cover different topical domains and both multi-class and multi-label CTA: Limaye [16], WikiTURL [6], and SOTAB V2 [14]. We evaluate our methods using two open-source models *Llama-3.1-8B-4-bit* and *Llama-3.1-70B-4-bit*, and two closed-source models *gpt-4o-mini-2024-07-18* and *gpt-4o-2024-03-15*. The main findings of our experiments are:

- **Knowledge generation prompting improves the performance of OpenAI models compared to five-shot prompting**: Generating label definitions using demonstrations from the training set improves the performance of the CTA task on the OpenAI models for 2 out of 3 datasets compared to using the same demonstrations for five-shot in-context learning. This

performance gain comes at the cost of a 1.3- to 6-fold increase in the number of tokens compared to five-shot prompting.
- **Self-refinement increases the performance of most models tested**: We explore refining the generated label definitions by using errors made when classifying the validation set. This refinement leads in most cases to an average increase of 3.9% in F1 score compared to non-refined definitions. However, the additional refinement step increases the amount of tokens that are required for generating the definitions by factor 40.
- **Fine-tuning LLMs for CTA is more token efficient for large use cases**: Fine-tuning brings an increase of at least 6.3% in F1 score in 7 out of 8 cases tested. While the token usage is high during the fine-tuning process, the number of tokens required during inference is 6 to 10 times lower in the fine-tuning scenario compared to the self-refinement scenario without fine-tuning. This means that for use cases requiring a large amount of tables to be annotated, fine-tuning is more efficient than self-refinement via prompting.
- **Combining fine-tuning and self-refinement is the best performing method for gpt-4o**: By generating and refining label definitions with the fine-tuned models, we find that using the fine-tuned *gpt-4o* model in combination with the refined definitions via prompting brings an additional 3% increase when compared to using the fine-tuned model with zero-shot prompting in the two datasets that we use in our fine-tuning experiments. This improvement comes at the cost of a 4- to 8-fold increase in token usage.

The paper is structured as follows: Sect. 2 discusses related work. In Sect. 3, we describe our experimental setup. In Sect. 4, we test zero-shot, few-shot and self-consistency prompting to establish baselines. In Sect. 5, we test knowledge generation prompting and self-refinement. In Sect. 6, we evaluate different fine-tuning setups. Section 7 summarizes our conclusions.

2 Related Work

We organize the related work section in three parts: table annotation, column type annotation, and LLM-based column type annotation.

Table Annotation. Numerous works [7,18,25] have been published that tackle different sub-tasks of table annotation such as CTA, column property annotation (CPA) and cell entity annotation (CEA). A line of these works [1,19,22] leverage knowledge bases (KBs) to discover table semantics by linking entities in a table to entities in KBs and determining the column type and relationships by looking at the type and properties of the entity in the KB. The Semantic Web Challenge on Tabular Data to Knowledge Graph Matching (SemTab) [11] is a challenge designed to yearly benchmark table annotation systems, like DAGOBAH [19] and MTab [22], on different datasets. Other works rely on building table representation models. In TURL [6] a TinyBERT model is further pre-trained on table corpora and learns cell representations which are then used on the tasks of CTA, CPA and CEA. TABBIE [10] uses two Transformers to learn row and column representations independently by further pre-training BERT with a cell corruption detection objective.

Column Type Annotation. Early CTA systems that use deep learning frameworks, like Sherlock [9], employ feature engineering to construct statistical and character-level features from column values. Later works, use PLMs in their methods. SeLab [24] fine-tunes a BERT model, using a column as an input sequence, while in Doduo [23] a BERT model is fined-tuned using multi-task learning for CTA and CPA and one table is treated as one input sequence. TorchicTab [5], uses Doduo as a base architecture and extends it by making it possible to include non-labeled columns as context and handle wide tables.

LLMs for CTA. In [13] prompt designs that include instructions and different formulations of the CTA task are explored using ChatGPT. Chorus [12] explores three table tasks including CTA and the idea of "anchoring" predictions, where out-of-vocabulary answers are embedded and mapped to the closest labels in the label space. In ArcheType [8] zero-shot and fine-tuning a Llama-7B model is explored on the task of CTA. The authors evaluate different ways of sampling rows, serializing tables and propose their own method of remapping out-of-vocabulary answers. In RACOON [27], a retrieval augmented method is explored where prompts are augmented with relevant information retrieved from a KB. In Table-GPT [15], the authors aim at creating a generalist model to increase the performance across multiple table tasks. They fine-tune ChatGPT on table structure understanding and on a variety of table tasks such as schema and entity matching and data imputation. In TableLlama [29] a Llama-7B model is fine-tuned on six table tasks, including CTA and CPA. In Jellyfish [28] a Llama-2-13B model is fine-tuned on four data pre-processing tasks and the model is tested on CTA. In contrast to these papers which aim at learning models that generalize to different tasks, we focus on adapting LLM-based methods to the way labels are used in the context of specific data sets. For this, we investigate two methods which have not yet been employed for the CTA task: knowledge generation prompting [17] and self-refinement [20].

3 Experimental Setup

This section introduces the datasets and models that we use in the experiments as well as the evaluation metrics. The code, prompts, data, and responses of the LLMs are available in the Github repository[1] accompanying this paper.

Datasets. In our experiments we use three datasets: SOTAB V2 CTA [14][2], the WikiTables-TURL [6] dataset, which we will refer as WikiTURL throughout the paper, and the Limaye [16] dataset. *SOTAB V2 CTA* is a large heterogeneous CTA dataset with tables that were collected from 44,268 different websites and are spread over 17 topical domains. Altogether, the SOTAB V2 dataset contains

[1] https://github.com/wbsg-uni-mannheim/TabAnnGPT.
[2] https://webdatacommons.org/structureddata/sotab/v2/.

Table 1. Datasets' statistics: number of tables in the training, validation and test splits, the average number of annotated columns per table (Avg. Columns), number of labels (Labels) and the number of annotated columns in the test set (Test Columns).

Dataset	Train	Validation	Test	Avg. Columns	Labels	Test Columns
SOTAB V2	44K	456	609	2.62	82	1,851
SOTAB V2-ds	698	199	239	2.80	50	824
WikiTURL	397K	4,8K	4,7K	1.60	255	13K
WikiTURL-ds	809	416	379	1.44	66	878
Limaye	105	-	107	1	26	107

over 40,000 training tables. For our experiments, we down-sample the dataset by choosing five domains out of the 17: book, recipe, restaurant, movie, and product. To work in the scenario with limited amounts of training data, we reduce the number of examples in the training set to 20 columns per label resulting in 698 training tables (see Table 1). *WikiTURL CTA* is a large dataset created by Deng et al. [6]. Similarly to SOTAB V2, we down-sample WikiTURL by limiting tables to five domains: film, music, book, food and broadcast and limiting the amount of training data to 20 columns per label resulting in 809 training tables. The last dataset we use is the *Limaye* dataset [16]. The tables are gathered from Wikipedia pages from domains such as book, person, film etc. and we use the version annotated by Efthymiou et al. [7]. We split the 212 total tables into a training and test set. The main statistics of all datasets are summarized in Table 1. For SOTAB V2 and WikiTURL, we show the statistics for both the full and down-sampled sets (*-ds* in the table). WikiTURL and Limaye include only textual columns, while in SOTAB V2 9% of the columns are numerical while the rest are textual.

Models. To test our methods we select two open-source models, *Llama-3.1-8B-4-bit*[3] and *Llama-3.1-70B-4-bit*[4] and two OpenAI models, *gpt-4o-mini-2024-07-18* and *gpt-4o-2024-03-15*. For the fine-tuning experiments with gpt-4o we use the available *gpt-4o-2024-08-06* model. For brevity, in the paper we will refer to these models as Llama-8B, Llama-70B, gpt-mini and gpt-4o respectively. For fine-tuning the open-source models we use the Unsloth[5] and Huggingface[6] libraries. We set the temperature of the models to 0 for OpenAI models and 0.001 for the Llama-3.1 models to reduce randomness.

Serialization. The tables in the prompts are serialized in markdown format. We include only the first five rows of a table in the prompts and restrict the number of words in cells to 20. For Limaye and SOTAB V2 where context columns (columns not annotated) are available, they are included in prompts and we

[3] https://huggingface.co/unsloth/Meta-Llama-3.1-8B-Instruct-bnb-4bit.
[4] https://huggingface.co/unsloth/Meta-Llama-3.1-70B-Instruct-bnb-4bit.
[5] https://docs.unsloth.ai/.
[6] https://github.com/huggingface/transformers.

ask for the classification only of the columns in the test set. In WikiTURL and Limaye, where column headers and table descriptions are available, we do not use this information as we focus on a scenario with missing metadata. We therefore substitute the column names to "Column 1", "Column 2", etc.

Evaluation Metrics. The CTA task is treated as a multi-class classification problem in SOTAB V2 and as a multi-label problem in WikiTURL and Limaye. All datasets have class imbalance and for this reason we use as our main metric the Micro-F1 score. Open-source models are run 3 times and we report the average metrics. Similarly, for our fine-tuning experiments, we run fine-tuning with open-source models three times with three different seeds and report the average metrics. We treat out-of-vocabulary answers as errors. In addition to F1, we report precision and recall for all experiments in the GitHub repository.

4 Zero-Shot, Few-Shot and Self-consistency Baselines

This section establishes baselines for the comparison of our experimental results. The first baseline is *zero-shot* prompting. A zero-shot prompt for the CTA task includes a *task description* where the label set is presented, *task instructions* on how to carry the annotation and the JSON format the model should respond with, and the *input table*. For Llama models, we do not include task instructions in the prompt as initial experiments showed that instructions harmed the performance of these models. An example of a zero-shot prompt is shown in Fig. 2. The second baseline is *few-shot* prompting, where we present to the model 5 demonstrations picked from the training set based on the similarity between test and training tables. To compute this similarity, we embed the test and training tables with the *text-embedding-3-small* model, without including the ground truth labels. We then compute the cosine similarity between the embedding vectors and choose for each test table the 5 most similar training tables to use as demonstrations to the LLM. Our third baseline is *self-consistency* introduced by Wang et al. [26]. This technique consists of prompting the model multiple times with different temperature values to generate different answers and use majority voting for selecting the final answer. In our experiments, we run the LLMs 3 times with temperatures: 0, 0.5 and 0.7.

Results. The results of the baseline experiments are shown in Table 2. Observing the WikiTURL label set, we notice that the labels follow a hierarchy, therefore we manually construct a label hierarchy without using the ground truth, and experiment by including this hierarchy information in the prompts (denoted as *-hier* in the table). We show a small part of the hierarchy in the task description in Fig. 2. We notice that by including the hierarchy information for all models we have large increases from 3.3–16.8% in F1 score in the zero-shot setting, with the larger increases (>10% F1) for Llama-8B and gpt-4o. Although the increase becomes smaller in the few-shot prompting case, the models benefit with an additional 1.3–3.2% F1. A general pattern that we observe is that in SOTAB V2 and WikiTURL, including 5 demonstrations improves the performance of the

 # System message (TASK DESCRIPTION)
Your task is to classify the columns of a given table with only one of the following classes that are separated with comma: tv character, written work, astronaut, award organization, politician, film genre, ...
The labels follow this hierarchy from lowest to highest: {'music producer': ['music producer', 'person'], 'music composer': ['music composer', 'person'], 'book': ['book', 'written work'], ... }

 # System message (TASK INSTRUCTIONS)
Your instructions are: 1. Look at the cell values in detail. The first row of the table corresponds to the column names. 2. For each column, select one or more label/s that best represents the meaning of all cells in the column. The column can have multiple labels that have the same semantic meaning. 3. Answer with the selected label/s for each column using the JSON format {column_name: [label/s]}. 4. Answer only with labels from the provided label set!

 # Human message (INPUT TABLE)
```
Classify these table columns: | Column 1  | Column 2 |
|:---------------------------|:-------------------------|
| " Achilles Last Stand "  | Jimmy Page , Robert Plant |
| " All My Love "          | John Paul Jones , Plant  |
```

 # AI message (LLM RESPONSE)
{ "Column 1": ['written work', 'music album'], "Column 2': ['music artist']}

Fig. 2. Example of a zero-shot prompt.

smaller Llama-8B model over the zero-shot scenario. On the other hand, when these demonstrations are fed to larger models, the F1 score is either similar or less than the zero-shot scenario. In contrast, in the Limaye dataset in all cases when we include 5-shots the F1 increases to around 90%. To understand the difference in this behavior, we calculate the average cosine similarity of the demonstrations we include in the prompts for all datasets. We find out that the average similarity of the test tables to the 5 selected demonstrations is higher in Limaye, with an average cosine similarity of 0.755 than in SOTAB V2 and WikiTURL where the similarity is 0.654 and 0.648 respectively, potentially being the reason for the smaller F1 increase. Regarding self-consistency, we notice that in most cases the results are similar to zero-shot prompting.

Costs. In the second half of Table 2, we report the total number of tokens and the cost required for annotating all tables in the test set using *gpt-4o-2024-08-06*[7] and the inference cost per single annotated column (Costs/Col.). From the perspective of the F1 score, self-consistency and zero-shot prompting have small differences between them, but on the token perspective, the inference tokens are 3 times higher for self-consistency, making this technique not efficient in both cost and F1 score. Similarly, in the 5-shot setup we have small changes in F1 score in 2 out of 3 datasets while the token usage increases by at least 1.3 times.

5 Self-refinement via Knowledge Generation Prompting

Liu et al. [17] introduce the concept of *generated knowledge prompting*, where knowledge facts are generated in an inital step and are then incorporated into

[7] $2.5/1M input tokens (as of January 2025) https://openai.com/api/pricing/.

Table 2. Baselines' Micro-F1 results and their inference costs for gpt-4o: number of inference tokens (Tokens) and costs (Cost), and annotation cost per column (Cost/Col.).

Dataset	Setup	Micro-F1 Results				Inference Costs		
		Ll-8B	Ll-70B	gpt-mini	gpt-4o	Tokens	Cost	Cost/Col.
SOTAB V2	0-shot	56.0	67.4	69.4	80.9	270K	$0.87	$0.001
	self-con.	55.2	67.5	70.6	80.0	810K	$2.61	$0.004
	5-shot	59.3	64.6	63.4	81.8	1,047K	$2.73	$0.004
Limaye	0-shot	66.0	76.7	76.5	82.6	33K	$0.09	$0.0008
	self-con.	66.7	77.8	80.8	84.3	99K	$0.28	$0.002
	5-shot	90.2	81.5	91.2	89.4	90K	$0.23	$0.002
WikiTURL	0-shot	15.0	54.0	53.3	55.3	–	–	–
	5-shot	40.4	54.8	56.0	61.5	–	–	–
	0-hier	31.8	57.3	61.0	70.0	476K	$1.29	$0.001
	self-con.	36.3	58.4	61.6	69.2	1,428K	$3.89	$0.003
	5-hier	43.6	56.1	58.5	63.7	625K	$1.63	$0.002

the prompt to support the LLM to complete its task. Inspired by this work, in this section we experiment with the idea of augmenting our prompts with LLM-generated label definitions and guiding the LLM towards more dataset-specific annotation. A label definition is one or more sentences that describe the meaning of a label in the dataset vocabulary and what values it can be used to annotate. The labels can have different, more specific meanings in different datasets influenced by the annotators' guidelines used to create the dataset. These specific meanings can be unknown to the LLMs. Motivated by the idea that we can guide the generation of the annotation based on the meaning of the labels, we test augmenting our prompts with LLM-generated label definitions.

We experiment with four types of label definitions: **(1) Initial Definitions**: The initial definitions serve as our baseline definitions and we generate them by prompting the LLM to give a definition for each label in the label set. They, therefore, contain the knowledge that the LLM has gained about the terms during its pre-training. An example of the prompt used to generate the initial definition for the label "BookName" is shown on the left side of Fig. 3. **(2) Demonstration Definitions**: With the idea to fit the definitions to the meaning that the labels have in a specific dataset, we generate definitions based on demonstrations from the training set. This means that when prompting the LLM, for each label we show 3 random demonstrations of the label and ask the model to generate the definition based on the shown demonstrations (shown on the left side of Fig. 3). These definitions reflect, therefore, not only the background knowledge of the LLM about the term, but also the specific usage of the label in the dataset. **(3) Error-based Comparative Definitions**: In this setup, motivated by the existence of labels that have close semantic meanings, we test the idea of gen-

Initial definitions

 # System message
Your task is to generate definitions about some terms that can help in the task of column type annotation.

 # Human message
Generate some definitions about the term 'BookName'.

Demonstration definitions

 # System message
Your task is to generate definitions about some terms that can help in the task of column type annotation.

 # Human message
Generate a definition about the term 'BookName' using the following examples:

BookName | BookFormat | ...
A Handbook for Morning Time | Paperback |...
...

Refined definitions

 # System message
You are a model that updates the definitions of labels based on errors that were done by a model.
The previous definition for label 'BookDescription' is the following: A "BookDescription" is a textual summary or overview of the content, ...

***** Correct way of using label 'BookDescription' *****
Column 1 | Column 2 | ... |
The Hollywood History Of The World | Quarto sized volume, illustrated throughout with hundreds of photographs...|
...
In the table above, **Column 2 is an example of label 'BookDescription'**

***** Incorrect ways of using the label 'BookDescription' *****
Column 1 | Column 2 | ... |
One More Chance | Talking about the book design, I find the flipping of the book to read the second novel a very wonderful...|...|
...
In the table above, **Column 2 was wrongly classified as 'BookDescription ' but the correct label is 'Review'**

 # Human message
Update or rewrite the old definition of the 'BookDescription' label so that the wrong classifications do not happen again. Reply only with the new definition!

Fig. 3. Prompts for initial, demonstration and refined definition generation.

erating comparative definitions that contain tips on how to distinguish different labels from each other. We generate the comparative definitions only for labels where errors were made when conducting CTA on the validation set. This setup consists of three steps: (i) Classification of the validation set using zero-shot prompting. (ii) Extract the columns that were wrongly annotated and group them under their correct label. (iii) For each label we feed the LLM the false positives and false negatives of the labels, 3 random demonstrations of the label from the training set in order to inform the model of the correct usage of the label, and we ask the model to generate tips on how to distinguish the label from the other labels that it was confused with. For example, if a column is labeled as *RecipeDescription* but its correct label was *Review*, we ask the model to generate tips of how to distinguish between the label *Review* and *RecipeDescription*. We show an example of the prompt for this generation in our repository. **(4). Error-based Refined Definitions.** In this setup we test the idea of the model self-refining the *demonstration* definitions based on errors of LLMs to further adapt the previously generated definitions to the examples of the dataset that the LLM annotates wrongly. It follows 3 steps: (i) First, the model classifies the validation set using prompts augmented with the *demonstration definitions.* (ii) Second, we gather the errors that were made for each label. (iii) Finally, for each label, we show an LLM the false positives and false negatives of the label, 3 random demonstrations from the training set to show the model the correct way of using the label, the demonstration definition for the label, and we ask the model to update the demonstration definition based on the information given (shown in Fig. 3 on the right).

Generation Details. We use *gpt-4o-2024-03-15* to generate and refine definitions. To test the definitions, we use our zero-shot prompts and add the label definitions in the *task description* after we present the label set. For the OpenAI models we include all the definitions in the prompt, while for the Llama models we include only the 10 most similar definitions to the test tables. The similarity is calculated by embedding the definitions as well as the test tables without the ground truth, and selecting for each test table 10 definitions that are most similar by cosine similarity. The model used for generating the embeddings is OpenAI's *text-embedding-3-small* model. This decision is influenced by initial experiments where the inclusion of all definitions in the prompt for the Llama models damaged the performance of CTA. When comparing the three sets of definitions, we notice that *initial* and *demonstration* definitions contain a descriptive sentence for the label and include a low amount of example values, while the *comparative* definitions explain a label mostly based on example values and comparisons to other labels making these definitions longer in length. All generated definitions are provided in our repository.

Table 3. Micro-F1 results across datasets and definition strategies, along with generation (Gen.) and inference (Inf.) costs and annotation cost per column for gpt-4o.

Dataset	Setup	Micro-F1 Results				Costs		
		Ll-8B	Ll-70B	gpt-mini	gpt-4o	Gen.	Inf.	Cost/Col.
SOTAB V2	initial	55.0	64.3	71.7	79.3	–	–	–
	demos.	58.3	70.4	72.9	82.6	\$0.07	\$3.72	\$0.006
	refined	**59.8**	**72.0**	**75.1**	**85.4**	\$3.50	\$4.27	\$0.007
	compar.	55.4	64.1	70.3	83.7	\$1.10	\$8.61	\$0.014
	*Δ*0-shot	+3.8	+4.6	+5.7	+4.5	–	–	–
	*Δ*5-shot	+0.5	+7.4	+11.7	+3.6	–	–	–
Limaye	initial	63.6	77.8	75.7	82.8	–	–	–
	demos.	66.1	81.6	77.7	86.8	\$0.01	\$0.54	\$0.005
	refined	**79.1**	**84.9**	<u>80.2</u>	**88.4**	\$0.53	\$0.65	\$0.006
	compar.	77.3	76.6	**80.5**	85.0	\$0.10	\$0.68	\$0.006
	*Δ*0-shot	+13.1	+8.2	+4.0	+5.8	–	–	–
	*Δ*5-shot	−11.1	+3.4	−10.7	−1.0	–	–	–
WikiTURL	initial	29.8	47.5	60.4	69.8	–	–	–
	demos.	35.3	**51.5**	61.6	72.2	\$0.04	\$7.43	\$0.010
	refined	**42.6**	49.3	**65.5**	**72.4**	\$8.36	\$12.1	\$0.016
	compar.	33.2	48.8	62.2	71.6	\$1.61	\$14.1	\$0.019
	*Δ*0-shot	+10.8	-5.8	+4.5	+2.4	–	–	–
	*Δ*5-shot	−1.0	-4.6	+7.0	+8.6	–	–	–

Results. The results of using our four types of definitions on the three datasets are listed in Table 3. The results show that the inclusion of the definitions generated only from the background knowledge of the model harms the performance of all the tested models in three datasets when compared to the zero-shot baseline. On the other hand, in most cases including the *demonstration* definitions in the prompts increase the F1 score by an average of 2.4% compared to 0-shot, while refining these definitions bring an additional average increase of 3.9% to the F1 score compared to the *demonstration* definitions. Overall, in 10 out of 12 cases, the *refined* definitions help the models reach the highest F1 score out of all other definitions. In contrast to our idea that *comparative* definitions can help the models distinguish better between the different labels, *comparative* definitions do not reach a higher F1 score than *refined* definitions. Comparing 5-shot prompting to the *refined* definitions (Δ5-shot in the tables), for the larger models *gpt-mini* and *gpt-4o*, the refined definitions bring an increase of at least 3.6% compared to few-shot prompting for SOTAB V2 and WikiTURL. For the Limaye dataset, this pattern can not be found as for 3 out of 4 models, the definitions are not better performing than few-shot prompting.

Error Analysis. To gain more insight into the differences between the methods, we conduct an error analysis of the results of zero-shot, five-shot prompting and error-based refined definitions using the *gpt-4o* model on the SOTAB V2 dataset. Overall, we notice that the errors follow a long-tail distribution, where at least 90% of the labels have 5 or less errors. The major error types (>10% of errors) are three: (i) Semantically similar labels: the errors in this category are columns that are annotated with a closely related label, such as boolean columns mistaken for numeric columns. (ii) Misusage of labels: in this category the model confuses the meaning of the label and uses it to annotate columns that are not related to the label. An example in this category is the model using the "*RestrictedDiet*" label to annotate columns that show the age rating of a movie. (iii) Disregarding context columns: the errors in this category can be solved by paying attention to the neighboring columns, for example columns showing weight information are followed by measurement columns which can distinguish them from distance columns. When we compare the 0 and 5-shot setups, even though overall the errors for each label decrease in the 5-shot case, the increase of errors in the "*Duration*" label (+14 errors) makes the final difference between both methods only 0.9% F1. The demonstrations in this case did not help the LLM as the embedding based search picked "*Time*" columns as demonstrations. When we compare five-shot prompting to using refined definitions, we notice that the latter improves 52% of the wrong annotations of five-shot prompting, most of which fall into the error category of semantically similar labels, aided by the definitions of the labels. However, using the refined definitions wrongfully changed 39% of the five-shot annotations, 10% of which are attributed to a wrongly generated definition for the "*Time*" label, while in 11 cases the model disregards the definitions. The full set of errors are reported in our repository.

Costs. We report the tokens of generating and using *demonstration*, *refined* and *comparative* definitions for the three datasets in the second half of Table 3 for *gpt-*

4o-2024-08-06. We report costs separated into generation and inference costs, the former including the cost used to generate the definitions and the classification of the validation set, and the latter including the cost for classifying the test sets. While the cost per annotated column is similar for *demonstration* and *refined* definitions, the latter brings an 2.8% increase in F1. This increase comes with a higher generation cost. On the other hand, *comparative* definitions have a high inference cost due to their length, making this method not useful in applications with large number of tables.

6 Fine-Tuning

In this section, we experiment with fine-tuning and investigate whether it is beneficial to use the previously generated definitions within the fine-tuning process. We compare integrating the definitions to the simple fine-tuning scenario, where we fine-tune the LLMs only on the CTA task and check if this simple fine-tuning is enough to adapt the model to the datasets' vocabularies.

We build two different training sets for fine-tuning: **1) Simple fine-tuning:** In the simple fine-tuning set, we fine-tune the models on the CTA task using our zero-shot prompts shown in Fig. 2. Based on initial experiments, we leave out the instructions when we fine-tune the models as they do not impact the results of CTA and keep the instructions only when fine-tuning the OpenAI models for SOTAB V2. **2) Multi-task fine-tuning:** In this set, we combine the tasks of CTA and definition generation. We combine the *simple* set with a *definition generation* set, where the model should learn to generate the *demonstration* definitions that we generated in Sect. 5. For example, for the SOTAB V2 dataset, this fine-tuning set has a total of 748 examples, 698 of which are the annotated tables of the CTA training set (shown in Table 1), while 50 are the definition generations which correspond to the total number of labels in the SOTAB V2 dataset. In our experiments, we also test a version of the multi-task training set where we include some demonstrations in the definition generation prompt. We name this variation as *multi-task with demonstrations*. For the CTA task, the model should generate the answer in the JSON format {column_name: label}, while definition generation is a text generation task.

To test the fine-tuned models, we use two prompting techniques: **zero-shot** and **knowledge generation** prompting. For the former, we use the zero-shot prompt in Fig. 2, while for the latter we use the methods described in Sect. 5 to generate *demonstration*, *comparative* and *refined* definitions. For this generation, we use the corresponding *gpt-4o* model fine-tuned on the best performing set to generate the definitions for each dataset. The motivation of testing knowledge prompting is to investigate whether augmenting prompts with generated definitions can help further adapt the fine-tuned models to the dataset vocabularies, or if the knowledge is already gained through the fine-tuning process.

Fine-Tuning Details. In our hyperparameter search we test two learning rates [2e-5, 1e-4] as well as three LoRA ranks and alphas [8, 16, 32]. Following this hyperparameter search, we fine-tune the Llama models using a learning rate of

Table 4. Fine-tuning results on SOTAB V2 and WikiTURL using zero and knowledge prompting. We report the differences of the best zero-shot prompting result to the best prompting with definitions results of fine-tuned models (Δ 0-shot fine-tuned).

Setup	SOTAB V2				WikiTURL			
	Ll-8B	Ll-70B	mini	gpt-4o	Ll-8B	Ll-70B	mini	gpt-4o
Zero-shot Prompting								
simple fine-tuning	77.9	86.4	87.0	87.8	62.5	64.3	**67.3**	71.1
multi-task fine-tuning	77.3	**87.8**	87.0	-	**64.1**	**66.4**	64.4	-
multi-task-with-demos	**80.7**	86.9	-	-	63.1	60.2	-	-
Δ 0-shot (not fine-tuned)	+24.7	+20.4	+17.6	+6.9	+32.3	+9.1	+6.3	+1.1
Prompting with Definitions								
FT + comparative defs	80.8	85.8	**89.1**	90.0	**70.1**	**65.6**	65.4	**74.1**
FT + demonstration defs	80.3	85.6	87.0	89.1	65.0	65.0	**66.4**	72.3
FT + refined defs	**81.3**	**86.1**	87.2	**91.8**	65.1	64.4	63.3	<u>74.0</u>
Δ 0-shot fine-tuned	+0.6	−1.7	+2.1	+4.0	+6.0	-0.8	-0.9	+3.0

1e-4, a batch size of 16 for the Llama-8B model and a batch size of 8 for the larger Llama-70B model. We train the models for 10 epochs using a maximum input length of 5020 and a LoRA rank and alpha of 32. The OpenAI models are fine-tuned using the default parameters for 3 epochs and a batch size of 1.

Results. The results of fine-tuning are reported in Table 4. In the first part of the table we present the results of fine-tuning LLMs using our three fine-tuning sets and testing them using zero-shot prompting. Fine-tuning all models brings a large increase in the metrics for both datasets, from 6.3% up to 32.3% in the F1 score. The largest increases (>20%) can be noticed for the Llama-8B model on both datasets. There is one exception for the WikiTURL dataset and the *gpt-4o* model, where the increase from the zero-shot scenario is only 1.1%. This small increase can be attributed to the multi-label problem, where in the training set we have different combinations of labels than in the test set. When fine-tuning the Llama models with the *multi-task* sets there is a small increase in F1 score compared to fine-tuning with the *simple* sets, on average a 2% F1 increase on both datasets. For the OpenAI models, due to limited resources for fine-tuning the models, we test on *gpt-mini* only the *multi-task* set that brought a small increase to the Llama models. We notice from these runs that this set does not improve but can also harm the performance of the model, as seen when fine-tuning *gpt-mini* with the WikiTURL dataset. We therefore only run simple fine-tuning with the large *gpt-4o* model. In the second part of Table 4, we report the results of using knowledge prompting on the highest performing fine-tuned model for each model and dataset. From the results, we notice that we have a positive impact on the large *gpt-4o* model: for both datasets we have an increase of at least 3% in F1 when we use the error-based definitions: *comparative* and *refined* definitions. Using these definitions on the smaller fine-tuned models, does

Table 5. Token usage and cost analysis for gpt-4o for three fine-tuning scenarios.

Dataset	Setup	FT	Gen.	Inf.	FT Cost	Gen. Cost	Inf. Cost
SOTAB V2	0-shot	1,895K	–	268K	$47.4	–	$1.28
	refined	1,895K	928K	1,023K	$47.4	$3.48	$3.90
	comp.	1,895K	385K	1,383K	$47.4	$1.45	$5.26
WikiTURL	0-shot	801K	–	473K	$20.0	–	$1.88
	refined	801K	2,912K	3,799K	$20.0	$10.9	$14.3
	comp.	801K	626K	4,457K	$20.0	$2.35	$16.8

not impact the performance of the models, and we conclude that the models have already gained some knowledge about the vocabulary from fine-tuning and the additional definitions do not impact the annotation. There is one exception for the small fine-tuned 8B model on the WikiTURL dataset, when paired with *comparative* definitions the model reaches the same performance as fine-tuning the large *gpt-4o* model with the simple fine-tuning set.

Error Analysis. To understand how the inclusion of definitions further help the fine-tuned model, we analyze the errors of the zero-shot and refined definitions in the fine-tuned scenario. As in the previous section, the errors follow a long-tail distribution with only 5 labels out of 50 having more than 5 errors. Using the definitions 49% of the wrong annotations in zero-shot are corrected and we notice an improvement of the long tail labels: in the zero-shot scenario there are 21 labels with no errors, and this number improves to 30 labels when *refined* definitions are used. Overall, these improvements in the number of errors brings the 4% increase to the SOTAB V2 dataset, bringing the F1 score to 91.8%.

Costs. In Table 5, we report the token usage and costs for testing the fine-tuned *gpt-4o-2024-08-06* model in a zero-shot scenario as well as in a knowledge prompting scenario with *refined* and *comparative* definitions. We report the generation (Gen.), inference (Inf.) and fine-tuning (FT) tokens alongside their cost. The generation tokens include the tokens used for generating/refining the definitions and classifying the validation set. The inference tokens include the tokens used to classify the test set, while the fine-tuning tokens include the total tokens used in fine-tuning. The cost of fine-tuning the model is $25/1M training tokens and the usage of the fine-tuned model is $3.75/1M tokens (as of January 2025).

Fine-Tuning or Knowledge Prompting Without Fine-Tuning? Comparing the cost of previous methods from Table 3 to the 0-shot fine-tuning scenario, we observe that using knowledge prompting with *refined* definitions in a non-fine-tuning scenario has a lower total cost for SOTAB V2 and WikiTURL than the fine-tuning scenario. However, the cost per column of the 0-shot fine-tuning scenario which is 0.002 for both datasets is lower than the cost per column of the *refined* definitions which stands at 0.007 and 0.016 for SOTAB V2 and WikiTURL respectively (from Table 3). We can conclude from this that in scenarios with small datasets such as SOTAB V2 and WikiTURL, knowledge prompting is more efficient cost-wise and performance-wise reaches a similar score to fine-tuning. In scenarios with a large amount of tables and columns, 0-shot fine-tuning becomes more efficient. For example, if the total amount of columns in a dataset exceeds 9400[8], the fine-tuning scenario becomes the cheaper of the two options.

Zero-Shot or Knowledge Prompting when Using Fine-Tuned Models? The highest F1 score in the fine-tuning scenario for both SOTAB V2 and WikiTURL are reached when the fine-tuned *gpt-4o* is tested using error-based definitions-augmented prompts. The sets of *comparative* and *refined* definitions have a small difference in F1 score between them, however from the inference cost we can conclude that *refined* definitions are the most efficient token-wise and more effective performance-wise between them. The cost per annotated column is also higher for *comparative* definitions, which means that in applications with large amount of tables, this method is more expensive.

7 Conclusion

In this paper, we evaluate knowledge generation and self-refinement strategies for LLM-based column type annotation. The methods that we test include the generation of label definitions with knowledge prompting and error-based self-refinement. We show that generated definitions increase the F1 score of the models in most cases by an average of 2.4% compared to zero-shot prompting, while using the error-based self-refinement method to enhance definitions brings an additional average increase of 3.9% in most cases. In our fine-tuning setup, we test integrating the label definitions into the fine-tuning process and observe that this integration results in an average increase of 2% F1 for the Llama models. We further conclude that fine-tuning the LLMs is more token efficient for use cases requiring a large amount of tables to be annotated (>9400 columns) compared to using knowledge prompting with refined definitions in the non-fine-tuning scenario. Finally, using the fine-tuned models in combination with the self-refinement method, increased the F1 score of two datasets on the *gpt-4o* model by at least 3% compared to zero-shot prompting the fine-tuned model.

References

1. Avogadro, R., Cremaschi, M., et al.: Mantistable V: a novel and efficient approach to semantic table interpretation. In: SemTab@ ISWC, pp. 79–91 (2021)
2. Chapman, A., Simperl, E., Koesten, L., Konstantinidis, G., et al.: Dataset search: a survey. VLDB J. **29**(1), 251–272 (2020)
3. Chen, J., Jiménez-Ruiz, E., Horrocks, I., Sutton, C.: Colnet: embedding the semantics of web tables for column type prediction. In: Proceedings of the AAAI Conference on Artificial Intelligence, vol. 33, pp. 29–36 (2019)

[8] Calculated using the Cost/Column and fine-tuning costs of SOTAB V2.

4. Christensen, M.P., Leventidis, A., Lissandrini, M., Rocco, L.D., et al.: Fantastic tables and where to find them: table search in semantic data lakes. In: Proceedings 28th International Conference on Extending Database Technology, EDBT 2025, Barcelona, Spain, 25–28 March 2025, pp. 397–410 (2025)
5. Dasoulas, I., Yang, D., Duan, X., Dimou, A.: Torchictab: semantic table annotation with wikidata and language models. In: Proceedings of the Semantic Web Challenge on Tabular Data to Knowledge Graph Matching (SemTab 2023) (2023)
6. Deng, X., Sun, H., Lees, A., et al.: TURL: table understanding through representation learning. Proc. VLDB Endow. **14**(3), 307–319 (2020)
7. Efthymiou, V., Hassanzadeh, O., Rodriguez-Muro, M., Christophides, V.: Matching web tables with knowledge base entities: from entity lookups to entity embeddings. In: d'Amato, C., et al. (eds.) ISWC 2017. LNCS, vol. 10587, pp. 260–277. Springer, Cham (2017). https://doi.org/10.1007/978-3-319-68288-4_16
8. Feuer, B., Liu, Y., Hegde, C., Freire, J.: Archetype: a novel framework for open-source column type annotation using large language models. arXiv preprint arXiv:2310.18208 (2023)
9. Hulsebos, M., Hu, K., Bakker, M., Zgraggen, E., et al.: Sherlock: a deep learning approach to semantic data type detection. In: Proceedings of the 25th ACM SIGKDD International Conference on Knowledge Discovery & Data Mining, pp. 1500–1508 (2019)
10. Iida, H., Thai, D., Manjunatha, V., Iyyer, M.: TABBIE: pretrained representations of tabular data. In: Proceedings of the 2021 Conference of the North American Chapter of the Association for Computational Linguistics: Human Language Technologies, pp. 3446–3456 (2021)
11. Jiménez-Ruiz, E., Hassanzadeh, O., Efthymiou, V., Chen, J., et al.: SemTab 2019: resources to benchmark tabular data to knowledge graph matching systems. In: Proceedings of The Semantic Web, pp. 514–530 (2020)
12. Kayali, M., Lykov, A., Fountalis, I., et al.: Chorus: foundation models for unified data discovery and exploration. arXiv preprint arXiv:2306.09610 (2023)
13. Korini, K., Bizer, C.: Column type annotation using chatgpt. In: Joint Proceedings of Workshops at the 49th International Conference on Very Large Data Bases (VLDB 2023) (2023)
14. Korini, K., Peeters, R., Bizer, C.: SOTAB: The WDC Schema.org Table Annotation Benchmark. Semantic Web Challenge on Tabular Data to Knowledge Graph Matching (SemTab), CEUR-WS.org (2022)
15. Li, P., He, Y., Yashar, D., Cui, W., et al.: Table-GPT: table-tuned GPT for diverse table tasks. arXiv preprint arXiv:2310.09263 (2023)
16. Limaye, G., Sarawagi, S., Chakrabarti, S.: Annotating and searching web tables using entities, types and relationships. Proc. VLDB Endow. **3**(1–2), 1338–1347 (2010)
17. Liu, J., Liu, A., Lu, X., Welleck, S., et al.: Generated knowledge prompting for commonsense reasoning. arXiv preprint arXiv:2110.08387 (2021)
18. Liu, J., Chabot, Y., Troncy, R., et al.: From tabular data to knowledge graphs: a survey of semantic table interpretation tasks and methods. J. Web Semant. **76**, 100761 (2023)
19. Liu, J., Troncy, R.: Dagobah: an end-to-end context-free tabular data semantic annotation system. SemTab@ ISWC (2019)
20. Madaan, A., Tandon, N., Gupta, P., Hallinan, S., et al.: Self-refine: iterative refinement with self-feedback. Adv. Neural. Inf. Process. Syst. **36**, 46534–46594 (2023)
21. Nargesian, F., Zhu, E., Miller, R.J., et al.: Data lake management: challenges and opportunities. Proc. VLDB Endow. **12**(12), 1986–1989 (2019)

22. Nguyen, P., Kertkeidkachorn, N., Ichise, R., et al.: Mtab: matching tabular data to knowledge graph using probability models. arXiv preprint arXiv:1910.00246 (2019)
23. Suhara, Y., Li, J., Li, Y., Zhang, D., et al.: Annotating columns with pre-trained language models. In: Proceedings of the 2022 International Conference on Management of Data, pp. 1493–1503 (2022)
24. Trabelsi, M., Cao, J., Heflin, J.: Selab: semantic labeling with bert. In: 2021 International Joint Conference on Neural Networks (IJCNN), pp. 1–8. IEEE (2021)
25. Venetis, P., Halevy, A., Madhavan, J., Pasca, M., et al.: Recovering semantics of tables on the web. Proc. VLDB Endow. **4**(9) (2011)
26. Wang, X., Wei, J., Schuurmans, D., Le, Q.V., et al.: Self-consistency improves chain of thought reasoning in language models. In: The Eleventh International Conference on Learning Representations (2023)
27. Wei, L.L., Xiao, G., Balazinska, M.: Racoon: an LLM-based framework for retrieval-augmented column type annotation with a knowledge graph. In: NeurIPS 2024 Third Table Representation Learning Workshop (2024)
28. Zhang, H., Dong, Y., Xiao, C., Oyamada, M.: Jellyfish: a large language model for data preprocessing. arXiv preprint arXiv:2312.01678 (2023)
29. Zhang, T., Yue, X., Li, Y., Sun, H.: Tablellama: towards open large generalist models for tables. In: Proceedings of the 2024 Conference of the North American Chapter of the Association for Computational Linguistics: Human Language Technologies, pp. 6024–6044 (2024)

Entity Resolution and Integration

LSBlock: A Hybrid Blocking System Combining Lexical and Semantic Similarity Search for Record Linkage

Dimitrios Karapiperis[1(✉)], Christos Tjortjis[1], and Vassilios Verykios[2]

[1] International Hellenic University, Thessaloniki, Greece
{dkarapiperis,c.tjortjis}@ihu.edu.gr
[2] Hellenic Open University, Patras, Greece
verykios@eap.gr

Abstract. Record linkage plays a crucial role in data integration by identifying and merging records that correspond to the same entity across different datasets. Traditional linkage methods are either lexically driven, relying on surface-level textual matching, or semantically driven, capturing contextual meaning. In this paper, we propose LSBlock, a hybrid blocking system that integrates lexical and semantic similarity search to enhance record linkage accuracy by improving recall while maintaining high precision. LSBlock employs minhash-based lexical similarity for efficient blocking key formulation and uses dense embeddings with cosine similarity to refine candidate matches. A Hierarchical Navigable Small-World (HNSW) index is leveraged for efficient approximate nearest neighbor searches. The retrieved results are further refined using two thresholds: one in the Jaccard space and another in the cosine similarity space, both derived after training a model using labeled matches. Our experimental evaluation on six real-world datasets demonstrates that LSBlock significantly outperforms state-of-the-art blocking techniques in terms of recall, precision, and F1-score, achieving a balanced trade-off between efficiency and effectiveness. LSBlock consistently achieves recall above 0.95 and precision above 0.65, when using structured data, outperforming its competitors by a large margin. These findings highlight the advantages of integrating lexical and semantic similarity, making LSBlock a robust solution for scalable record linkage.

1 Introduction

The explosion of data generated by modern applications has led to an increasing need for efficient data processing techniques. Among these, record linkage, which is the process of identifying and linking records that correspond to the same entity across different datasets, plays a critical role in numerous domains such as healthcare, fraud detection, census, and social media. In traditional settings, record linkage is usually performed in a batch processing mode whose outcome is the set of potentially matching record pairs.

P. K. Chrysanthis et al. (Eds.): ADBIS 2025, LNCS 16043, pp. 131–146, 2026.
https://doi.org/10.1007/978-3-032-05281-0_9

A record linkage task comprises two distinct phases: blocking and matching. The former aims to group records in blocks that share some common attributes, termed as blocking keys, which are likely to indicate a potential match. Dates of birth, or zip codes, or surnames are commonly used as blocking keys to partition records into blocks. The purpose of blocking is to mitigate the quadratic complexity of the number of record pairs, each of which originate from different data sets, that are formulated in the blocks. Unlike exact blocking, where the blocking keys should be identical, approximate blocking allows for flexibility by grouping records that are similar, accommodating minor variations such as typos, formatting differences, or slight inconsistencies. This makes it particularly useful for handling noisy or inconsistent data. During matching, these record pairs need to be resolved to determine their matching status.

Regarding blocking, numerous methods have been explored in the literature, including token blocking [20], suffix arrays [30], TF-IDF [5,25], minhash blocking [12], and Locality-Sensitive Hashing (LSH) blocking [13], which have played a dominant role in record linkage over the past decade. Schema-agnostic approaches have also emerged, which incorporate all attribute values for blocking, usually by assuming all their tokens, to enhance its effectiveness [9,21,24,32]. Although, this might be the only resort given unstructured data, in the realm of schema-aware datasets often results in low quality blocks, populated with extremely large numbers of records. To this end, several metablocking techniques [22,23] have been devised to filter out superfluous comparisons that incur high computational costs. Even constructing these low-quality blocks may be infeasible given the vast volumes of records today.

All of the above methods assume only lexical similarity, which focuses on the surface-level similarity of attribute values. It is particularly effective for identifying straightforward matches or resolving minor variations, such as typographical errors or formatting differences (e.g., *'John Smith'* vs. *'Jon Smith'*). However, relying solely on lexical similarity can lead to false positives when unrelated records coincidentally appear similar and false negatives when the same entity is represented with significantly different terms (e.g., nicknames or abbreviations). On the other hand, semantic similarity addresses these challenges by focusing on the underlying meaning of the records, enabling the identification of relationships between records even when the textual representation varies significantly, e.g., *'Proceedings of the VLDB Endowment'* and *'PVLDB'*. Pre-trained embeddings encapsulate semantic and contextual information from large corpora into dense numeric vectors. The emergence of these embeddings has revolutionized the field of record linkage, as they offer a compact, continuous representation of data attributes, capturing rich semantic and contextual information beyond lexical patterns. Static embeddings, such as Word2Vec [18], GloVe [26], and FastText [1], offer efficient representations for individual tokens by leveraging co-occurrence statistics. However, they lack context-awareness, as the same word is always mapped to the same vector, regardless of its surrounding words. Pre-trained language models, like BERT-based models [7] generate dynamic embeddings that address this limitation by creating token representations that adapt to their context. Sentence embeddings, such as SentenceBERT [27], extend this concept to sentence-level tasks, efficiently encoding entire text spans into fixed-size vectors. Several studies, including [3,4,8,14,19,29,33,34], have explored the use of static and dynamic embeddings for record linkage.

However, given the fundamental differences in objectives, it is not safe to assume that language models trained or fine-tuned for semantic similarity tasks will perform equally well in record linkage tasks. Pre-trained models typically focus on sentence-level semantics to derive holistic semantic similarity, which may overlook critical lexical differences that are essential for effective record linkage. For example, consider the following patient records from different hospitals: *Paul Jones, 12/05/1985, 123 Elm St, Type 1 Diabetes, New York'* and *John Smith, 12/05/1985, 123 Elm Street, Type 1 Diabetes, NY'.* A semantic similarity model or a schema-agnostic approach might rank these as highly similar due to their shared context and significant token overlap. However, a simple blocking approach using the name as the blocking key would place these records in separate blocks, preventing them from being considered as potential matches. Similarly, the pair *Paul Jones, 12/05/1985, 123 Elm Street, Diabetes Mellitus, NY* and *P. Jones, 12/05/1985, 125 Elm Street, Type 1 Diabetes, NJ*, detected using similar blocking keys, should not be matched due to the clinically different diseases and distinct states. Therefore, using schema-aware data, to achieve optimal performance, both in formulating high-quality blocks and achieving accurate matching, it is often necessary to combine both lexical and semantic similarity techniques.

In record linkage, the concept of *discriminative power* refers to an attribute's (or set of attributes) ability to uniquely distinguish between matching and non-matching records. Attributes with high discriminative power, such as passport numbers or author names combined with the title of a publication, provide strong indicators of identity due to their uniqueness, whereas common attributes like gender or country of residence have low discriminative power. Given large volumes of records, high-quality blocks is a key feature towards performing successful record linkage tasks. To this end, using attributes, with high discriminative power, to play the role of blocking keys improves precision by reducing false matches and increases recall by bringing together potential matches.

In this paper, we propose a hybrid system that operates using both lexical and semantic similarity search components, which operate at different levels, to perform record linkage tasks. Specifically, we use lexical similarity to formulate high-quality blocks using attributes that exhibit high discriminative power. We then filter out candidate matches using their semantic similarity to capture subtleties in meaning, phrasing, and contextual overlap between the pairs of records that have been formulated. We derive the blocking keys, using the SHAP [16] framework, whose lexical similarity relies on the Jaccard distances of their minhash representations. The semantic similarity of records is based on the cosine similarity of their embeddings using a pre-trained language model. Fundamental aspects of our proposed system are the two thresholds applied in the Jaccard space and the cosine similarity space, both of which are derived using simple neural network models. In addition to LSBlock's full version, which utilizes both components, it can also operate solely with its semantic component in scenarios where all attributes are weak or missing, such as in semi-structured or unstructured data. We conduct a detailed experimental evaluation using six real-world benchmark datasets for record linkage that provides valuable insights and conclusions. We argue that combining both components ensures that we can capture both the surface-level and deep-level similarities between records that helps improve matching accuracy, especially when dealing with noisy or inconsistent data sources. To the best of our knowledge, this is the first proposal to combine lexical and semantic similarity search, highlighting both the strengths and limitations of this composite approach.

The remainder of this paper is organized as follows. Section 2 discusses related work in both the lexical and semantic domain. Section 3 presents the proposed system. Section 4 provides a detailed experimental evaluation of our proposal along our competitors, using real-world datasets. Finally, Sect. 5 concludes with a discussion of our main findings and our future work.

2 Related Work

In the lexical spectrum, several techniques are used to efficiently group similar records based on textual attributes. Token blocking [5] splits text into tokens and groups records that share common tokens, though it can lead to large candidate sets. Suffix arrays [30] index suffixes of text attributes, enabling fast approximate string matching by identifying shared prefixes. The Bag-of-Words model represents text as an unordered set of words, and blocking is done based on common words, ignoring word order. TF-IDF [5] (Term Frequency-Inverse Document Frequency) weighs words based on their importance across the entire dataset, highlighting rare or significant terms for more precise blocking. Paulsen et al. in [25], very recently, introduced Sparkly, built on top of Apache Lucene.[1] Sparkly indexes records by tokenizing them into 3-grams and utilizes Lucene for retrieval. For each query record, it identifies the top-k most similar indexed records using a modified Okapi BM25 similarity function based on TF-IDF weights. Minhash blocking [12] relies on the estimation of the Jaccard similarity between sets by creating compact representations, commonly known as signatures, while Locality-Sensitive Hashing (LSH) blocking [13] builds on these signatures, or on binary vectors by grouping similar records into the same hash buckets with high probability. The most appealing feature of LSH is its theoretical guarantees of accuracy within the specific spaces it operates in, such as Hamming space for binary vectors and Jaccard space for minhash signatures.

The first approach to leveraging dense vectors for semantic similarity search was DeepER [8], which utilized GloVe [26] for the embedding task and Cosine LSH for indexing and querying. AutoBlock [34] extended DeepER by incorporating FastText [1] embeddings combined with cross-polytope LSH. However, it treated blocking as a classification task, requiring substantial labeled data to

[1] https://lucene.apache.org.

train its bidirectional-LSTM. DeepBlocker [29] generalized its blocking mechanism to support any language model. It leverages FastText embeddings with an autoencoder, although Word2Vec [18] and GloVe are also supported. It also considers transformer-based models with Byte Pair Encoding, offering slightly better recall at the cost of higher runtimes. DeepMatcher [19], provides a framework for deep learning-based matching algorithms that supports static embeddings such as GloVe and FastText. Recent research focuses on BERT-based models due to their dynamic and context-aware representations. EMTransformer [3] pioneers the use of BERT and its main variants—XLNet [31], RoBERTa [15], and DistilBERT [28]—in matching tasks. GNEM [4] extends EMTransformer by incorporating a graph-based method in order to capture relationships between all candidate pairs. DITTO [14] improves upon EMTransformer by combining BERT-based models with domain-specific information and data augmentation to create synthetic training data. A more recent study by Zeakis et al. [33] demonstrated the effectiveness of SentenceBERT models, achieving notable results even without fine-tuning the top attention layers.

3 LSBlock

The following describes the operation of LSBlock, which employs a hybrid blocking technique combining minhash vectors for lexical similarity of blocking keys and dense embeddings for semantic similarity of records. This approach enables the fast retrieval of candidate records, followed by a refinement step to identify the most relevant matches to a query record. Given the nature of the dataset, e.g., unstructured data, LSBlock can adapt by dropping the lexical component to perform semantic-based linkage.

3.1 Formulation of the Blocking Keys

For a random sample of labeled matches and non-matches of each paired dataset, we computed the Jaccard distances of the minhash vectors between their attributes. These distances, for each pair, were fed into a neural network model, which was trained in order to predict whether a pair is a match or a non-match. The model essentially learned the best combination of attributes' importance for detecting matches. Given this trained predictive model, the contribution of each attribute to a specific prediction was estimated using the Shapley Additive Explanations (SHAP) [16], which is a unified framework based on Shapley values, that explains the output of any machine learning model. SHAP computes the contribution of each distance (feature) to the model's prediction for each individual record. It then aggregates these contributions to provide global attributes importance. A high Shapley value for an attribute suggests that the corresponding distances strongly influence the model's prediction, aligning with the theoretical notion of high discriminative power. Conversely, attributes with low discriminative power typically receive lower Shapley values, indicating minimal influence.

3.2 Creation of the Blocking Vectors

Minhash [2] is a probabilistic technique designed to estimate the similarity between sets, often used for tasks such as near-duplicate detection. For lexical similarity, strings are represented as sets of q-grams[2] (e.g., bigrams or trigrams), which are then used by minhash to estimate their similarity according to the Jaccard metric. Minhash applies a hash function to the elements of a set to map these elements to numerical values. Subsequently, it picks the minimum value as the result of its operation. The probability of the Minhash values of two sets to be the same is equal to the Jaccard similarity of these sets. We implement the minhash technique using MurmurHash3[3], which is a non-cryptographic hash function, to map the q-grams[4] of a string to numerical hashes, in a well-distributed manner, and return the minimum of these hashes. Two sets of q-grams that correspond to a pair of strings is considered as similar if their Jaccard distance meets the specified threshold.

Although, we rely on the theoretical framework of [12] to derive the optimal number m of hashes, we adopt a different path for building our blocking vectors. Specifically, instead of using a single hash to play the role of the blocking vector, we concatenate m such hashes generated for a blocking key. The dimensionality m of these minhash vectors is derived according to the concept of *collisions* described in [12]. A collision occurs when a pair of these vectors exhibits the same minhash value at the same position. In summary, if we assume a pair of such blocking vector, we expect a certain number of collisions, which depend on their Jaccard similarity. We ensure that any deviation from the expected number of collisions remains within an allowable error margin, with a specified confidence, by repeating the minhash technique m times.

3.3 Hierarchical Navigable Small World

The Hierarchical Navigable Small World (HNSW) [17] is an efficient proximity graph-based algorithm for approximate nearest neighbor search. It constructs a multi-layer graph structure, where each layer contains a subset of connections with varying densities. The top-most layer has the sparsest connections, while the bottom layer includes all records inserted and is the densest.

The search process begins at the top layer and iteratively progresses to lower layers. At each layer, the algorithm employs a greedy traversal strategy to move closer to the nearest neighbors of the query record by selecting candidates in the current layer. This hierarchical structure significantly reduces the search space, allowing the algorithm to achieve logarithmic scaling with respect to the number of records that it contains.

[2] The set of q-grams of a string consists of its substrings of length q.

[3] https://github.com/aappleby/smhasher/blob/master/src/MurmurHash3.cpp.

[4] The set of q-grams of a record consists of its substrings of length q.

3.4 Setting the Thresholds

The chosen thresholds play an important role in the operation of LSBlock, because they essentially specify the block that contains the similar vectors to a query. Choosing an appropriate threshold depends on balancing precision and recall. A lower threshold improves precision by excluding dissimilar pairs but may reduce recall by missing true matches. Conversely, a higher threshold increases recall but risks admitting more false positives. Traditional rule-based approaches rely on manually defined distance thresholds, which may not generalize well across different datasets. To determine the optimal thresholds, we first computed pairwise Jaccard distances (and inner product similarities) between the minhash vectors of the blocking keys and the dense vectors of entire records for randomly selected labeled record pairs. We then trained two simple neural network models to predict match probabilities based on these Jaccard distances and inner product similarities. The optimal thresholds were determined by computing the precision-recall curve from the model's output probabilities, evaluating precision and recall at different thresholds, and selecting the threshold that maximized the F1-score. The thresholds that maximized the F1-score were selected, ensuring the best trade-off between false positives and false negatives.

The effectiveness of both thresholds, θ for the Jaccard space and ϕ for the cosine similarity space, depends on the quality of the minhash vectors and dense vectors, respectively. Higher-dimensional minhash vectors (e.g., $m = 120$ hashes) provide better approximations of Jaccard similarity. Similarly, using an effective model to embed records and capture their contextual relationships is essential for accurate semantic similarity estimation. These considerations ensure that θ aligns with the theoretical foundations of minhash, while ϕ meets the practical requirements for effective semantic similarity search.

3.5 Operations of LSBlock

We showcase the full version of our algorithm, utilizing both the lexical and semantic components to demonstrate its effectiveness. LSBlock leverages an HNSW index B to efficiently retrieve lexically similar blocking keys (or semantically similar dense vectors) to a given query. To further refine the search, given that the lexical component has not been dropped, LSBlock utilizes dense embeddings of both the retrieved records and the query to ensure that the final matches are not only lexically similar but also semantically meaningful. The dense vectors of each blocked record are stored in dictionary T, which may reside either in main memory or in secondary storage.

Algorithm 1, which presents the full version of LSBlock, resolves a query record q and aims to return a set mp that contains the Ids of the records that match q. The process begins with the extraction of a blocking key from the query record (bk), as shown in line 1. Using the blocking key, the algorithm generates its minhash vector mv by iteratively applying minhash functions (lines 3–4). Moreover, the dense vector dv of q is obtained by embedding q into a high-dimensional vector space using a pre-trained SentenceBERT model, which will

Algorithm 1. Resolving a query record q using the full version of LSBlock.

Input: Query record q, dimensionality m of minhash vectors, parameter k, HNSW index B (built on blocking vectors), dictionary T (maps record Ids to dense vectors of the blocked records), Jaccard distance threshold θ, similarity threshold ϕ, and an empty set mp for storing the Ids of matching records.
Output: The set mp containing Ids of records matching q.

1: $bk \leftarrow extract\_bkey(q)$ ▷ Extract blocking key from q
2: $mv \leftarrow []$
3: **for** $j = 1$ to m **do** ▷ Generate minhash vector for bk
4: $\quad mv.append(MinHash(bk, j))$
5: $dv \leftarrow generate\_embedding(q)$
6: $sim\_vector\_ids,\ distances \leftarrow B.search(mv, topk = k)$ ▷ Find the blocked records in B for the blocking key bk
7: **for each** $vector\_id, distance \in (sim\_vector\_ids,\ distances)$ **do**
8: $\quad$ **if** $distance < \theta$ **then**
9: $\quad\quad dvr \leftarrow T.get(vector\_id)$ ▷ Retrieve dense vector from T for each blocked record
10: $\quad\quad sim \leftarrow inner\_product(dv, dvr)$
11: $\quad\quad$ **if** $sim > \phi$ **then**
12: $\quad\quad\quad mp.add(match.Id)$ ▷ Add matching record Id to the result set
13: **return** mp

be used to capture semantic similarities. The vector mv is then used to query B, which, in turn, returns a list of record Ids (sim_vector_ids), and their distances, of the top-k similar vectors found, in the Jaccard space, to mv (Line 6). For each record Id in sim_vector_ids, if its distance from mv meets θ (Line 8), then the algorithm retrieves the corresponding normalized dense vectors from dictionary T (line 7). These vectors, generated using a pre-trained SentenceBERT model, are compared with dv using their inner product similarity (Line 10). Fine-tuning the pre-trained model used on domain-specific data will allow it to better capture specialized terminology, which may lead to higher recall and precision rates. The Ids of these semantically similar records to q are added to the set of results mp (line 12), ensuring that the final output contains records whose blocking keys are lexically similar and, in addition, are semantically relevant to the query. Once all records from sim_vector_ids have been processed, the algorithm returns mp as the set of matching record Ids. Dropping the lexical component results in indexing the dense vectors and comparing only the retrieved vectors with the query vector using ϕ.

Constructing the minhash vector requires $O(m)$ operations, where m is the number of minhash functions. Generating the dense embedding of the query record using a pre-trained model has a complexity of approximately $O(dL)$, where d is the embedding dimension and L is the length of the query record in terms of tokens. Searching the HNSW index for the top-k similar blocked records has an approximate complexity of $O(\log(n))$ for each query, where n is the number of blocked records. Retrieving the dense vector for each of the k

retrieved candidates from the dictionary takes $O(1)$ per lookup, resulting in $O(k)$ in total. Computing the inner product similarity for each of these candidates has a complexity of $O(d)$, leading to an overall similarity computation cost of $O(kd)$. Thus, the total complexity of the LSBlock algorithm is approximately $O(m+dL+\log(n)+kd)$. For sufficiently large n, the overall complexity simplifies to $O(\log(n))$.

4 Experimental Evaluation

We evaluated LSBlock, referred to as LSB, against the state-of-the-art system from the deep learning domain DeepBlocker, for blocking, combined with Ditto, for matching, [14,29] (DPD), and Sparkly [25] (SPK), as the top performing competitor using token-based similarity. DPD first transforms each record into dense embeddings using a pre-trained model. It, then, generates candidate pairs using efficient similarity-based methods, like Locality Sensitive Hashing [10]. DPD does not require labeled training data and leverages self-supervised learning to enhance blocking effectiveness. Sparkly uses Apache Lucene's full-text search capabilities to efficiently identify candidate pairs of records that are likely to match. To handle variations in text, Sparkly tokenizes textual attributes into 3-grams and indexes them in Lucene. It employs a two-phase approach: first, it uses Lucene to perform a coarse-grained search to identify potential matches, and then it applies a more refined similarity computation to these candidates to determine actual matches.

Datasets. The DBLP-Scholar paired dataset[5] contains bibliographic records from different sources and are commonly used the record linkage literature. This dataset includes variations in author names, titles, and venues, making them challenging for entity matching. The DBLP-Scholar is a non deduplicated dataset, also termed as *dirty* in the literature. The Abt-Buy paired dataset (see footnote 5) focuses on product matching containing structured product descriptions with differences in naming conventions and missing attributes. Similarly, the Amazon-Google paired dataset (see footnote 5) aligns products from Amazon and Google Shopping, exhibiting heterogeneity in product names, descriptions, and specifications. The Fodor's-Zagat's restaurants paired dataset[6] contains structured attributes such as restaurant name, address, city, phone number, and cuisine type. The attribute values exhibit differences in formatting as well as abbreviations. Each such paired dataset includes the labeled matches, which serve as a gold standard for evaluating record linkage algorithms. We also used the semi-synthetic voters dataset,(see footnote 5) with $10M$ records, which includes demographic information. This dataset is particularly useful for evaluating scalable techniques due to its large size. These datasets are widely used as benchmarks in the literature [6,9,11,12,14,19,24,29,33].

[5] https://old.dbs.uni-leipzig.de/research/projects/object_matching/benchmark_datasets_for_entity_resolution.

[6] https://www.cs.utexas.edu/~ml/riddle/data.html.

Metrics. Recall, precision, and the F1-Score are key metrics for assessing the effectiveness of record linkage methods. Recall quantifies the fraction of true matches that were successfully retrieved, whereas precision measures the fraction of correctly identified matches among all retrieved pairs. The F1-score is a weighted harmonic mean of precision and recall, that places greater emphasis on recall, making it particularly useful in scenarios where missing relevant matches is more critical than retrieving false positives.

Implementation Details. We performed the experiments using a virtual machine with a Xeon CPU, lacking a GPU, and 48GB of main memory. The accompanying software was implemented using Python ver 3.10. The neural networks used to estimate the discriminative power of attributes as well as θ and ϕ were simple feed-forward models for binary classification, implemented using PyTorch. They consisted of an input layer with one feature, followed by a fully connected hidden layer with 16 neurons using the ReLU activation function. The output layer contained a single neuron with a sigmoid activation, producing probabilities between 0 and 1. The networks were trained using the binary cross-entropy loss function for 10 epochs, using a random sample of matches and non-matches. For instantiating the index of LSBlock,[7] we used the implementation of HNSW offered by FAISS.[8].

The dimensionality of the minhash vectors was set to $m = 120$ hashes, as it is the smallest integer greater than 116 that satisfies the theoretical guarantees outlined in [12] while also being exactly divisible by 8, as required by FAISS. To generate the embeddings for each dataset, which were also used by DPD, we utilized the `all-MiniLM-L6-v2` SentenceBERT model from the Sentence Transformers library.[9] This model, based on a distilled version of BERT, efficiently encodes text into a 384-dimensional dense vector space, making it well-suited for semantic similarity tasks. We fine-tuned the model on each dataset using a labeled random sample of pairs and contrastive loss for 15 epochs. For both LSBlock and SPK, which rely on top-k retrieval, we set the number of returned records to $k = 5$ for each submitted query.

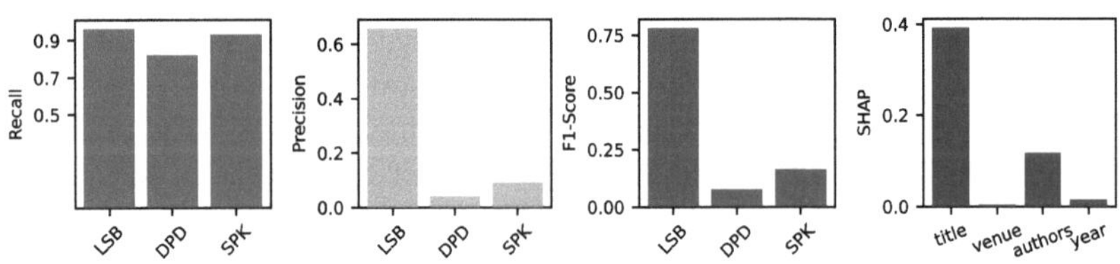

Fig. 1. DBLP-Scholar.

---

[7] The implementation of LSBlock can be found at https://github.com/dimkar121/LSBlock.

[8] https://github.com/facebookresearch/faiss.

[9] https://huggingface.co/sentence-transformers.

DBLP-Scholar (Fig. 1). The SHAP values indicated that the *title* attribute was the most influential, playing a key role in distinguishing records. The *author names* attribute also contributes to disambiguation, though to a lesser extent, while *venue* and *year* have minimal impact. We combined the *title* and the *author names* to formulate the blocking key, which helped maintain high recall while reducing false positives, ensuring a highly effective blocking strategy that resulted in a recall of 0.96 and precision of 0.66. LSB outperformed DPD and SPK by 16% and 4%, respectively, in terms of recall and by an order of magnitude in terms of precision. The high discriminative power of the blocking key in conjunction with the effective approximate blocking of the minhash vectors, are the two key factors behind this outstanding performance. LSB found matches with a large token overlap, resulting in high lexical similarity among their blocking keys but with significant lexical differences in the venue attribute. For example, *R Agrawal, Srikant R., Fast Algorithms for Mining Association Rules, Conference on Very Large Databases, Santiago, Chile'* paired with *R Agrawal, R Srikant, Fast Algorithms for Mining Association Rules in Large Databases, VLDB* and *U Maheshwari, B Liskov, Partitioned Garbage Collection or a Large Object Store, ACM SIGMOD International Conference on Management of Data'* paired with *U Maheshwari, B Liskov, Partitioned Garbage Collection of a Large Object Store, SIGMOD Conference'*. These record pairs would have been classified as non-matches if LSB relied solely on lexical similarity. In contrast, pairs such as *R Yerneni, C Li, JD Ullman, H Garcia-Molina, Optimizing Large Join Queries in Mediation Systems, LECTURE NOTES IN COMPUTER SCIENCE* and *R Yerneni, C Li, H Garcia-Molina, J Ullman, Computing Capabilities of Mediators, SIGMOD Conference'*, which are not matches despite their high lexical similarity in blocking keys, were ruled out by matching their corresponding dense embeddings.

We have to stress that DPD fails in terms of precision because the record pairs created, although exhibit some semantic relationship, are not real matches. For instance, *'W Kim, Editorial: Charter and scope, ACM Transactions on Database Systems (TODS)'* and *'W Chen, Declarative Updates of Relational Databases, ACM Trans. Database Syst.'* were identified as a match mainly due to the venue attribute. However, the venue is clearly not a discriminatory attribute to rely for finding matches. Another false positive was the following pair: *'T Zhang, R Ramakrishnan, M Livny, Birch: an efficient data clustering method for very large databases, Proc. ACM SIGMOD Intl. Conf. on Management of Data'* paired with *'S Guha, R Rastogi, K Shim, CURE: An Efficient Clustering Algorithm for Large Databases, SIGMOD Conference'*. This pair although exhibits some semantic similarity in the title and venue attributes is not an actual match, which can be easily identified by the set of the corresponding authors. Sparkly excels in terms of recall but in the absence of a threshold, during matching, approximately only one out of ten record pairs was correctly identified as a match.

Fodor's-Zagat's (Fig. 2). The restaurants dataset lacks meaningful contextual relationships among the matches, exhibiting only lexical inconsistencies in

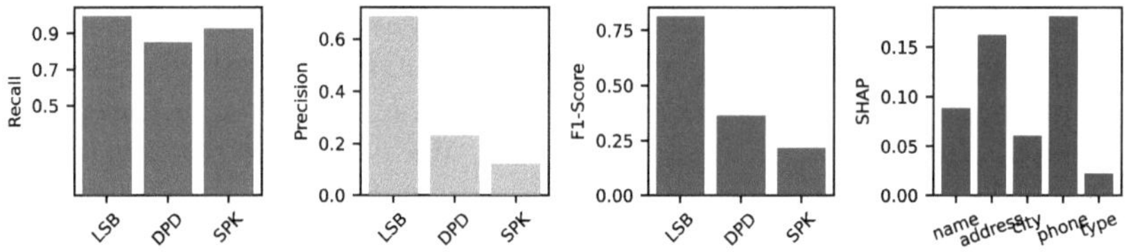

Fig. 2. Fodor's-Zagat's.

attribute values. LSB used the *restaurant name* and *phone* as the blocking key, resulting in a recall of 0.99 and precision of 0.68. We did not use *address* for the formulation of the blocking key, which also exhibited a relatively high SHAP value, because addresses can have high cardinality (too many unique values), that may cause overblocking with too many small blocks, missing true matches. During the experiments, the only match missed by LSB was *'mifune japan center kintetsu building, 1737 post st., san francisco, asian, 415/922-0337'* and *'mifune, 1737 post st., san francisco, japanese, 415-922-0337'* due to the large lexical variations between the blocking keys. SPK also demonstrated high recall with the same blocking key, whereas DPD exhibited significantly lower recall, trailing LSB and SPK by 16% and 9%, respectively. In terms of precision, LSB outperformed both DPD and SPK by a large margin.

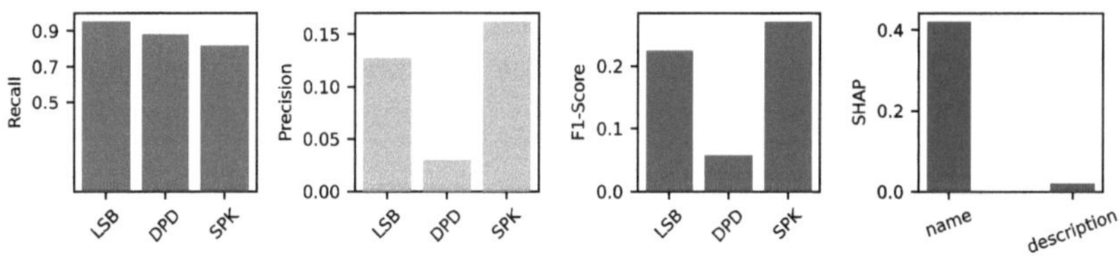

Fig. 3. Abt-Buy.

Abt-Buy (Fig. 3). This dataset comprises product names along with their descriptions, that closely resemble to semi-structured data. The SHAP values indicate that *name* is by far the most discriminative attribute. This suggests that product names contain crucial identifiers, such as brand names, model numbers, and key specifications, which are essential for distinguishing between different items. The low importance of descriptions suggests that the corresponding distances in the Jaccard space had minimal contribution towards the formulation of matches. On the other hand, descriptions excel in the vector space due to their rich contextual information they convey. LSB demonstrated moderate recall and precision, namely 0.85 and 0.26, respectively. DPD surpassed LSB by 0.03 in terms of recall, at the cost of significantly reduced precision, which was one order of magnitude lower than LSB. The initial similarity search of LSB in the Jaccard space missed some matches, which exceeded the distance threshold, such as *'Sony NP-FH100 Lithium Ion Camcorder Battery, Lithium Ion (Li-Ion) - 8.4V DC - Photo Battery'* and *'Sony NP-FH100 InfoLithium H Series Standard Rechargeable Battery Pack, Sony NP-FH100 InfoLithium...'*. In contrast, DPD failed to match pairs such as *'adobe photoshop cs3 (mac os x v.10.4.8) - create powerful images with the professional standardadobe ...'* and *'adobe photoshop cs3 [mac] - adobe photoshop cs3 software...'* due to semantic differences in the corresponding descriptions. LSB outperformed SPK by 5% in terms of recall, but the gap was much larger in terms of precision. Given the unstructured nature of the dataset, with the lengthy descriptions, we removed the lexical component of LSB and relied solely on its semantic engine. As a result, recall increased to 0.95, while precision dropped to 0.12, as shown in Fig. 3. This indicates a significant improvement over DPD, with an 8.8% increase in recall and an order-of-magnitude increase in precision, respectively. By removing the lexical component, the key difference between LSB and DPD lied in the use of the threshold, specified in the cosine similarity space, to filter the results returned by the index.

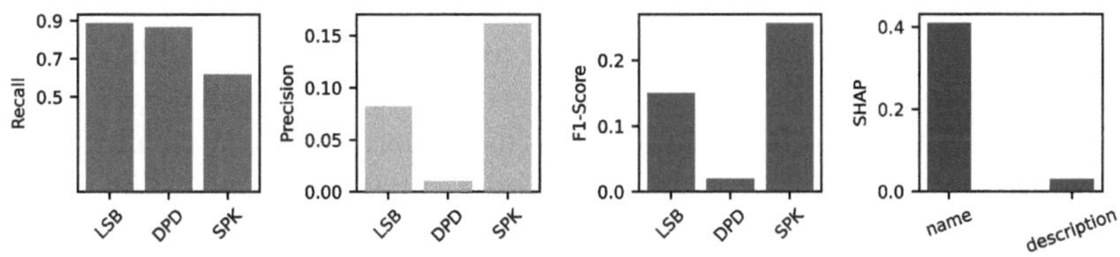

Fig. 4. Amazon-Google.

Amazon-Google (Fig. 4). The key characteristic of this dataset is its extensive product descriptions, which provide contextual information beyond simple string matching, making it particularly well-suited for deep learning models. The SHAP values are consistent with the Abt-Buy dataset, showing that *name* remains the dominant attribute, while *description* contributes minimally. We observe that DPD surpassed LSB by 18% in terms of recall, but DPD's precision was considerably lower than LSB's, indicating a trade-off between recall and precision. LSB missed pairs with low lexical similarity between the blocking keys, *'encore software 10538 - adventure workshop preschool ...'* and *'adventure workshop preschool (jewel case) adventure workshop 7 preschool - 1st grade features three great award...'*. DPD, on the other hand, missed pairs with short or missing descriptions in either record, such as *'adobe photoshop cs3 - create powerful images with the professional standardadobe...'* and *'adobe photoshop cs3 [mac] - adobe photoshop cs3 software...'*. SPK, which is based exclusively on lexical similarity, demonstrated generally low performance. Using exclusively LSB's semantic engine led to an increase in recall to 0.88, , while precision dropped to 0.08, surpassing DPD by 2.3% and an order of magnitude, respectively, which can be attributed to LSB's fine-tuned threshold.

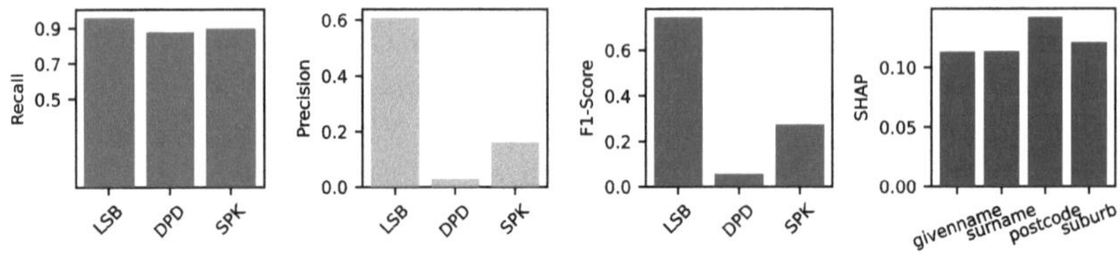

Fig. 5. Voters.

Voters (Fig. 5). Given its size of $10M$ records, this dataset is well-suited for evaluating LSB and its competitors in a large-scale linkage task. It consists of four attributes: *surname*, *given name*, *suburb*, and *postcode*. The SHAP values suggest that *postcode* is the most influential attribute, indicating its high discriminative power. This can be attributed to the structured nature of postcodes, which exhibit lower variability compared to names. From a pure SHAP value perspective, *postcode* and *suburb* seems the best option for formulating the blocking key. Blocking, though, depends on practical constraints; since postcodes are more precise, adding a suburb will not add much extra filtering power. Therefore, we chose *surname* to complement *postcode* to further refine the formulated blocks. Blocking all records and saving the index to disk took approximately 7 minutes. We then used chunks of $500K$ query records to perform the matching process. The average query resolution time per record was 0.0011 seconds. The combination of *given name* and *surname* exhibited high discriminative power, resulting in recall and precision values of 0.95 and 0.60, respectively. The records were embedded into the vector space without any added value for the linkage task due to the lack of any meaningful context conveyed by the records. LSB outperformed DPD by 8% in recall and by an order of magnitude in both precision and query time. SPK achieved high recall and fast query times, but had relatively low precision due to the unfiltered records being returned.

5 Conclusions

In this paper, we introduced LSBlock, a novel hybrid blocking system that combines lexical and semantic similarity to enhance record linkage accuracy. Leveraging HNSW indexing, LSBlock achieves logarithmic complexity in search, making it well-suited for large-scale datasets. By utilizing minhash-based blocking keys and dense embeddings, LSBlock effectively balances precision and recall by filtering the top-k candidate matches retrieved from the HNSW index. To this end, LSBlock uses a pair of thresholds: one for filtering the minhash vectors that correspond to the blocking keys and another for filtering the dense embeddings of records that correspond to the minhash vectors that have passed the first filtering stage. The use of these thresholds is a key feature of LSBlock because they essentially define the blocks, which are not explicitly implemented, since HNSW is a graph-based data structure.

References

1. Bojanowski, P., Grave, E., Joulin, A., Mikolov, T.: Enriching word vectors with subword information. Trans. Assoc. Comput. Linguist. **5**, 135–146 (2017)
2. Broder, A.Z., Charikar, M., Frieze, A., Mitzenmacher, M.: Minwise independent permutations. In: ACM STOC, pp. 327–336 (1998)
3. Brunner, U., Stockinger, K.: Entity matching with transformer architectures - a step forward in data integration. In: EDBT, pp. 463–473 (2020)
4. Chen, R., Shen, Y., Zhang, D.: GNEM: a generic one-to-set neural entity matching framework. In: WWW, pp. 1686–1694 (2020)
5. Christen, P.: Data Matching - Concepts and Techniques for Record Linkage, Entity Resolution, and Duplicate Detection. Data-Centric Systems and Applications. Springer (2012)
6. Christophides, V., Efthymiou, V., Stefanidis, K.: Entity Resolution in the Web of Data. Morgan & Claypool Publishers (2015)
7. Devlin, J., Chang., M., Lee, K., Toutanova, K.: BERT: pre-training of deep bidirectional transformers for language understanding. In: Conference of the North American Chapter of the Association for Computational Linguistics: Human Language Technologies, pp. 4171–4186 (2019)
8. Ebraheem, M., Thirumuruganathan, S., Joty, S.R., Ouzzani, M., Tang, N.: Distributed representations of tuples for entity resolution. In: PVLDB, pp. 1454–1467 (2018)
9. Gagliardelli, L., Papadakis, G., Simonini, G., Bergamaschi, S., Palpanas, T.: GSM: a generalized approach to supervised meta-blocking for scalable entity resolution. Inf. Syst. **120** (2024)
10. Gionis, A., Indyk, P., Motwani, R.: Similarity search in high dimensions via hashing. In: VLDB, pp. 518–529 (1999)
11. Karapiperis, D., Tjortjis, C., Verykios, V.: A randomized blocking structure for streaming record linkage. In: PVLDB, pp. 2783–2791 (2023)
12. Karapiperis, D., Tjortjis, C., Verykios, V.: A suite of efficient randomized algorithms for streaming record linkage. TKDE **36**(7), 2803–2813 (2024)
13. Karapiperis, D., Verykios, V.: An LSH-based blocking approach with a homomorphic matching technique for privacy-preserving record linkage. TKDE **27**(4), 909–921 (2015)
14. Li, Y., Li, J., Suhara, Y., Doan, A., Tan, W.C.: Deep entity matching with pre-trained language models. In: PVLDB, pp. 50–60 (2020)
15. Liu, Y., et al.: Roberta: a robustly optimized bert pretraining approach. arXiv preprint arXiv:1907.11692 (2019)
16. Lundberg, S., Lee, S.: A unified approach to interpreting model predictions. In: NIPS, pp. 4768–4777 (2017)
17. Malkov, Y., Yashunin, D.: Efficient and robust approximate nearest neighbor search using hierarchical navigable small world graphs. IEEE Trans. Pattern Anal. Mach. Intell. **42**(4), 824–836 (2018)
18. Mikolov, T., Chen, K., Corrado, G., Dean, J.: Efficient estimation of word representations in vector space. arXiv preprint arxiv:1301.3781 (2013)
19. Mudgal, S., et al.: Deep learning for entity matching: a design space exploration. In: SIGMOD, pp. 19–34 (2018)
20. O'hare, K., Jurek-Loughrey, A., Campos, C.D.: High-value token-blocking: efficient blocking method for record linkage. TKDD **16**(2), 1–17 (2021)

21. Papadakis, G., Alexiou, G., Papastefanatos, G., Koutrika, G.: Schema-agnostic vs schema-based configurations for blocking methods on homogeneous data. In: PVLDB, pp. 315–323 (2015)
22. Papadakis, G., Koutrika, G., Palpanas, T., Nejdl, W.: Meta-blocking: taking entity resolution to the next level. TKDE **26**(8), 1946–1960 (2014)
23. Papadakis, G., Papastefanatos, G., Koutrika, G.: Supervised meta-blocking. In: PVLDB, pp. 1929–1940 (2014)
24. Papadakis, G., Svirsky, J., Gal, A., Palpanas, T.: Comparative analysis of approximate blocking techniques for entity resolution. In: PVLDB, pp. 684–695 (2016)
25. Paulsen, D., Govind, Y., Doan, A.: Sparkly: a simple yet surprisingly strong TF/IDF blocker for entity matching. In: PVLDB, pp. 1507–1519 (2023)
26. Pennington, J., Socher, R., Manning, C.D.: Glove: global vectors for word representation. In: Conference on Empirical Methods in Natural Language Processing, pp. 1532–1543 (2014)
27. Reimers, N., Gurevych, I.: Sentence-bert: sentence embeddings using siamese bert-networks. In: EMNLP-IJCNLP, pp. 3980–3990 (2019)
28. Sanh, V., Debut, L., Chaumond, J., Wolf, T.: Distilbert, a distilled version of bert: smaller, faster, cheaper and lighter. arXiv preprint arXiv:1910.01108 (2019)
29. Thirumuruganathan, S., et al.: Deep learning for blocking in entity matching: a design space exploration. In: PVLDB, pp. 2459–2472 (2021)
30. de Vries, T., Ke, H., Chawla, S., Christen, P.: Robust record linkage blocking using suffix arrays and bloom filters. TKDD **5**(2), 1–27 (2011)
31. Yang, Z., Dai, Z., Yang, Y., Carbonell, J., Salakhutdinov, R.R., Le, Q.V.: Xlnet: generalized autoregressive pretraining for language understanding. In: Advances in Neural Information Processing Systems, vol. 32 (2019)
32. Zeakis, A., Papadakis, G., Skoutas, D., Koubarakis, M.: Pre-trained embeddings for entity resolution an experimental analysis. In: PVLDB, pp. 2225–2238 (2023)
33. Zeakis, A., Papadakis, G., Skoutas, D., Koubarakis, M.: Pre-trained embeddings for entity resolution: an experimental analysis. PVLDB **16**(9), 2225–2238 (2023)
34. Zhang, W., Wei, H., Sisman, B., Dong, X.L., Faloutsos, C., Page, D.: Autoblock: a hands-off blocking framework for entity matching. In: WSDM, pp. 744–752 (2020)

Evaluating Quality of Disparate Data Sources: A Discord-Driven Approach

Yeasmin Ara Akter[1,2(✉)], Alberto Abelló[1], Petar Jovanovic[1], Tomer Sagi[2], and Katja Hose[2,3]

[1] Universitat Politècnica de Catalunya, 08034 Barcelona, Spain
yeasmin.ara.akter@upc.edu
[2] Aalborg University, 9220 Aalborg Øst, Denmark
[3] Technische Universität Wien, 1040 Vienna, Austria

Abstract. Among other measures of data quality, determining the reliability of conflicting values from different sources is especially challenging. Traditional data fusion approaches often infer correct values in simple cases, but struggle to handle variations in data granularity (such as differences in temporal, spatial, or categorical aggregations) and offer limited insight into the nature of disagreements. Thus, we propose a new source evaluation approach for numerical attributes that measures discordance (i.e., the extent to which sources differ from each other). Unlike existing methods that focus solely on point estimation, we allow both fine-grained and coarse-grained analysis, allowing more sophisticated data quality assessments. We employ a linear programming solver that transparently adapts to any data alignment expressed in a set of operators resembling relational algebra. Extensive experiments on real-world datasets demonstrate that our method generalizes existing truth discovery techniques measuring differences with Mean Absolute Error (MAE), Root Mean Square Error (RMSE), and can adapt to diverse and complex scenarios.

Keywords: Data Fusion · Truth Discovery · Discordance · Linear Programming

1 Introduction

By current estimates[1], over 400 million TB of data are produced every day. This growth, inevitably leads to an increase in the number of cases where multiple data sources collect the same or complementary information about different real-world events. Thus, applications often require a consolidated platform for users to integrate data from different sources. However, the significant research conducted over the years in the field of data integration is still not enough.

Data integration processes address two major challenges: semantic heterogeneity (differences in how data are structured and represented across sources) and conflicting data (inconsistencies in the actual values reported for the same

[1] https://www.statista.com/statistics/871513/worldwide-data-created.

P. K. Chrysanthis et al. (Eds.): ADBIS 2025, LNCS 16043, pp. 147–163, 2026.
https://doi.org/10.1007/978-3-032-05281-0_10

entity). Semantic heterogeneity can appear both at the schema and at the instance level; however, conflicting data appear at the attribute value level. Consequently, data integration goes through three steps [6]: (i) **schema alignment** identifies semantic relationships between entity classes and attributes in different schemata; (ii) **entity resolution** identifies distinct instances that correspond to the same real-world entity; and (iii) **data fusion** aims to resolve conflicting attribute values among those instances to produce accurate and reliable information.

Our work specifically contributes to the field of data fusion to evaluate value conflicts. Thus, we focus on analyzing how different sources report diverse values for the same attribute instance (e.g., the temperature of the same place and at the same time reported by different satellites).

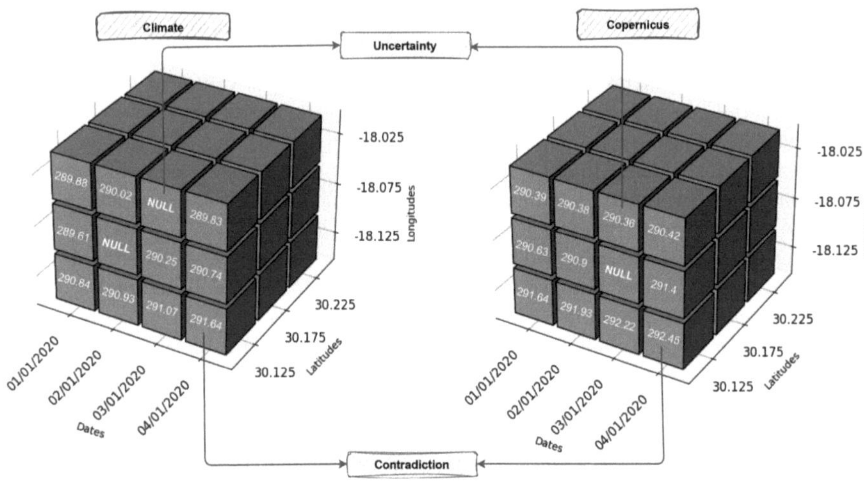

Fig. 1. Comparison of SST measurements from Climate and Copernicus.

As an application domain, Sea Surface Temperature (SST) measured by satellites represents important information in environmental monitoring, and is essential for predicting climate patterns and weather conditions. However, inconsistencies in the SST values reported hinder their use. In general, the value conflicts between sources may arise due to differences in non-simultaneous orbital passes over the same location (e.g., one satellite passing at 1:30 AM and another at 2:15 PM), spatial resolutions (e.g., 5 km vs. 4 km grid cells), sensor characteristics (e.g., thermal infrared sensor vs. microwave radiometer), and atmospheric interference from clouds and aerosols (e.g., 30% cloud cover vs. clear sky conditions). Our case study is illustrated in Fig. 1 for only two satellite sources (we will consider four in Sect. 5), namely Climate and Copernicus. Matching measurements of identical geographical coordinates (latitude and longitude grid) on the same date, we highlight two of the most critical issues: *measurement gaps* where sources have missing values (i.e., NULLs), and *temperature inconsistencies* where two sources show different temperature values for the same fact (e.g., 291.64 K vs. 292.45 K). These value conflicts raise the fundamental questions of which source provides higher-quality data and to which factor we can attribute the discrepancies. Furthermore, as mentioned above, differences in spatial resolution between satellites add more complexity to the analysis due to mismatched measurement alignments (e.g., for the same area of $20\,\text{km} \times 20\,\text{km}$, we can find a grid of 16 cells of size $5\,\text{km} \times 5\,\text{km}$, which needs to be mapped against another grid with 25 cells of $4\,\text{km} \times 4\,\text{km}$).

Early data fusion approaches relied on rule-based methods that use simple conflict handling functions such as minimum/maximum [3], average [14], and majority voting [4]. However, these methods have significant limitations. Firstly, they lack mechanisms to handle missing data. Second, they cannot process measurements in incompatible units (e.g., Kelvin vs. Celsius), thus requiring a preprocessing step for unit conversion. Third, they fail to consider the inherent relationships between numerical values (e.g., three SST readings 293.45 K, 292.99 K, and 285.66 K from different sensors would be treated as categorical, ignoring the numerical distance between values even though 293.45K and 292.99 K could be considered relatively similar, while 285.66K is clearly different). Moreover, these approaches do not consider uncertainty encoding (e.g., representing uncertainty with either NULL or symbolic variables), as is broadly considered in the incomplete database research field [12].

Hence, our goal is to create a unified framework that encodes uncertainty to evaluate data quality by measuring the discordance among the different sources available (i.e., the extent to which they differ from each other). We achieve this by building on Eris [1], which provides foundations for measuring discordance in multidimensional sources using linear expressions of symbolic variables to effectively encode uncertainty. These symbolic variables are then integrated with Linear Programming (LP) optimization techniques using OSQP [9] to measure the discordance. However, Eris used a uniform approach to variable generation which did not allow to dynamically evaluate discordance at varying levels of granularity. In addition, it lacked advanced optimization features for LP formulation, which limited its ability to produce high-quality results when dealing with large LP problems, for which OSQP has limited capacity.

Hence, we added to Eris the ability to dynamically generate symbolic variables. Now, the user can express the source alignment using an algebraic expression (that can include unit conversions, aggregations, etc.), and symbolic variables representing uncertainty (or potential errors) in the values can be introduced at any time in the query. This feature allows not only flexible and precise comparisons of sources at varying levels of granularity, but also to attribute errors to different analytical dimensions (e.g., the amount of error could depend on time, latitude, or longitude). This way, we address a critical gap in previous uncertainty encoding and data fusion methods and provide a scalable and adaptable solution for complex, multi-source, and multidimensional scenarios. Besides this, for efficiently dealing with large LP problems while ensuring the high accuracy of the discordance evaluation, we propose a subproblem decomposition approach. Our **contributions** can be summarized as follows:

1. We enable the dynamic introduction of symbolic variables for conflicting data, attributing discordance to any analytical dimension at any level of granularity.
2. We propose to use a sub-problem decomposition approach to handle large-scale datasets that challenge the computational capabilities of OSQP, which enables effective processing and improves the accuracy of the result.
3. We demonstrate the effectiveness of our method through a comprehensive case study of SST that showcases multiple real-world challenges. We show that Eris can deal with temporal variations (different collection times), spatial discrepancies (varying grid resolutions), measurement unit inconsistencies (Kelvin vs. Celsius), and direct value conflicts between satellite sources. In addition, we compare our approach against state-of-the-art fusion techniques to highlight its generality over existing techniques.

The remainder of the paper is organized as follows. In Sect. 2, we review related work. Section 3 formally defines the problem of measuring discordance. In Sect. 4, we describe our methodology in detail, including the techniques for uncertainty representation, data alignment, and optimization. Section 5 reports on our experimental evaluation and discusses the results, while Sect. 6 concludes the paper.

2 Related Work

We divide the related work into two key relevant aspects: (i) uncertainty handling in incomplete databases and (ii) data fusion methods.

2.1 Uncertainty Handling in Incomplete Databases

A common challenge in data fusion is uncertainty, which arises in multi-source settings when one source provides a NULL value. In such cases, it becomes unclear whether the NULL represents missing information or if the event truly did not occur.

In the domain of incomplete databases, uncertainty is typically addressed using the Open World Assumption (OWA), which assumes that missing data indicates unknown information rather than non-existence. To encode this uncertainty, researchers often use symbolic variables that represent whether a tuple should be included or excluded in the result of a query [12] and allow to reason over them.

M-table was proposed as a symbolic representation to handle the uncertainty of missing values in databases [21]. The approach uses positive relational algebra operators (e.g., $\sigma, \pi, \cup, \times, \ltimes, \rho$) to derive possible answers. Although this approach addresses unknown and partially known values, it suffers from high computational costs and cannot effectively handle aggregation operations (e.g., min, max, avg, count). AU-DB [8] introduces an attribute-annotated uncertain database that handles both tuple and attribute-level uncertainty using ETL heuristics and bag semantics. While pioneering multi-aggregation uncertainty computation, it is limited to ordinal data, excluding nominal and numerical data types. Lenses [24] introduces an on-demand ETL approach using virtual C-tables, which encode uncertainty through boolean expressions and probability estimates. Its successor, Mimir [19], enhances the system by optimizing VC-tables through query partitioning and inline evaluation. However, these approaches, while effective for uncertainty representation, lack mechanisms to analyze source disagreement and struggle with numerical data complexity.

2.2 Data Fusion Methods

Data Fusion research addresses the challenge of resolving contradictions in multi-source settings by inferring the most likely truth from conflicting information [5]. Over time, the field has evolved significantly, proposing iterative techniques such as TRUTHFINDER [25], which use Bayesian networks to jointly evaluate data consistency and source quality. The iterative process assumes that the higher the quality of the source, the more accurate its data. This key principle has influenced many subsequent and complex methods in the field. Nevertheless, most algorithms still focus exclusively on categorical data and are not suitable for numerical data. In the following, we categorize and analyze these algorithms based on data types being either categorical or numerical.

Categorical Data Fusion: For resolving conflicts in categorical data, several advanced methods have been proposed. A multi-truth fusion method DART [16] enhances the basic framework of TRUTHFINDER [25] by incorporating weighted voting and iterative Bayesian models, using domain expertise scores to determine source weights. STORM [2] advances the field by introducing copy-based source authority measurement and clustering techniques to handle source discrepancies. Methods based on confidence intervals emerge through works such as CATD [14], MTD-VCI [7], HCDT [18], and ETCIBoot [23], which innovated by resolving conflicts within ranges rather than point estimates. FaitCrowd [17] proposes a probabilistic approach to jointly estimate topic-specific expertise and true answers while modeling question content and answers.

Numerical Data Fusion: To resolve numerical data conflicts, a weighted mean with probability derivation can be used [13], which works well for a limited number of sources but lacks support for missing value imputation and query aggregation. The Gaussian truth model (GTM) [27] introduced a probabilistic framework using Gaussian distributions, where true values serve as mean parameters and source quality is reflected in the variance. Dynamic data fusion evolved through EvolvT [28], which integrates Kalman Filtering and Hidden Markov Models to capture temporal truth evolution. The FTS method [26] makes substantial advances by incorporating the Hub Authority method and considering both reported and missing values. They evaluate source quality through silent rate, true rate, and false rate metrics. CRH [15] presents a versatile optimization

Table 1. Notation used

Notation	Description
$K \rhd V$	Finite map signature with key K and value V attributes
$R : K \rhd V$	Relation R with signature $K \rhd V$
Σ	Schema with relations and signatures
X	Set of symbolic variables
W	Weights associated with X
ϕ	Conjunction of linear integrity constraints
$LLUN$	Internally consistent state
$h : \mathbb{R}^X \rightarrow \mathbb{R}$	Valuation function mapping uncertainty variables to real values
$\delta : h \rightarrow \mathbb{R}$	Discordance metric/LP cost function

Table 2. Operators

Symbol	Operation
σ	Select
π	Project
$\hat{\pi}$	Project Away
$\bowtie$	Join
ρ	Renaming
$\setminus$	Difference
γ	Aggregation
$\cup$	Union
$\uplus$	Discriminated Union
$\varepsilon_{B:=e}$	Derivation
κ	Coalescing

framework using normalized absolute deviation as the loss function, capable of handling both numerical and categorical data using appropriate loss functions. KDEm [22] uses kernel density estimation to resolve conflicting information in data. Although these methods perform advanced numerical data fusion, they struggle with integrating missing values and heterogeneous granularities.

In general, incomplete database approaches effectively handle uncertainty through symbolic representations, but lack a comprehensive analysis of source disagreements. Conversely, data fusion methods excel at identifying source agreement patterns but struggle with data uncertainty and granularity differences. Our work bridges previous approaches by introducing a unified framework that combines uncertainty handling and advanced source quality evaluation using LP.

3 Problem Definition

In this section, we present the formal background of our approach and state the problem definition. Table 1 summarizes the key mathematical notation used throughout the paper.

To systematically measure discordance and allow precise comparisons among data sources, for each relation R we use a finite map signature, denoted as $K \rhd V$, which defines that a set of key attributes K functionally determines a set of value attributes V. A schema Σ represents a collection of such relations with their finite map signatures. We assume that all the input datasets comply with such a structured representation. Furthermore, using an alignment algebraic query (using operations listed in Table 2), we generate a set of symbolic variables X to encode the uncertainties (a.k.a. errors) associated with the values V. Operations make variables in X form linear expressions for each attribute that are eventually combined with the same attribute in other tuples sharing keys to define a linear integrity constraint (LIC). The conjunction of all LICs (ϕ) asserts the integrity among the sources to detect potential conflicts. To represent an ideal internally consistent state implied by these constraints, we also generate LLUN variables [10] for every key, which correspond to an interpretation of the data that satisfies ϕ. Then, to resolve the uncertainties of the values V, we introduce a valuation function $h : X \to \mathbb{R}$ which maps symbolic variables to real values based on the imposed constraints. Finally, to quantify the level of discordance between sources, we define the discordance metric δ (a.k.a. cost function in the associated LP), which measures the deviation of observed data from the LLUNs. The discordance metric is a function of the values associated to variables (X) and their corresponding weights ($W \subset \mathbb{N}$), where a weight $w_x \in W$ of a variable $x \in X$ corresponds to the number of times x appears in the constraint set ϕ. This metric helps assess relative data quality by measuring discordance among data sources without using any external ground truth.

Problem Definition: *Given the schema Σ containing finite maps with signatures of the form $K \triangleright V$ (being K descriptive keys used to identify and group data, and V numerical values to be analyzed) and an alignment/mapping between them, the objective is to find the values of the dynamically created symbolic variables X (representing uncertainty/errors) that minimize the discordance/cost function δ, subject to integrity constraints ϕ corresponding to that mapping.*

Example 1. Following the example in Fig. 1, we store the Climate and Copernicus data sources in a relation $R : \{s, d, la, lo\} \triangleright \{SST\}$. We then create variables for each source and date, and introduce them using linear expressions. Among many possibilities, given a value $v_{[s,d,la,lo]}$, we take $v_{[s,d,la,lo]} + x_s + x_d$ generating variables x_s and x_d representing the uncertainty per source and date respectively, with $s \in \{\text{"Climate"}, \text{"Copernicus"}\}$ and $d \in \{01/01/2020, \ldots, 04/01/2020\}$. The corresponding LICs in ϕ would then be of the form $v_{[\text{"Climate"},d,la,lo]} + x_{Climate} + x_d = v_{[\text{"Copernicus"},d,la,lo]} + x_{Copernicus} + x_d$. Thus, we look for the values of x_s and x_d that minimize δ under those constraints. One of the several discordance functions we might use is the weighted sum of squared errors $\delta = \sum_{x \in X} w_x \cdot x^2$.

4 Data Sources Evaluation

In this section, we describe our approach, which comprises two main stages (sketched in Fig. 2): (1) process alignment query, and (2) compute discordance.

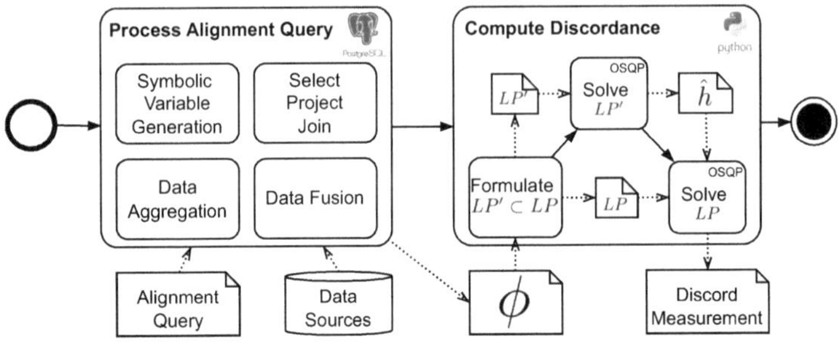

Fig. 2. Discordance Measurement Approach

4.1 Process Alignment Query

This stage (commonly referred to as *data alignment*) transforms an algebraic alignment query into SQL. The query serves four purposes: (i) apply restricted versions of relational operations (i.e., any operator from Table 2 except γ and κ), (ii) perform aggregation (γ) to align granularities in the sources; (iii) generate symbolic variables dynamically by introducing linear expressions (ε), and (iv) execute data fusion to unify records referring to the same real-world entity (κ). These operations can be freely combined in any order to structure the data.

Since the use of restricted relational algebraic operations has already been extensively discussed in [1], we do not elaborate on these details here. Instead, our focus is on symbolic variable generation and its interaction with aggregation.

Symbolic Variable Generation: Symbolic variables can be generated in any relation at any moment (i.e., before or after applying any sequence of algebraic operations including aggregation), which allows uncertainty to be captured at different levels of granularity, either at the atomic data point or after aggregation into consolidated values, providing flexibility in how it is modeled.

Definition 1. ***Dynamically Generating Symbolic Variables:*** *Given a linear formula $f{:}(X,\mathbb{R}){\rightarrow}\mathbb{R}$ and a relation $R : K{\triangleright}V$, where a value attribute $B_i \in V$ has some associated uncertainty, we generate a new attribute B_i' representing the uncertainty of B_i. Thus, given a subset of key attributes $\{A_1,\ldots,A_n\} \subseteq K$, we generate a set of symbolic variables $X_{\{A_1,\ldots,A_n\}} \subseteq X$, containing variables of the form $x_{t[A_1,\ldots,A_n]}$, being $t \in R$. Then, $\forall t \in R, t[B_i'] = f(x_{t[A_1,\ldots,A_n]}, B_i)$. Notice that we can only guarantee that $\mid R \mid = \mid X_{\{A_1,\ldots,A_n\}} \mid$ if $\{A_1,\ldots,A_n\} = K$ (i.e., in general, variables can be reused in different tuples).*

Specifically, we introduce variables in $R : K \triangleright V$ applying a derivation operation $\epsilon_{B:=e}$, by using in the expression e the function "New(str, K')", where str is a string used as the root name for all the rows, and $K' \subseteq K$.

Table 3. Example of symbolic variable generation

Query	Result signature
$R' = \varepsilon_{SST':=SST+\text{New}(``x",\{s\})+\text{New}(``x",\{d\})}(R)$	$R' : \{s, d, la, lo\} \triangleright \{SST, SST'\}$
Values of $R'[SST']$ are of the form $SST + x_s + x_d$	$X = \{x_{R[s]}\} \cup \{x_{R[d]}\}$

Example 2. Taking again the data from Fig. 1, we can create the variables as exemplified in Table 3. This specifies the derivation of a new SST value attribute with just two new variables. Both will have names starting with "x", concatenated to some value in the corresponding tuple. For the first one, the value comes from the source attribute and for the second from the date, getting among other tuples:

$$[``Climate", 04/01/2020, 30.125, -18.125, 291.64, 291.64 + x_{Climate} + x_{04/01/2020}]$$

The underlying assumption is that the error in the measurement can be attributed to both the corresponding satellite hardware, and the date of the measurement. If the value v is null (i.e., missing or unknown), any imputation method could have been applied to the corresponding generated variable. However, implementing such imputation process is beyond the scope of this work.

Data Fusion: After the uncertainty variables are introduced, we match data using the coalescing operator κ, which merges records that refer to the same real-world entity (i.e., having the same keys) into a single tuple. Thus, when sources provide different values for the same entity attribute, the coalescing operator resolves these conflicts by introducing a new variable (called LLUN as in [10]) to represent the coincidence in reality. Thus, after this, for each source and LLUN, we get a LIC corresponding to the equality between the LLUN and the data provided by the source (including uncertainty variables), which asserts that all reported values for a given key correspond to the same real world value represented by the LLUN. Such structured alignment creates the conjunction of linear integrity constraints ϕ.

Example 3. Continuing with the example in Fig. 1, we want to measure discordance imputing it independently per source and date, by creating in our alignment query the variables as in Table 3. Then, this query would perform a fusion on these symbolically represented data by linking each pair of tuples (one from Climate and one from Copernicus) whenever they share the same K (i.e., identical date, latitude and longitude). Thus, we introduce a new variable $LLUN_{R[d,la,lo]}$ and its corresponding constraints:

$$LLUN_{04/01/2020,30.125,-18.125} = 291.64 + x_{Climate} + x_{04/01/2020}$$
$$LLUN_{04/01/2020,30.125,-18.125} = 292.45 + x_{Copernicus} + x_{04/01/2020}$$

4.2 Compute Discordance

We further compute discordance by formulating an LP from ϕ and solving it.

Formulate LP: Once the data are aligned, we obtain ϕ, a comprehensive system of LICs, which we translate into a standard linear (or convex quadratic) program by combining all equality constraints and adding bounds for the variables to ensure that the solution stays within a realistic range (e.g., a boundary is set from 291.64 to 292.45 for the variable LLUN if it tries to resolve the conflict between these two values). Each coalesced tuple contributes one or more equations of the form $\mathbf{A}\,\mathbf{X} = \mathbf{B}$, where $\mathbf{X}$ is the vector of symbolic variables, and $\mathbf{B}$ is a vector of constants (either reported atomic values or aggregated ones). Finally, we set the cost function δ that measures discordance. In our case, we use the weighted sum of absolute values or squares of the error variables, where the weights (W) reflect the importance of each variable for the overall cost depending on their frequency of occurrence (e.g., the same date variable with four different measurements would have weight four).

Solve LP: OSQP [20] is used as LP solver, which generates a mapping h for both the uncertainty variables and LLUNs. However, solving a large constraint system using OSQP easily falls in local optimums [9]. To address this, we take a *subproblem decomposition* approach, and solve first a smaller problem LP' for $\phi' \subset \phi$ (e.g., we used only 20% of the constraints in our experiments) initializing the variables to zero, as OSQP(LP', δ, $\emptyset$). This subproblem gives an approximate solution $\hat{h}$ which serves as a warm-start for solving the full problem OSQP(LP, δ, $\hat{h}$), that yields the final solution h resulting in a discordance of $\delta(h)$.

5 Experimental Evaluation

In this section, we present how the experiments are designed and conducted to evaluate the discordance among sources. We analyze:

1. **Data Source Quality**: Showing the capabilities and usefulness of our approach in a real use case to determine whether discordance between different sources of satellite data arises from specific spatial or temporal differences.
2. **Discordance Metric Quality**: Scrutinizing the quality of the LP results of the current implementation compared to the original Eris implementation [1].
3. **SotA Metrics**: Comparing the effectiveness of Eris against the current State-of-the-Art methods on a public dataset.

5.1 Data Source Quality

Our satellite case study uses real environmental data obtained from four different sources: Copernicus,[2] Climate Change Service,[3] Moderate Resolution Imaging Spectro Radiometer (MODIS),[4] and the National Oceanic and Atmospheric

[2] https://data.marine.copernicus.eu.
[3] https://cds.climate.copernicus.eu.
[4] https://oceandata.sci.gsfc.nasa.gov.

Administration (NOAA) PathFinder.[5] Notably, while the other three sources rely on independent sensors, Copernicus aggregates data from Climate Change Service and NOAA PathFinder for a specified time range. Among all data they provide, we select the Mediterranean sea as the area of interest, relying on its coverage by all the data sources in the year 2020, to get the following number of tuples: Copernicus (Co - 25,376,654 rows), PathFinder (Pa - 63,396,861 rows), Climate (Cli - 9,420,718 rows), and Modisaqua (Mo - 35,591,624 rows).

As in the running example, the environmental variable analized is SST, and the heterogeneities in terms of resolution, units, and sensor technologies among different sources are summarized in Table 4. Hence, our goal is to evaluate the discordance of SST reported by these sources.

Figure 3 shows raw monthly nighttime SST for each source (Climate, Copernicus, and Pathfinder in Kelvin, and Modisaqua in Celsius), revealing significant variations in the reported values. After aligning both scales, all of them follow a similar seasonal pattern with SST rise until the middle of the year and then fall. However, Copernicus and Climate provide almost the same values along the year, while Pathfinder systematically reports the lowest SST, and Modisaqua falls somehow in between. Differences in trends likely come from the different data collection time, satellites, and sensors, but we would like to quantify this.

Thus, although a visual inspection of raw data can be insightful, we want to go further and assess the quality of the sources using two indicators: (i) *Discordance*, expressing δ as the weighted mean of: a) absolute error values $\frac{\Sigma w_i \cdot |x_i|}{n}$ (a.k.a. MAE), and b) squared error values $\sqrt{\frac{\Sigma w_i \cdot x_i^2}{n}}$ (a.k.a. RMSE), where w_i represents the weight of variable x_i, and n the overall number of variables; and (ii) *Deviation*, indicating how the different dimensions contribute to the errors by showing separately the values of the different variables, depending on whether they were generated based on either source, month, latitude, or longitude. These indicators allow us to capture both the magnitude and the origin of disagreement among sources.

Table 4. Data Source Description

Source Name	Reso.	Units	Sensors
Cli	0.05°	Kelvin	SLSTRB
Mo	0.04°	Celsius	MODIS-Aqua
Pa	0.04°	Kelvin	AVHRR
Co	0.05°	Kelvin	[1981-2016] ESA CCI [2017-present] C3S [1981-2014] Pathfinder

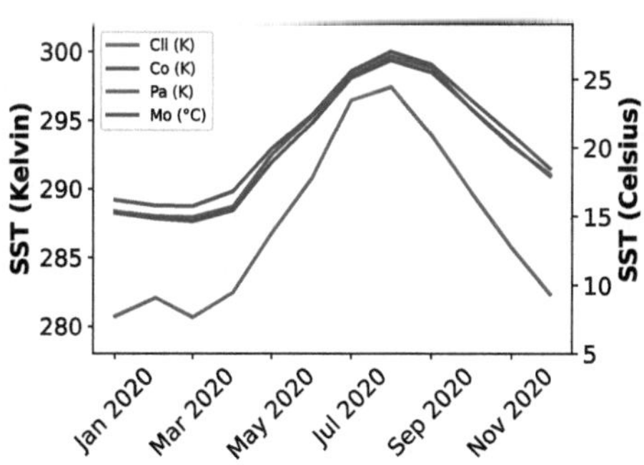

Fig. 3. Monthly SST Distribution

[5] https://www.ncei.noaa.gov.

Firstly, our queries transform Celsius into Kelvin, and to align spatial resolutions across sources (i.e., 0.04° vs. 0.05°), we aggregate the data to a common resolution of 1° for both latitude and longitude. Then, to investigate the internal consistency of each source, we create different variables and at different levels of granularity depending on the question we want to answer (Sect. 4 describes how the queries are generated and formulated as an LP).

Experiment 1: Pairwise Source Discordance Analysis. To evaluate the consistency of the SST measurements, we first analyzed the pairwise discordance between all source combinations, aiming to identify which sources align closely and which diverge. To this end, we generated a different query per month and combination of sources, which averages values per month and source, generating one variable for each source and matching both values. Each query was then configured to use MAE and RMSE as discordance metrics.

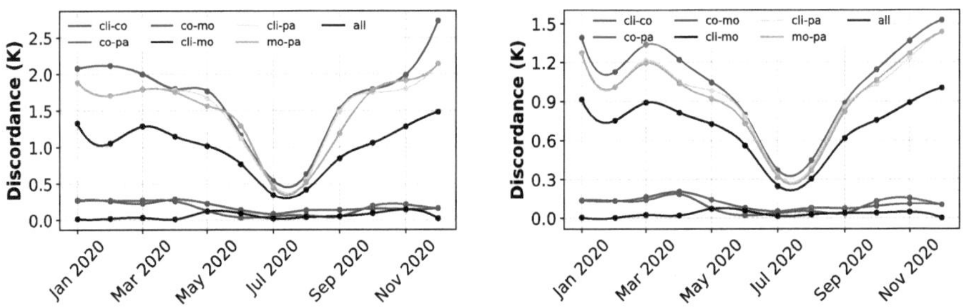

(a) Discordance per month using MAE (b) Discordance per month using RMSE

Fig. 4. Discordance using Different Metrics

Figure 4a and 4b present the monthly trends in pairwise discordance for all combinations of the four sources, along with an aggregated line labeled `all`, representing the combined discordance of all sources. As seen in pairs `cli-pa`, `co-pa`, and `mo-pa`, PathFinder clearly exhibits the highest discordance with all other sources with some fluctuation through the year. Actually, if we compare these pair-wise lines against the one corresponding to all sources together, we see that they follow the same trend, indicating a clear domination of Pathfinder discordances in the total. This is consistent with the domain knowledge, as it collects temperature data very early in the morning (2–5 A.M.), when the ocean surface is typically at its coldest. In contrast, the other sources all gather data earlier (around 10:30 P.M. to 12 A.M.), when it is a bit warmer, resulting in higher estimation of SST and lower deviations between them.

Both MAE and RMSE show similar trends. Seasonally, we observe decreases in discordance during the warmer months (May–August), maybe due to more stable ocean conditions, better satellite visibility (due to less cloud cover), and fewer abrupt temperature changes, which improve agreement among sources.

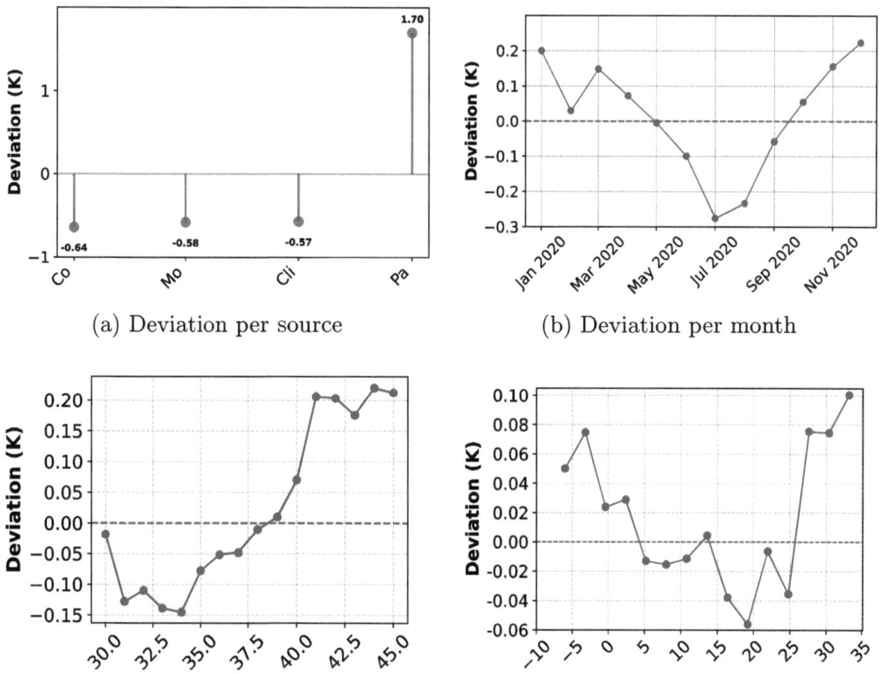

(a) Deviation per source

(b) Deviation per month

(c) Deviation per latitude

(d) Deviation per longitude

Fig. 5. Deviation of SST imputing differences to different analysis dimensions

Overall, the experiment highlights that PathFinder displays the greatest variability, especially during non-summer periods. Using MAE or RMSE does not make any difference in this conclusion, though RMSE shows more consistent trends in the lines involving Pathfinder.

Experiment 2: Temporal and Spatial Deviation Analysis. In this experiment, we aimed to dig dipper and identify the main causes of discordance. For this, we aggregated data per source, month, latitude, and longitude; generated independent variables for each of them; and match values for the same month, latitude and longitude. Then, we can visualize separately the values of the different variables corresponding to the source (Fig. 5a), the month of the year (Fig. 5b), the latitude (Fig. 5c), or the longitude (Fig. 5d).

Figure 5a shows the deviation per source and confirms that Pathfinder stands out with the highest deviation (1.70 K). Figure 5b corresponds to how SST deviation varies over the months of the year. A clear seasonal pattern emerges, indicating that deviation is negative (i.e., overestimation) and larger during the warmer months (May to September), while it appears to be positive (i.e., underestimation) but smaller during the cooler months (January-April and October–December). Figure 5c plots the deviation of SST across latitudinal bands. In the south (30°N to 36°N), there is a negative deviation, but as we move toward the north (40°N to 46°N), the deviation increases to become positive and reaches higher absolute values, which can be attributed to more unstable weather conditions. Finally, Fig. 5d represents how SST values deviate across longitudinal bands. In the western Mediterranean (−6° to 5°E), sources underestimate the SST. Moving eastward (10°E to 25°E), the deviation decreases until in the eastern Mediterranean (25°E to 36°E), deviation varies more widely, indicating underestimations again. These patterns suggest more complex coastal dynamics in the extremes and more stability in the wider center of the sea.

Overall, we conclude that the main factor introducing discrepancies in the measurements is whether the source is Pathfinder or not (by one order of magnitude), but month, latitude, and longitude also substantially affect the quality.

5.2 Discordance Metric Quality

In this section, we compare the performance of the new Eris version (Eris_v2), with the original one [1] (Eris_v1). Given that OSQP can incur faulty results (i.e., local optima) [11], we need to study how much this happens. At this point, it is important to note that, when measuring discordance between only two sources (e.g., SST 291.64 and 292.45), the value that minimizes the error at every point must be in between the given ones (e.g., 292.045), hence one must be overestimating (i.e., −0,405) and the other underestimating (i.e., +0,405). However, this is not necessarily so when reusing the same variable at many points, and trying to minimize all at once, as we do in general. Any possibility where both have the same sign will not minimize neither MAE nor RMSE, and thus, for the purpose of this experiment, we count it as incorrect. Aggregating this for all pairs of variables (one from each source) gives us an indicator of the result's correctness.

Figure 6 depicts the ratio of the quality of the results in both systems. While the accuracy of the original system deteriorates as the number of LICs and variables increases, Eris maintains a consistently high quality after implementing the subproblem decomposition approach, even in complex scenarios. Despite both performing perfectly for small problems, for higher numbers of LICs and variables, the original system achieves only around 10–20% correct estimations, whereas the new version maintains over 65–95% accuracy (up to 6.9 times more), demonstrating its robustness and scalability.

In terms of execution time, we ran each experiment 20 times and took the mean time to solve the LP. In the original system, every call to OSQP took 0.1 s, while with the new implementation, solving the same problem takes 0.005 s more (this difference is found not significant according to a t-test). We have to add the resolution of the subproblem LP', though, which takes only 0.024 s (for the 20% of LICs we used). Hence, introducing a subproblem-solving phase does not substantially increase the overall execution time while achieving significantly better solution quality.

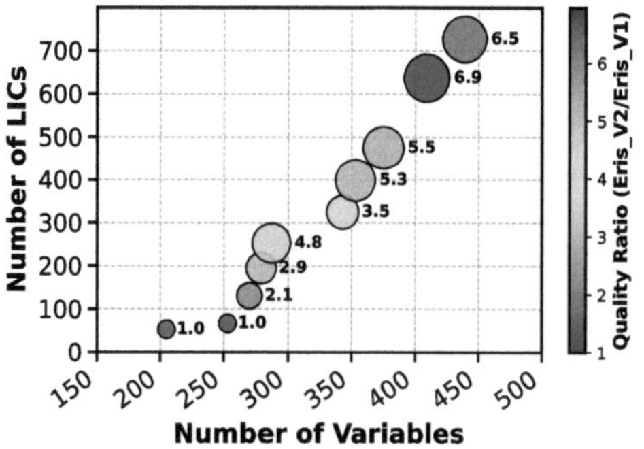

Fig. 6. Comparison of Quality of the Results

Table 5. Comparison with SotA

Method	MAE	RMSE
Mean	200,926	1,136,605
Median	11,075	129,850
TruthFinder [25]	1,535	8,888
GTM [27]	84,393	2,020,279
CATD [14]	99,599	567,960
CRH [15]	152,257	770,056
KDEm [22]	1,546	8,883
Eris	1,547	8,843

5.3 SotA Metrics

In this section, we want to show that Eris, beside the scenario exemplified above, can also provide the same features as competing approaches. To this end, we took the same dataset as in [22] (a.k.a. Population), which contains Wikipedia edit history related to city populations across different years. We cleaned the dataset using the transformations proposed by [22], resulting in 4,008 claims (a.k.a. values) for 1,124 entities (a.k.a. cities) from 2,344 sources (a.k.a. editors). We then annotated its schema according to our $K \triangleright V$ finite map signature, where entities and sources are treated as keys, and population claims as values. Finally, the query takes the values obtained for the LLUNs and automatically compares them against the ground truth (which was already in the database).

Table 5 summarizes the results of every system using MAE and RMSE, showing that Eris achieves the near-best MAE, but the best RMSE (which means it is never far from the truth), outperforming or closely matching the top-performing methods such as KDEm and TruthFinder. In contrast, traditional imputation strategies like Mean and Median, or other methods like GTM, CATD, and CRH, perform significantly worse, with MAE values being a hundred times worse and RMSE values a thousand times worse in some cases. Importantly, both MAE and RMSE were measured against the same ground truth in all these cases.

6 Conclusions

We presented an enhanced version of Eris for measuring discordance across multiple, multidimensional data sources. Our main contribution is the dynamic generation of symbolic variables representing uncertainty in the sources, enabling flexible, fine-grained comparisons across independent dimensions like time, latitude, and longitude. We also introduced a subproblem decomposition strategy to mitigate the limitations of LP-solvers such as OSQP in complex problems.

Experimental results show that (1) the approach is useful to interpret complex scenarios like satellite data, (2) our system achieves significantly higher accuracy than the baseline, while maintaining comparable execution time, and

(3) the truth values obtained as by-product of the quality measurement are in the level of the SotA.

Overall, the improved Eris offers a robust and scalable solution for reliable source quality assessment in complex, real-world settings. As future work, we plan to incorporate data provenance techniques to trace data transformations and better capture the propagation of quality and trust.

Acknowledgements. Y. A. Akter is funded by the EC Horizon 2020 research and innovation programme (DEDS: grant agreement No 955895). A. Abelló and P. Jovanovic are funded by the Spanish Ministerio de Ciencia e Innovación under project PID2020-117191RB-I00/AEI/10.13039/501100011033 (DOGO4ML) and the EC Horizon Europe programme (ExtremeXP: grant agreement No 101093164).

References

1. Abelló, A., Cheney, J.: Eris: efficiently measuring discord in multidimensional sources. VLDB J. **33** (2024)
2. Azzalini, F., Piantella, D., Rabosio, E., Tanca, L.: Enhancing domain-aware multi-truth data fusion using copy-based source authority and value similarity. VLDB J. **32**(3) (2023)
3. Bimonte, S., Ren, L., Koueya, N.: A linear programming-based framework for handling missing data in multi-granular data warehouses. Data Knowl. Eng. **128**, 101832 (2020)
4. Chen, J., Liu, Y., Xiang, Y., Sood, K.: RPPTD: robust privacy-preserving truth discovery scheme. IEEE Syst. J. **16**(3) (2021)
5. Dayal, U.: Processing queries over generalization hierarchies in a multidatabase system. In: Proceedings of 9th International Conference on Very Large Data Bases (VLDB). Morgan Kaufmann (1983)
6. Dong, X.L., Naumann, F.: Data fusion: resolving data conflicts for integration. Proc. VLDB Endow. **2**(2) (2009)
7. Fang, X., Shen, C., Sheng, Q.Z., Sun, G., Tang, Y., Zhuo, H.: A multi-truth discovery approach based on confidence interval estimation of truths. In: Proceedings of 19th International Conference on Advanced Data Mining and Applications (2023)
8. Feng, S., Glavic, B., Huber, A., Kennedy, O.A.: Efficient uncertainty tracking for complex queries with attribute-level bounds. In: Proceedings of ACM International Conference on Management of Data (SIGMOD) (2021)
9. Fercoq, O.: A generic coordinate descent solver for non-smooth convex optimisation. Optim. Methods Softw. **36**(6) (2021)
10. Geerts, F., Mecca, G., Papotti, P., Santoro, D.: The llunatic data-cleaning framework. Proc. VLDB Endow. **6**(9) (2013)
11. Ghadimi, E., Teixeira, A., Shames, I., Johansson, M.: Optimal parameter selection for the alternating direction method of multipliers (ADMM): quadratic problems. IEEE Trans. Autom. Control **60**(3) (2014)
12. Jaradat, A., Safieddine, F., Deraman, A., Ali, O., Al-Ahmad, A., Alzoubi, Y.I.: A probabilistic data fusion modeling approach for extracting true values from uncertain and conflicting attributes. Big Data Cogn. Comput. **6**(4) (2022)
13. Jiang, Z.: A decision-theoretic framework for numerical attribute value reconciliation. IEEE Trans. Knowl. Data Eng. **24**(7) (2012)

14. Li, Q., et al.: A confidence-aware approach for truth discovery on long-tail data. Proc. VLDB Endow. **8**(4) (2014)
15. Li, Y., et al.: Conflicts to harmony: a framework for resolving conflicts in heterogeneous data by truth discovery. IEEE Trans. Knowl. Data Eng. **28**(8) (2016)
16. Lin, X., Chen, L.: Domain-aware multi-truth discovery from conflicting sources. Proc. VLDB Endow. **11**(5) (2018)
17. Ma, F., et al.: Faitcrowd: fine grained truth discovery for crowdsourced data aggregation. In: Proceedings of 21th ACM International Conference on Knowledge Discovery and Data Mining (SIGKDD) (2015)
18. Marshall, J., Syed, M., Wang, D.: Hardness-aware truth discovery in social sensing applications. In: Proceedings of International Conference on Distributed Computing in Sensor Systems (DCOSS). IEEE (2016)
19. Nandi, A., et al.: Mimir: bringing CTables into practice. CoRR abs/1601.00073 (2016)
20. Stellato, B., Banjac, G., Goulart, P., Bemporad, A., Boyd, S.: OSQP: an operator splitting solver for quadratic programs. Math. Program. Comput. **12**(4) (2020)
21. Sundarmurthy, B., Koutris, P., Lang, W., Naughton, J.F., Tannen, V.: m-tables: Representing missing data. In: Proceedings of 20th International Conference on Database Theory (2017)
22. Wan, M., Chen, X., Kaplan, L., Han, J., Gao, J., Zhao, B.: From truth discovery to trustworthy opinion discovery: an uncertainty-aware quantitative modeling approach. In: Proceedings of 22nd ACM International Conference on Knowledge Discovery and Data Mining (SIGKDD) (2016)
23. Xiao, H., et al.: Towards confidence interval estimation in truth discovery. IEEE Trans. Knowl. Data Eng. **31**(3) (2018)
24. Yang, Y., Meneghetti, N., Fehling, R., Liu, Z.H., Kennedy, O.: Lenses: an on-demand approach to ETL. Proc. VLDB Endow. **8**(12) (2015)
25. Yin, X., Han, J., Yu, P.S.: Truth discovery with multiple conflicting information providers on the web. IEEE Trans. Knowl. Data Eng. **20**(6) (2008)
26. Zhang, J., Wang, S., Wu, G., Zhang, L.: A effective truth discovery algorithm with multi-source sparse data. In: Shi, Y., et al. (eds.) ICCS 2018. LNCS, vol. 10862, pp. 434–442. Springer, Cham (2018). https://doi.org/10.1007/978-3-319-93713-7_37
27. Zhao, B., Han, J.: A probabilistic model for estimating real-valued truth from conflicting sources. In: Proceedings of International Workshop on Quality in Databases (QDB), vol. 1817 (2012)
28. Zhi, S., Yang, F., Zhu, Z., Li, Q., Wang, Z., Han, J.: Dynamic truth discovery on numerical data. In: IEEE International Conference on Data Mining, (ICDM), pp. 817–826. IEEE Computer Society (2018)

Pasteur: Scaling Privacy-Aware Data Synthesis

Antheas Kapenekakis[1(✉)], Daniele Dell'Aglio[1], Martin Bøgsted[2], Minos Garofalakis[3], and Katja Hose[4]

[1] Aalborg University, Aalborg, Denmark
antheas@cs.aau.dk
[2] Aalborg University Hospital, Aalborg, Denmark
[3] Athena Research Center, Athens, Greece
[4] TU Wien, Vienna, Austria

Abstract. Privacy-aware data synthesis is a field aiming to liberate data access through the generation of synthetic data which mirrors the original without resulting in privacy exposure. State-of-the-art algorithms for structured data perform well in datasets with tables of a few million rows but result in prohibitive runtimes when scaling to hundreds of millions of rows. In addition, due to the sensitive nature of data, practitioners are often limited to a single server environment. In this paper, we present the framework Pasteur, which aims to scale privacy-aware data synthesis linearly under a single server environment. Pasteur achieves this through a parallelization approach tailored for synthesis, optimized memory representations, and an accelerated marginal calculation algorithm (bottleneck in a class of privacy-aware algorithms). We show Pasteur performing pre-processing, synthesis, and evaluation of a tabular dataset with 1 billion rows (200 GB) in 1 h on a 16 core CPU server.

1 Introduction

Data synthesis aims to promote responsible data sharing through anonymization [11]. However, state-of-the-art algorithms have focused on structured data of limited size and complexity, i.e., tables with up to 5 million rows (less than 100 MB) [3,4,13,14,20,23]. Hence, these datasets are small enough to fit conveniently in main memory and can be processed swiftly without the need for parallel or efficient processing. However, real-world datasets may contain dozens of tables with a larger degree of complexity and billions of rows. This is exacerbated with sensitive data, as GDPR resources are limited, often to a small percentage of a cluster's resources and to a single machine, due to regulatory requirements.

Therefore, in this paper we attempt to address the challenge of scaling synthesis linearly to larger-than-memory datasets when confined to a single server environment. To solve this challenge, we present Pasteur, a system for scalable privacy-aware data synthesis for datasets far larger than main memory.

P. K. Chrysanthis et al. (Eds.): ADBIS 2025, LNCS 16043, pp. 164–180, 2026.
https://doi.org/10.1007/978-3-032-05281-0_11

We focus on scaling privacy-aware Probabilistic Graphical Model (PGM) algorithms [3,12,13,22], since recent benchmarks [18] show them performing well in terms of privacy versus utility.

In summary, the main contributions of this paper are: 1. Improved memory usage by information pruning and optimized memory formats. 2. A parallelization approach that maintains a constant scaling factor equal to the number of cores, with stable memory use regardless of dataset size. 3. A theoretically optimal Single Input Multiple Data (SIMD) marginal calculation algorithm for PGMs offering a 150x speedup over existing implementations in single core containing multiple parallelization strategies to ensure it scales linearly across all CPU cores for tables from 1M to 1B rows and beyond.

The remainder of this paper is structured as follows: Section 2 provides preliminaries and the problem definition, Sect. 3 presents the Pasteur system with its improved memory usage and parallelization. Afterwards, Sect. 4 presents the marginal calculation algorithm; the first part covers the algorithm and the second part its parallelization. Our evaluation results are discussed in Sect. 5. Finally, Sect. 6 covers related work and Sect. 7 concludes the paper.

2 Problem Definition and Analysis

Let D be a sensitive relational dataset consisting of multiple tables $T \in D$ with foreign key dependencies. Each row in T describes a protected entity such as patients. The goal of privacy-aware data synthesis is to produce a dataset S that approximates D without incurring privacy exposure of the entities of D. For privacy-aware data synthesis, recent work has focused on Probabilistic Graphical Model (PGM) algorithms [3,12,13,22] as they have compelling performance [18] when privacy is required, so in this work we optimize those. When working with sensitive data, analysts are often limited to a single-machine environment, typically a server or laptop, with a predefined number w of CPU cores and amount M of main memory. The problem we aim at studying is:

Problem 1. How can we maximize the usage of w available CPU cores to efficiently perform PGM-based data synthesis with datasets D, where the size of a dataset D in memory $|D|$ is up to $|D| >> M$.

To solve this problem, we identify three challenges. Firstly, the memory footprint of data needs to be optimized, as a smaller footprint allows for higher throughput, leading to faster synthesis. Therefore, the first challenge is *identifying and removing unnecessary information* in the data and optimizing data encodings to compress the data both in main memory and disk. Once the data is in an optimal format, the next challenge is to optimize parallelization. That means, we aim *to maximize the usage of w cores*, while *memory usage remains stable under M*, but close to it to minimize swapping. Finally, we aim to optimize how synthesis algorithms handle data themselves. We focus on PGM algorithms, which only access input data through a single operation: marginal calculation. Therefore, the third challenge lies in *how to optimize marginal calculation.* By achieving these challenges, the PGM algorithms we surveyed scale to larger-than-memory datasets without any changes.

Here, it is worth to present an overview of marginal calculation to complement the complexity analysis in Sect. 4. Let $T \in D$ be a table with N rows and C columns associated to discrete variables $T_1, \ldots, T_C$. Let $supp(T_j)$ be the support of T_j, i.e., the possible values that T_j can take. Given a subset of the column indices $\{j_1, \ldots, j_r\} \subseteq \{1, \ldots, C\}$, we define an r-dimensional marginal of T as a set of r columns $T' = \{T_{j_1}, \ldots, T_{j_r}\} \subseteq T$, where $E' = supp(T_{j_1}) \times \cdots \times supp(T_{j_r})$ is the set of possible values of the rows in T'. We define the marginal frequency distribution of T' as:

$$f(e_1, \ldots, e_r) = \frac{1}{N} \sum_{i=1}^{N} 1[t_{i,j_1} = e_1, \ldots, t_{i,t_r} = e_r], \ \forall (e_1, \ldots, e_r) \in E',$$

where $1[\cdot]$ denotes the indicator function. The goal of marginal calculation is computing the marginal frequency distribution of T'. This is performed in three steps. Firstly, we linearize the column values into a unique value per combination:

$$s_i = t_{i,j_1} + \sum_{k=2}^{r} t_{i,j_k} \prod_{l=1}^{k-1} |supp(T_{j_l})|, \forall i \in \{1, \ldots, N\}.$$

Secondly, we create the marginal frequency distribution of the linearized values:

$$h_s = \sum_{i=1}^{N} 1(s_i = s), \ \forall s \in \{1, \ldots, |E'|\}.$$

Thirdly, we reshape h_s to the r-dimensional distribution $T' = [T_{j_1}, \ldots, T_{j_r}]$.

For a collection of R marginals, we obtain the following complexity analysis. Linearization is an arithmetic operation that scales with the number of marginals R and, the number of columns per marginal r, and the number of rows N in the table the marginal is calculated for: $O(R \cdot N \cdot r)$. The second step accesses one vector of size N per marginal: $O(R \cdot N)$. Finally, post-processing depends on the marginal domain size $|E'|$ and the number of marginals R: $O(R \cdot |E'|)$. $|E'|$ is in the thousands for all marginals, as that is required for them to be statistically significant, which results in $N \gg |E'|$. We may treat r as a constant, as it is a small single digit number for all marginals. Given the above, the dominant term $O(R \cdot N)$. Since $O(R \cdot N)$ is a very large term, if the synthesis algorithm requests a large R, it is important this operation is performed efficiently. In addition, the size of a table T affects its ideal placement in memory (single copy, multiple copies, or swapping), necessitating varying memory placement depending on the magnitude on N.

3 Pasteur System and Principles

In this section, we cover a solution to the two first challenges, defining the architecture system: *identifying and removing unnecessary information* in the data and optimizing data encodings to compress the data both in main memory and disk, and *optimizing for parallelization* in a single node with memory pressure.

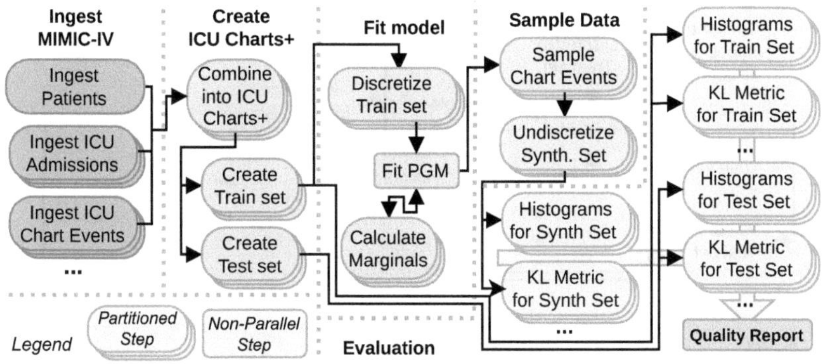

Fig. 1. Example Pipeline for the ICU Charts Dataset.

3.1 Data Representation and Preprocessing

Improved memory density acts as a multiplier for the efficiency of data synthesis. This is for two reasons: first, it allows holding more data in memory at a time, minimizing swapping, and second, denser representations enable Single Instruction Multiple Data (SIMD) to handle more data per CPU instruction. Following, we walk through the steps of optimizing a dataset by using as an example the dataset MIMIC-IV [9], a large medical dataset (110 GB uncompressed) sourced from a US hospital and provided as a set of 27 tables $T \in D$ in CSV files. In it, we find text columns containing medicine, dosages, doctor's notes, dates in US format, and numerical columns with various ranges.

Pruning Columns. We distinguish columns in two types: learnable and non-learnable. Learnable columns store the information that should be learned and synthesized, such as age and diagnosis. Non-learnable columns are personal information with little or no value in the context of data synthesis, such as patient name and email. We prune non-learnable columns during ingestion, to avoid overhead, and when one is used as a join attribute, we replace it with an integer.

Encoding. Many datasets come with data in string format. For example, MIMIC-IV is available as a CSV file, where dates and categorical values are encoded as strings. We transform these values to reduce the memory footprint and improve performance. Specifically, dates are converted to Unix timestamps (int64). Categorical columns are converted to integers through dictionaries to store the original value. For integers, when they represent categorical values, the smallest bit width possible is chosen. Finally, numerical columns are converted to floating point values, with a preference for float32. The end result is a compacted dataset that can be used directly for synthesis.

Partitioning. Despite the optimized dataset being smaller, it might still not fit in main memory. For such cases, we designed a partitioning strategy for efficient swapping between disk and main memory. Here, we leverage the properties of

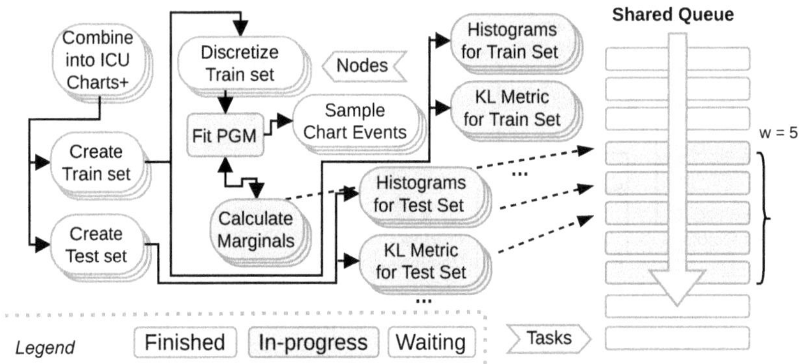

Fig. 2. The architecture applied to the example in Fig. 1

synthesis: it consists of two types of operations: processing steps (e.g., discretizing, encoding) and model fitting (e.g., sampling marginals) on a set of entities (e.g., patients). In these operations, synthesis entities (e.g., patients, customers) are treated independently (e.g., during preprocessing each patient is encoded independently, in fitting models batches may be used). Therefore, it is possible to shard the input dataset using synthesis entities as a key.

Conclusion. Pruning, partitioning, and encoding form a foundation for processing arbitrarily large datasets with maximal throughput. Following, we build upon it with a parallelization approach tailored to running on single nodes.

3.2 Parallelization Approach

Data synthesis experiments are processing heavy: dataset is re-ingested every time to account for hyper-parameters, output dataset has to be processed and evaluated. We observe that this results in many independent processing steps that can be executed simultaneously (evaluation, preprocessing model fitting; inter-task parallelism). In addition, each step is parallelizable, as the data is partitioned (intra-task parallelism).

Therefore, in the architecture of Pasteur, we represent synthesis pipelines as Directed Acyclic Graphs (DAGs) G where nodes correspond to processing steps and edges to artifact dependencies. Figure 1 shows a Pasteur's synthesis pipeline with the MIMIC-ICU dataset. The parallelization challenge can then be reformulated into parallelizing a synthesis DAG G with a constant parallelization factor, close to w, while memory use is under M.

We note that due to the memory constraint, naive parallelization with fixed factors for inter-task (w_1) and intra-task (w_2) would result in a varying parallelization factor $w_{\max} = w_1 \cdot w_2$ and memory use, which is not optimal on a single server (either average CPU use is low or there is a risk of memory overflow). Therefore, we propose a new parallelization approach where each synthesis step schedules sub-steps, which act as the atoms of computation, where each sub-step operates on a partition of the dataset.

In this architecture, there is a shared queue which is bound to process up to w sub-steps at a time. Steps in G launch as light threads that schedule sub-steps to run per partition on the queue. Each sub-step loads, processes, and saves a partition to disk, avoiding Inter Process Communication (IPC). Since each sub-step operates on a similar sized partition, they have similar memory footprint proportional to that. Therefore, given a predetermined partition size, w parallel tasks can be bound to a memory footprint under M. This architecture is shown in Fig. 2. The benefit of this approach is that it allows processing multiple heterogeneous and arbitrary operations simultaneously, while remaining agnostic to them, with a stable factor close to w and bounded memory use under M, for a wide range of w and $|D|$ values.

Partition Sizing. The final part is determining the number of partitions N_P to split the dataset to. Assuming each partition has size $|P|$, we have $|D| = N_P \cdot |P|$. Each sub-step execution uses a similar amount of memory $M_{\text{sub-step}}$, where $w \cdot M_{\text{worker}} < M$. In addition, there is certain overhead for sub-step operations, therefore we define c as a constant representing memory headroom (e.g., 2–5): $M_{\text{worker}} = c \cdot |P|$. From combining the previous we get $w \cdot c \cdot |P| < M$. Finally, we derive the partition number needs to meet the following $N_P > \frac{|D|}{M} \cdot c \cdot w$.

This is a lower bound. For the optimal number, we insert two additional constraints. First, since sub-steps are scheduled in groups of size N_P (one for each partition) and have similar execution times (similar operation and size), there is a preference for making N_P a multiple of w. Then, a smaller N_P minimizes scheduling overhead by maximizing memory use. Applying both constraints results in the following: $N_P = \left\lceil \frac{|D|}{M} \cdot c \right\rceil \cdot w$, where $\lceil \cdot \rceil$ is the ceiling operator.

4 Accelerated Marginal Calculation

As we analyzed in the third challenge of Sect. 2, marginal calculation is a crucial operation in PGM algorithms, and the only function which affects their scaling related to dataset size. Various PGM works [3,13] use the NumPy function "histogramdd" for marginal calculation. This is sufficient for small datasets but features a large overhead due to not being specialized for marginal calculation, as it is a floating point multidimensional histogram function (e.g., unnecessary binning, 32 bit widths). Therefore, in this section, we optimize this operation in two steps: first, we derive an optimized SIMD variant of the marginal calculation algorithm (Sect. 4.1) under a single core in-memory. Then, we integrate this approach to the parallelization architecture of the system (Sect. 4.2), to provide linear scaling across a wide range of dataset sizes.

4.1 Optimizing the Computation of Histograms

To optimize linearization, we begin with a vectorized approach. Then, we further optimize it with a custom Operator Fusing SIMD algorithm, which applies both SIMD and operator fusing optimizations.

Vectorized Approach. First, we apply the marginal calculation steps using vectorized functions, resulting in the following complexity: when linearizing the column values of an r-dimensional marginal, we perform one multiplication for each column apart from the first one ($r-1$ reads, $r-1$ writes) and then sum the results (r reads, $r-1$ writes). For N rows, we have a memory access complexity of $O(N(4 \cdot r - 3))$. Finally, counting is performed by iterating over the final index (1 read) and increasing a marginal vector (1 read, 1 write) for $O(3N)$. The total memory access complexity is $O(4 \cdot N \cdot r)$. In addition, through using vectorization we benefit from SIMD. SIMD allows processing multiple rows at a time which, when combined with the efficient memory structure introduced in Sect. 3.1, results in a considerable theoretical speedup (16–32x for 8–16bit values over 256bit SIMD). However, due to the way the linearization and counting steps interact (loup dependencies, mixed integer widths), operator fusing, another crucial optimization is not applied automatically.

Operator Fusing SIMD Approach. Linearization creates multiple intermediary multiplication results for calculating the vector s_i, which is then used for the marginal frequency distribution calculation. Both s_i and other intermediary multiplication results are not used after the calculation, and saving them to memory consumes main memory bandwidth. Ideally, we would use the concept of operator fusing, where all intermediary outputs are kept in registers, and only h_s is written to memory. Since h_s has a cardinality up to thousands (see Sect. 2), this results in it residing in the CPU cache, eliminating all memory writes.

While operator fusing and Single Input Multiple Data (SIMD) are widespread techniques, marginal calculations break automatic optimizations that apply in compilers in two ways: by mixing integer widths during linearizing (columns use different widths depending on cardinality) and introducing a loop-driven dependency during the creation of the frequency vector. To embed both SIMD and Operator Fusing at the same time, we create a custom algorithm, shown in Algorithm 1. We use 256 bit SIMD instructions as a base. Since the majority of marginals have a size smaller than 65535, we present a version for 16 bit marginals in the paper, which can be trivially extended to 32 bits.

In an initial step, the algorithm separates variables based on their cardinality into unsigned 8 bit and 16 bit groups and calculate a multiplier for each based on cardinality. Then, a partially unrolled loop is used to handle 16 (16 byte) rows at a time. 8 bit variable groupings are first expanded to 16 bit. Then, the columns are linearized using 16 bit vectorized instructions (1 read per variable and row), by multiplying the values with a precalculated multiplier. The final result is used to increment the histogram with individual instructions, as SIMD does not apply due to potentially incrementing the same memory location. We compile different versions of the algorithm for combinations of n_8 and n_{16}, which removes the inner for loops that would be used for linearization.

Therefore, during execution, the algorithm runs a single for loop that processes 16 rows per iteration, with no inner-for loops. It executes a set of predefined SIMD operations back-to-back which result in creating the linearized vector, then 16 individual writes increment the histogram residing in CPU cache.

Algorithm 1. 256 bit marginal calculation for 16 bit marginals.

procedure MARGINAL_16BIT_256BIT(N, out, n_8, mul_8, arr_8, n_{16}, mul_{16}, arr_{16})
  for $i \leftarrow 1$ **to** $N - 16$ **step** 16 **do**
    $r_{idx,256} = 0$
    for $j \leftarrow 1$ **to** n_8 **do** ▷ Unrolled loop
      $r_{128} \leftarrow arr_8[j][i : i + 16]$ ▷ $16 * 8 = 128$
      $r_{256} \leftarrow$ EXPAND(r_{128})
      $r_{256} = mul_8[j] * r_{256}$
      $r_{idx,256} = r_{idx,256} + r_{256}$
    end for
    for $j \leftarrow 1$ **to** n_{16} **do** ▷ Unrolled loop
      $r_{256} \leftarrow arr_{16}[j][i : i + 16]$
      $r_{256} = mul_{16}[j] * r_{256}$
      $r_{idx,256} = r_{idx,256} + r_{256}$
    end for
    for $j \leftarrow 1$ **to** 16 **do** ▷ Unrolled loop
      INCREMENT($out[r_{idx,256}[j]]$)
    end for
  end for
  ... ▷ Handle last $N \mod 16$ rows.
end procedure

Incrementing and SIMD use different ALU units, causing the out-of-order execution of the CPU to partially overlaps them. Since the resulting marginal frequency distribution fits in CPU cache, and all operands are stored in registers, the algorithm performs $O(rN)$ memory accesses, which is 4 times lower than the vectorized approach. Moreover, if the dataset is smaller than the processor's cache (e.g., a couple thousand rows), we remove all memory accesses after calculating the first marginal as the dataset is cached.

4.2 Parallelizing Marginals

The prior approach calculates a single marginal on a single core in memory. Given a collection of marginals to compute, we need to parallelize and scale it to w cores, for a wide range of dataset sizes. Here, memory bandwidth becomes a concern. We have three memory tiers ordered by bandwidth: CPU Cache, RAM, and disk. Given a certain dataset size, we want to remain to the fastest tier possible, and ensure we saturate its bandwidth. Therefore, we propose three memory strategies, where each is optimal for a specific dataset size range, as we investigate in Sect. 5.

Using Shared Memory. In this approach, we place the whole dataset in a shared memory pool and let cores share it during marginal calculation. Then, we evenly distribute the marginal collection to each core. If the dataset is small enough to fit in the CPU L3 cache (e.g., up to 100k rows), this removes most memory accesses with both the marginal vector and the dataset staying in it. If it is larger, however, cache invalidation and limited memory channel bandwidth harm performance, making this approach non-optimal.

Using Unique Copy. Here, we give each CPU performing marginal calculations its own copy of the data and a subset of marginals to calculate. The random placement ensures that the w cores utilize all DRAM channels and their access patterns do not cross, which would cause cache retrieval invalidation. The downside of this approach is that for w cores, the system memory used is $(w+1) \cdot |D|$, so it is only applicable if $(w+1) \cdot |D| < M$.

Using Unique Partition. Finally, for datasets that do not fit in memory, we partition the dataset as in Challenge 2, and give each core a partition and the whole marginal batch, having each core calculate sub-marginals for its partition. In the end, the sub-marginals are merged to create the final dataset marginal. If $N_P > w$, after finishing, each core receives another partition. This approach is only viable if there are enough rows in each partition, otherwise it results in slower performance. The overall computation scales with $O(R \cdot |D|)$ where merging sub-marginals scales with $O(R \cdot |r| \cdot N_P)$. If $|D| \gg r \cdot N_P$, which is equivalent to $|P| \gg r$, the overhead of merging sub-marginals is theoretically amortized. This approach scales infinitely, as cores can load partitions from disk. Moreover, if $|D| < M$, storing the partitions in memory can remove disk overhead for subsequent marginal batches.

5 Evaluation

In this section, we present the results of our evaluation of Pasteur; we evaluate scalability, with special focus on memory usage, parallelization, computing marginals, and current state of the art. For these experiments, we use the dataset Adult [6] and two datasets derived from MIMIC-IV [9]. Adult is a 55k row table containing US census data. Then from MIMIC-IV, we derive an Admissions dataset, which is the combination of MIMIC's Admissions and Patients tables, and the ICU Charts dataset, which combines the ICU chartevents, ICU stays and Patients tables. To obtain a challenging large dataset with ca. 1B rows, we duplicate the ICU Charts dataset rows three times and shuffle. After preparing the datasets, we obtain the statistics sketched in Table 1.

State-of-the-art PGM algorithms for data synthesis typically require input data in a single table with categorical attributes only. Therefore, we then transform numerical attributes and dates into categorical ones (shown under Categorical), by binning the numerical attributes and extracting information from dates (e.g., season, time of day). Scripts for creating these datasets from the original artifacts are available with the Pasteur source code online at https://github.com/pasteur-dev/pasteur.

Regarding the experimental environment, we run the in-memory marginal experiment on an AMD EPYC VM with a T4 GPU (as JAX requires a GPU), using a single core. We run the parallelization comparison with Dask on the same machine, using 20 virtual cores and 64 GB, to study the behavior of the

Table 1. Datasets used in experiments

Dataset	Rows	Original Form			Categorical Form
		Date	Number	Discrete	Discrete
Adult	50 k	0	5	10	15
MIMIC IV Admissions	450 k	4	0	10	19
MIMIC IV ICU Charts	$\leq$ 1 B	4	1	5	15

systems under increased memory pressure. For the rest of the experiments, we use an AMD EPYC 16c/32t 256 GB RAM bare-metal server with a SATA SSD. Both machines use Ubuntu 20.04.

5.1 Memory Representation

To evaluate memory consumption (Challenge 1), we use ICU Charts before (Original Form) and after discretizing all attributes (Categorical Form). For an in-memory representation, we use Python/Pandas because of its widespread use by most implementations of state-of-the-art synthesis algorithms. Since there is a need for swapping to disk, we compare popular file formats.

In Table 3, we see that using Pandas to load the data into main memory without providing specific type information (Naive) results in a very high memory consumption (777 GB and 136 GB respectively) because suboptimal formats are used (strings, float64, and int64). Both numbers are extrapolated from 15M sample. When applying the optimizations from Sect. 3.1 (Optimized, i.e., pruning non-learnable columns, dictionary encoding, shrinking the integer width), we considerably reduce memory footprint to 49 GB and 24 GB, respectively.

Then, we compare popular file formats for disk storage: CSV, CSV with GZIP compression, and Parquet. We report storage space on disk as well as the time to create those files (write time). We see that the most space-optimal storage format is Parquet, which requires only 19 GB (12 GB) instead of 197 GB (41 GB) in CSV format. CSV with GZIP achieves a similar compression ratio, with a consumption of 31 GB (14 GB). However, unlike Parquet, the time required for compressing and writing to disk is very high (3h and 2h respectively). Therefore, according to our measurements, Parquet files combine an efficient compression ratio with efficient disk writes. Moreover, when we partition the data, we can even parallelize the write process, achieving 1 m to write 19 GB with 60 partitions and 16 hyper-threaded cores (w is 32).

5.2 Parallelization Approach

To evaluate parallelization, we compare Pasteur against an in-memory baseline with Pandas (no parallelization) and Dask [5], a state-of-the-art distributed replacement for Pandas, shown in Fig. 3.

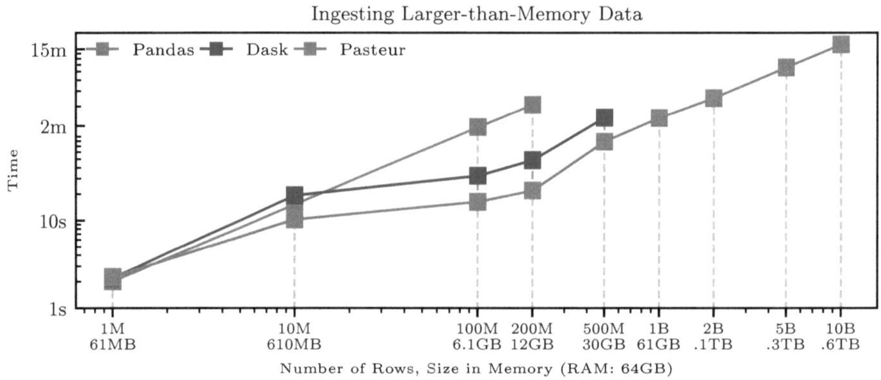

Fig. 3. Comparisons of Parallelization Approaches.

We use the transformation that loads the ICU Charts dataset, converts it from the Original format to Categorical, and saves it. Our approach is intertwined with our system, so we create a micro-benchmark with Pandas that does a similar transformation, such that at 1M rows it has a similar footprint, where we have a 10 % slowdown over the Pandas baseline while using Pandas, for fairness. Then, as Dask is a replacement for Pandas, we swap it out.

Results. Pandas is a single-core in-memory baseline that scales linearly up to 200M rows (12 GB) in memory and then runs out of memory, as it requires a certain overhead for data processing. Dask performs better and can run partially on larger-than-memory data, as processing 500M rows (30 GB) required more than 64 GB of RAM for Pandas. However, when the dataset is larger than memory, Dask creates memory spikes that cause it to crash. Moreover, its eager spill-to-disk query architecture introduces overhead over Pandas (at 10M rows, it is 1.5x slower) and it only starts to parallelize effectively only above 10M rows.

Our approach uses a custom partition size and begins to parallelize before 10M rows. From 10M-200M rows, there is not enough work to parallelize across 40 workers which causes execution time to flatten out. Since our approach does not introduce overhead over Pandas, it is 1.5x faster than Dask from 10M to 500M rows. Following, it continues to parallelize linearly regardless of the dataset size. We stop the experiment after 10B rows, which have a memory size of 600 GB, a magnitude larger than system memory.

5.3 Marginal Calculation

To evaluate our marginal approach, we use a collection of marginals collected while running PGMs on the ICU Charts dataset. We examine the influence of the number of rows and the marginal domain size $|E'|$ on runtime. In Table 2, we report the average time to calculate a marginal with one core in memory across several implementations. We consider the following baselines. Two baselines are based on popular frameworks: one with Pandas' value_counts [15] and one with

Table 2. Marginal Calculation Times

Rows	$\|E'\|$	Arbitrary (SOTA)		Vectorized			C-based	Pasteur	
		Pandas	**NumPy**		JAX			**AVX2 (CPU)**	
		vc[1]	**hdd**[2]	Custom	CPU	GPU		uint32	**uint16**
20k	2^{12}	2.39 ms	2.99 ms	99 μs	211 μs	312 μs	118 μs	27.9 μs	**18.7** μs
500k	2^{12}	51.9 ms	76.2 ms	1.65 ms	4.13 ms	538 μs	2.76 ms	613 μs	**383** μs
1M	2^{12}	87.0 ms	144 ms	4.02 ms	8.33 ms	736 μs	5.47 ms	1.20 ms	**742** μs
2M	2^{12}	167 ms	291 ms	10.3 ms	16.8 ms	1.07 ms	10.7 ms	2.40 ms	**1.55 ms**
5M	2^{12}	424 ms	716 ms	28.9 ms	48.6 ms	1.37 ms	26.9 ms	6.15 ms	**4.03 ms**
10M	2^{12}	785 ms	1.44 s	66 ms	94.2 ms	3.77 ms	53.8 ms	12.2 ms	**7.77 ms**
5M	2^{8}	313 ms	449 ms	20.8 ms	47.4 ms	2.68 ms	19.5 ms	5.11 ms	**3.29 ms**
5M	2^{12}	412 ms	727 ms	47.6 ms	6.19 ms	3.91 ms	26.9 ms	6.19 ms	**3.91 ms**
5M	2^{16}	556 ms	956 ms	40.2 ms	47.5 ms	2.21 ms	32.3 ms	11.3 ms	**8.15 ms**
5M	2^{20}	618 ms	1.24 s	71.9 ms	41.9 ms	2.2 ms	38.1 ms	**20.9 ms**	–
5M	2^{24}	642 ms	OVF[3]	79.1 ms	OVF[1]	OVF[1]	42.5 ms	**21.8 ms**	–
5 M	2^{28}	966 ms	OVF[3]	127 ms	OVF[1]	OVF[1]	107 ms	**61.2 ms**	–

[1]Pandas value_counts function, [2]NumPy histogramdd function
[3]Memory Overflow (allocates 300 GB-1.3 TB of heap)

NumPy's histogramdd [3,12,13]. We built baselines that exploit the vectorized approach (as presented in Sect. 4) using two standard libraries: NumPy and JAX [2] – both environments lack support of operator fusing. For JAX, we report results of two versions: CPU and GPU. We also implement a baseline for marginal calculation in C using a standard for loop without SIMD. The last columns present Pasteur, which includes the Operator Fusing SIMD approach presented in Sect. 4.1 in full, using the AVX2 instruction set, both for 16 bit and 32 bit marginal domain sizes $|E'|$. The best CPU results are bolded and the best GPU results are underlined.

The results show that our proposed algorithm in 16 bits achieves the best performance, with up to 150x speedup in comparison to state-of-the art baseline implementations. In addition, by extending it to 32 bits, given a 1.5x slowdown, it can handle occurances with $|E'|$ larger than 16 bits. The vectorization approach shows with a speed-up of 20x on average in comparison to histogramdd, due to lower overhead. However, without operator fusing, it does not achieve the results of Pasteur. The JAX implementation using a GPU shows the fastest runtime. However, in the context of using a full GPU and being limited to its VRAM, compared to using a single CPU core, it has a higher cost.

Parallelization. We evaluate parallelizing the Operator Fusing SIMD approach using the three memory strategies (Shared Memory, Unique Copy, Unique Partition) proposed in Sect. 4.2, under 32 logical cores (i.e. 16 hyper-threaded cores). Table 4 reports the results for varying numbers of input rows; showing marginal throughput per second as well as relative speed-up when running on 32

Table 3. Memory Format performance

	Format	ICU Charts Dataset (1B)			
		Original		Categorical	
		Space	t^1	Space	t^1
Mem	Naive	777 GB	–	136 GB	–
	Optimized	49 GB	–	24 GB	–
File	.csv	197 GB	17 m	41 GB	11 m
	.csv.gz	31 GB	3h	14 GB	2h
	.pq	19 GB	10 m	12 GB	8 m
.pq partition		19 GB	1 m	12 GB	1 m

[1]Save (write) time

Table 4. Parallelization Strategies (rows/s)

Rows	Single Core	Parallelized 16c/32t					
		Shared Memory		Unique Copy		Unique Partition	
20k[1]	16.8k	**27k**	**2x**	26.6k	1.6x	–	
100k[1]	7.1k	**23k**	**3x**	23.3k	3.3x	–	
500k	1.6k	**14k**	**9x**	13.2k	8.3x	–	
1M	754	**9k**	**12x**	8.9k	12x	–	
5M	160	935	5.8x	**2.3k**	**14x**	–	
10M	81	413	5.0x	**1.2k**	**15x**	470	6x
50M	16.0	122	7.6x	**223**	**14x**	179	11x
100M	8.0	98.6	12x	–		**104**	**13x**
500M	1.6	18.6	12x	–		**25**	**16x**
1B	0.8	10.1	13x	–		**12**	**16x**

[1]Single thread orchestration limit

logical cores versus a single core. If it is not feasible to use one of the approaches, we put "-" in the table. Specifically, Unique Partition is not applicable for low row numbers because there are not enough for multiple partitions and Unique copy requires a copy of the dataset for each core, and there is not enough memory for it beyond 50M rows.

These results confirm our theoretical analysis. Shared Memory works well for smaller numbers of rows, whereas for higher numbers ($> 1M$), after the dataset stops fitting in the CPU cache, total memory bandwidth is limited. By giving each core its own copy of the dataset, Unique Copy maximizes memory bandwidth and is faster. Due to the additional memory required, it can only be applied up to 50M rows. Afterwards, Unique Partition performs well, as the dataset size increases to amortize the cost of merging the partition marginals. By varying the approach based on dataset size (bolded results) it is possible to achieve linear parallelization (13x-16x for 16 hyper-threaded cores) for tables beyond 5M rows with great parallelization down to 500k rows. If the dataset is trivial ($\leq$ 100k rows), the single thread that orchestrates the marginals and distributes them to cores becomes the overhead.

5.4 Synthesis Pipelines

With all system components in place, we perform an experiment with end-to-end synthesis pipelines while using three state-of-the-art PGM algorithms: AIM [13], MST [12], PrivBayes [22] and three datasets (Adult, Admissions, ICU Charts). For the ICU Charts 3 dataset, we evaluate it by subsampling it with different row counts and use 60 partitions starting at 100M rows. In other datasets, we use a single partition. For all algorithms, we use hyper-parameters recommended by

Table 5. Performance of the Synthesis Pipeline

Dataset	Rows	ϵ	**Ingest**		PrivBayes		MST		AIM		**PrivBayesHV**	
			pre	**data**	time	N	time	N	time	N	**time**	**N**
Adult	50 k	1	**1 m**	**1 m**	1 m	168	3 m	120	13 m	120	**1 m**	**10.5 k**
MIMIC												
Admit.	550 k	1	**16 m**	**30 s**	1 m	8k	5 m	210	22 m	210	**9 m**	**1.3 M**
ICU Charts	1M	1		**3 m**	1 m	257	4 m	136	187 m	136	**1 m**	**11.3 k**
	10 M	0.1		**5 m**	2 m	302	5 m	136	218 m	136	**3 m**	**17.7k**
	100 M	0.01	**16 m**	**6 m**	6 m	306	8 m	136	9 m	136	**11 m**	**16.3 k**
	500 M	0.002		**13 m**	17 m	307	27 m	136	23 m	136	**37 m**	**15.8k**
	1 B	0.001		**21 m**	30 m	300	42 m	136	30 m	136	**72 m**	**16.0 k**
1 core →	500 M			**139 m**	180 m (11x)		197 m (7.3x)		145 m (6.2x)		**196 m (13.6x)**	

their authors (e.g., AIM uses all two-way column combinations as a workload) and only vary the privacy budget ϵ.

A pipeline in this experiment consists of three parts: pre-ingestion (transform raw data into a set of tables in Parquet format that are readily available), ingestion (create the dataset table and convert it to categorical), and synthesis (fit, sample the PGM algorithm, and evaluate the resulting data). For evaluation metrics, as a dummy workload, we calculate per-column histograms, KL divergence for each column pair, and χ^2 tests for each column. These are parallelizable and contribute to the pipeline runtime. Since our optimizations do not change the output of the PGM algorithms, we omit qualitative results.

To ensure comparability of the evaluation results between different row counts on the ICU Charts dataset, we keep the ratio $\epsilon \cdot N$ constant, where N is the row number of the table and ϵ is the privacy budget hyperparameter of differential privacy, the mechanism used to ensure privacy. This is because the complexity of all three surveyed algorithms scales with that ratio. PrivBayes uses noise reversly proportional to ϵ and keeps the signal-to-noise ratio of marginals constant, where the signal is proportional to N. AIM and MST instead work by adding noise to denormalized marginals (sum to N) in reverse proportion to ϵ; where a larger ϵ equates to less noise and a larger signal-to-noise ratio. Therefore, when we vary the number of rows in the ICU Charts dataset, we update ϵ such that $\epsilon \cdot N$ remains constant.

Our evaluation results are illustrated in Table 5, which shows the execution time in minutes and the number of calculated marginals. The runtime of the approaches increases sub-linearly with the growing number of rows, e.g., AIM requires 30 min for 1B rows and 23 min for 500M rows. The results meet our expectations; as other than marginals, we do not parallelize the algorithms themselves, which have a fixed runtime. There is a small exception with AIM using the ICU Charts dataset at 1M and 10M rows. The ICU Charts dataset is noisy by its nature and especially with a low row count, it causes AIM, an iterative algorithm, to try to overfit it and repeatedly perform measurements (we re-ran the experiment multiple times with different dataset samples and obtained similar results).

Finally, we evaluate the influence of parallelization, by executing the ICU Charts dataset with 500M rows using a single core and compare it to multicore. In dataset ingestion, the peak RAM use is 70 GB for multicore (32 logical cores) and 25 GB for single core. For the rest of the steps, the peak RAM use is 85 GB for multicore and 10 GB for single core (all algorithms). PrivBayes has great parallelization for multicore (11–14x) as it contains a small fixed overhead. AIM and MST have a larger fixed overhead per execution, where it is large enough to result in non-linear parallelization (7.3x, 6.2x respectively).

6 Related Work

Synthesis Algorithms. Data Synthesis algorithms fit into two broad categories: Neural Network (NN) and Marginal based. Notable Neural Network algorithms are PATE-GAN [10], CT-GAN [19], DP-GAN [8], and ADS-GAN [20]. For tabular data, a class of Probabilistic Graphical Model (PGM) based algorithms has been shown perform well when privacy is required [18]. Notable algorithms are: MST [12], PrivBayes [22], AIM [13], PrivMRF [3], which are enhanced through inference, shown in PrivPGM [14]. For privacy, the mechanism Differential Privacy [7] (DP) provides strong guarantees.

Data Synthesis Systems. Synthcity [17] and Synthetic Data Vault (SDV) [16] are collections of algorithms and evaluation metrics for data synthesis. Neither attempts to perform scaling nor larger-than-memory synthesis.

Machine Learning Orchestration. In Machine Learning Operations (MLOps), multiple systems have been proposed to parallelize machine learning workflows using DAG pipelines in an inter-task context: Prefect, Kubeflow, Apache Airflow, Luigi, and Kedro [1]. For intra-task parallelization, Dask [5] and Spark [21] have been proposed to parallelize computations themselves. The previous solutions focus on cluster environment scaling, causing a performance overhead (shown in Sect. 5) when the data fits in a single server. Our approach is tailored to a single server environment, which is where a large segment of medical data processing occurs due to GDPR limitations. In those environments, the low overhead of our approach (launching processes versus containers or cross-server communication) makes it optimal.

7 Conclusion

Privacy-aware data synthesis has become an important research area in fields such as Medicine. Yet available systems and frameworks suffer from limited scalability and do not scale to the current needs of real-world medical applications and datasets. Hence, in this paper, we proposed, Pasteur, a system that achieves linear scaling and stable memory usage, by employing parallelization, optimized data representations, and an optimized method of computing marginals (the core bottleneck in PGM algorithms). Our experiments show that Pasteur supports data sets of up to 1 billion on a single CPU in the period of one hour.

Acknowledgements. This project received funding from the European Union's Horizon 2020 research and innovation programme under Marie Skłodowska-Curie (grant No 955895), the Poul Due Jensens Fond (Grundfos Foundation), and the Novo Nordisk Foundation (grant number NNF23OC0083510).

References

1. Alam, S., et al.: Kedro (2022)
2. Bradbury, J., et al.: JAX: composable transformations of Python+NumPy programs (2018)
3. Cai, K., Wei, J., Lei, X., Xiao, X.: Data synthesis via differentially private markov random fields. VLDB (2021)
4. Cai, K., Xiao, X., Cormode, G.: PrivLava: synthesizing relational data with foreign keys under differential privacy. Proc. ACM Manag. Data **1**(2), 1–25 (2023)
5. Dask Development Team: Dask: Library for dynamic task scheduling (2016)
6. Dua, D., Graff, C.: UCI machine learning repository (2017)
7. Dwork, C.: Differential privacy (2006)
8. Ho, S., Qu, Y., Gu, B., Gao, L., Li, J., Xiang, Y.: DP-GAN: differentially private consecutive data publishing using generative adversarial nets (2021)
9. Johnson, A., Bulgarelli, L., Pollard, T., Horng, S., Celi, L.A., Mark, R.: MIMIC-IV
10. Jordon, J., Yoon, J., van der Schaar, M.: PATE-GAN: generating synthetic data with differential privacy guarantees (2018)
11. Liu, B., Ding, M., Shaham, S., Rahayu, W., Farokhi, F., Lin, Z.: When machine learning meets privacy: a survey and outlook. ACM Comput., Surv. (2022)
12. McKenna, R., Miklau, G., Sheldon, D.: Winning the NIST contest: a scalable and general approach to differentially private synthetic data (2021)
13. McKenna, R., Mullins, B., Sheldon, D., Miklau, G.: AIM: an adaptive and iterative mechanism for differentially private synthetic data (2022)
14. Mckenna, R., Sheldon, D., Miklau, G.: Graphical-model based estimation and inference for differential privacy. In: Proceedings of the 36th International Conference on Machine Learning, pp. 4435–4444. PMLR (2019)
15. McKinney, W.: Data structures for statistical computing in Python. In: van der Walt, S., Millman, J. (eds.) Proceedings of the 9th Python in Science Conference, pp. 56–61 (2010)
16. Patki, N., Wedge, R., Veeramachaneni, K.: The synthetic data vault (2016)
17. Qian, Z., Cebere, B.C., van der Schaar, M.: Synthcity: facilitating innovative use cases of synthetic data in different data modalities (2023)
18. Tao, Y., McKenna, R., Hay, M., Machanavajjhala, A., Miklau, G.: Benchmarking differentially private synthetic data generation algorithms (2021)
19. Xu, L., Skoularidou, M., Cuesta-Infante, A., Veeramachaneni, K.: Modeling tabular data using conditional GAN (2019)
20. Yoon, J., Drumright, L.N., Schaar, M.: Anonymization through data synthesis using generative adversarial networks (ADS-GAN). IEEE J. Biomed. Health Inform. **24**(8), 2378–2388 (2020)

21. Zaharia, M., et al.: Apache spark: a unified engine for big data processing (2016)
22. Zhang, J., Cormode, G., Procopiuc, C.M., Srivastava, D., Xiao, X.: PrivBayes: private data release via Bayesian networks. ACM Trans. Database Syst. **42**(4), 1–41 (2017)
23. Zhang, Z., et al.: PrivSyn: differentially private data synthesis

Generating Semantically Enriched Mobility Data from Travel Diaries

Maja Schneider[1(✉)], Charini Nanayakkara[2], Matthias Mohn[1], Peter Christen[2], and Erhard Rahm[1]

[1] Center for Scalable Data Analytics and Artificial Intelligence (ScaDS.AI) Dresden/Leipzig, Dresden/Leipzig, Germany
mschneider@informatik.uni-leipzig.de

[2] School of Computing, The Australian National University, Canberra, Australia

Abstract. Human mobility data is valuable for many applications, but it can pose a significant privacy risk for individuals. A person's daily movements are closely linked to their sociodemographic characteristics and their points of interest (POI), which can reveal sensitive information about them, such as their religion or educational background. Researching and mitigating these risks requires realistic, semantically rich data, which is often unavailable or lacks semantic features. We introduce ASTRA, an agenda-based approach for generating synthetic mobility data with semantic attributes. Using real travel surveys and census data, ASTRA simulates artificial agents with sociodemographic features that follow daily activity agendas. It maps activities to semantically similar POIs, and projects them onto real-world locations within a user-defined geographical region. To model movement, ASTRA extends the exploration and preferential return (EPR) model with spatio-temporal and semantic constraints as imposed by an agenda. Our evaluation against real check-in data and the EPR baseline model shows that ASTRA can generate realistic mobility data at scale, preserving important characteristics of human movement.

Keywords: Synthetic Data · Human Mobility · Points of Interest · Agenda-Based Modeling

1 Introduction

Driven by the extensive use of location-based services on mobile devices, location data of app users is becoming increasingly relevant. Such data can give valuable insights into human mobility patterns and enable a range of applications, such as urban planning or personal recommendations [26]. However, the collection of their location data bears potentially serious privacy risks for an individual [23]. Human mobility is highly correlated with an individual's sociodemographic characteristics and their *agenda*, which is the schedule of their daily activities [14,22]. These attributes determine a person's points of interest (POI)

P. K. Chrysanthis et al. (Eds.): ADBIS 2025, LNCS 16043, pp. 181–197, 2026.
https://doi.org/10.1007/978-3-032-05281-0_12

which are the places they stay at or visit, for example their home, work, or places of leisure activities. POIs are especially privacy-critical because they can reveal sensitive information about a person, such as their educational background, age or gender [30]. To understand the involved privacy risk and mitigate such risk with location privacy-preserving methods, researchers require suitable data to study questions such as: *Does an individual's geographical trajectory reveal their demographic attributes or even allow to re-identify them? Is there an increased privacy risk for certain demographic groups?*

When data is collected for research purposes, it is usually limited to small regions or certain groups of people, such as students, and can therefore be biased [2,17,27]. A solution to the lack of available data is to generate synthetic mobility data [13]. However, existing methods usually do not include semantic features, such as the location and label of a POI or the sociodemographic details of a person. In addition, real and synthetic mobility data is often modified to protect private user attributes, making such data unsuitable for privacy research. To be useful for privacy research, synthetic human mobility data needs to fulfill certain requirements: The data needs to be *realistic* to reflect actual human movement behavior. It needs to be *representative* for a society, for example a city population, in order to avoid bias and allow for generalizable and fair analyses. To understand privacy implications, it needs to be *explainable* with regard to the privacy-sensitive factors that influence mobility, such as the sociodemographic attributes of a person and the purpose of their movement as represented by their POIs. Finally, to understand privacy concerns in different geographical contexts, the synthesis of mobility data must be *flexible* with respect to the geographical simulation region.

Existing approaches to synthesizing human mobility data generally do not meet all of the above criteria, and in particular lack explainability [13]. To fill this gap, we present ASTRA, our approach to generate **A**genda-based **S**emantic **TRA**jectories. Agenda-based modeling is based on the assumption that a person's trajectory is not random but highly influenced by the activities they pursue and their social characteristics [12,29]. These factors determine which POIs a person visits, while travel can be seen as a by-product of these visits. To model the different sociodemographic groups in a society, ASTRA samples a population of artificial mobile agents with certain sociodemographic features and a daily *agenda* from a data set of real travel diaries. We define an agenda to be a sequence of activities that can be divided into alternating segments of consecutive travel activities or non-travel activities. A non-travel segment refers to an agent's POI and all of its activities are assumed to be carried out at the same location. For example, a non-travel segment with the activities *sleeping*, *eating*, and *reading* can be linked to the label of a POI where these activities can be executed, such as an *apartment*, *hotel*, or *tent*. Travel segments connect two non-travel segments and indicate the main mode of transportation (like *walking* or *driving*) that the agent uses to move between the respective locations.

ASTRA uses an embedding approach [21] to map the activities of non-travel segments to semantically similar POI labels. The segments are then projected

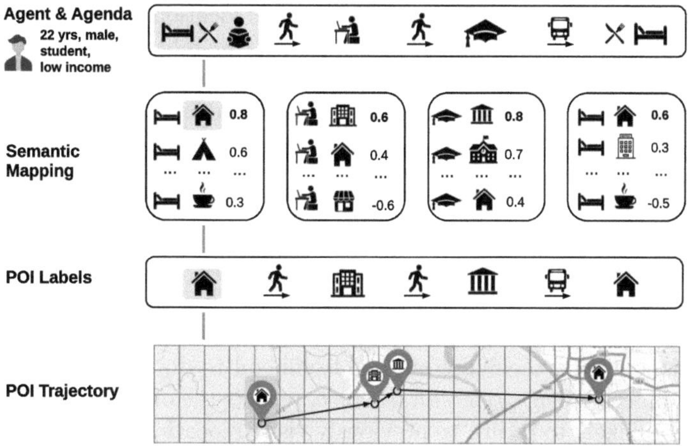

Fig. 1. An agent's agenda is mapped by ASTRA to a sequence of POI labels using a semantic similarity score. We use Cosine similarity, ranging from –1 (low similarity) to 1 (high similarity). The POI label sequence is further processed into a POI trajectory by selecting the GPS coordinates of actual POIs on a geographical map.

to the GPS coordinates of actual POIs with the respective label on a real map, creating a *POI trajectory*. To select a POI, ASTRA uses an extension of the exploration and preferential return model (EPR) [24] that takes into account spatio-temporal travel constraints and individual and collective semantic constraints, imposed by the agent's agenda and the map environment. Figure 1 illustrates the mapping from an agent's agenda to a POI trajectory. The example highlights how certain POIs can potentially disclose private attributes, such as the educational background. Note that while the aim of ASTRA is to synthesize realistic trajectory data to allow researchers to study privacy, its intention is not to reconstruct an actual person's trajectory or to re-identify them.

Our contributions are as follows: (1) We extend the EPR model [24] by three spatio-temporal and semantic constraints to reflect that human mobility is driven by people's daily activities. (2) We present ASTRA, a new approach for synthesizing human mobility data that includes not only GPS locations, but also semantic features, such as the labels of POIs and the sociodemographic features of the data producers. The POI trajectories generated with ASTRA are realistic, representative, explainable, and flexible. (3) We evaluate ASTRA against a real data set of POI trajectories and against the baseline EPR model. Our results show that ASTRA can generate mobility data that is realistic with respect to spatial, temporal, and semantic measures of mobility.

2 Background and Related Work

Traditional mobility models focus on reproducing spatial and temporal aspects of human mobility. Many approaches create synthetic trajectories by sampling certain features of a trajectory, such as start and end locations, speed, or trip length, from distributions that are extracted from real mobility data sets [11,15,18]. Agenda-based models in turn rely on travel survey data to extract semantic mobility patterns and demographics, assigning them to artificial agents that move in a predefined geographical area in a rule-based manner [1,25,29]. Transitions between consecutive trajectory points are often sampled based on Markov models [3,20] or random walk models [6,7].

To overcome the limitations in existing random walk methods to not sufficiently reproduce certain aspects of human mobility, the exploration and preferential return model (EPR) [24] introduced two fundamental mechanisms to describe human mobility. The first, *exploration*, describes that individuals visit new locations during their day but the likelihood to do so decreases over time. The second mechanism, *preferential return*, reflects the observation that individuals are more likely to return to previously visited locations where the likelihood to choose a location is proportional to its visitation frequency.

Several extensions have been proposed to account for inaccuracies in EPR modeling. Empirical observations indicate that the likelihood to choose a location as a return does not only depend on its visitation frequency. Instead, individuals tend to return to recently visited locations, even if these have not been visited frequently. The concept of *recency* [4] thus introduces a rank-based approach that considers for each potential return location a frequency rank and a recency rank that are considered with a certain likelihood when choosing a return location.

The original EPR model also does not account for the observation that human mobility is influenced not only by individual motivations but by collective behavior, without which some mobility metrics cannot be properly modeled. The density-EPR [19] model incorporates such collective forces during the exploration phase by introducing the *gravity model*. Instead of following an equal distribution, a destination location is selected using a normalized gravity score S_{grv} which indicates a transition probability between two locations. This score is calculated based on the relevance of start and destination locations and their geographical distance from each other. The relevance of a location can, for example, be measured by its population density or another measure of popularity.

The density-EPR approach however ignores the semantic relevance of a location in the context of a typical activity schedule of a person with certain sociodemographic characteristics. Therefore, ASTRA incorporates such knowledge by combining density-EPR with an agenda-based modeling approach.

3 Agenda-Based Semantic Trajectory Generation

We now describe our approach for generating agenda-based semantic trajectories, ASTRA. Our goal is to generate realistic and representative POI trajectories for a population of artificial agents in an arbitrary simulation region. A POI trajectory consists of the GPS locations and labels of the POIs that an agent visits throughout a day. A key aspect of ASTRA is that it associates agents and their POI trajectories with corresponding sociodemographic features. We first

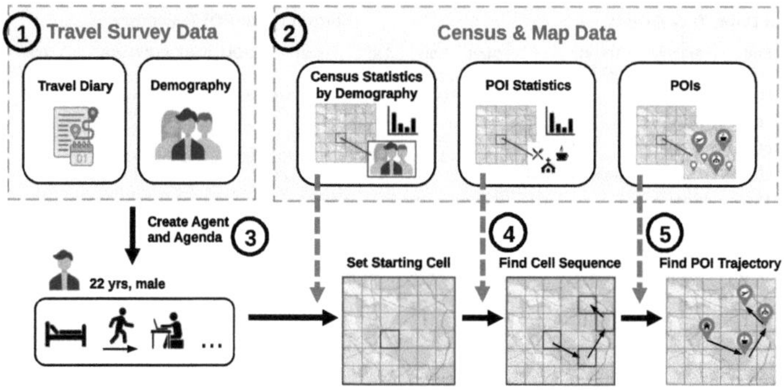

Fig. 2. The data processing pipeline of ASTRA.

give an overview of ASTRA before we discuss its individual steps in the following subsections.

Figure 2 outlines the data flow in ASTRA. (1) We use daily travel diaries from a travel survey data set, which includes sociodemographic attributes. (2) We then define the geographical simulation region, which we partition with a spatial tessellation into cells to reduce complexity and obtain census and POI data for each cell. (3) We create a user-defined population of artificial agents and assign each agent a starting cell, age and gender, proportionally to the actual population distribution in the cell. Each agent is randomly assigned a real travel diary and sociodemographic attributes from a person with matching age and gender. Diaries are preprocessed into agendas consisting of travel segments and non-travel segments. (4) ASTRA maps non-travel segments to cells according to the labels of their available POIs, while taking into account the spatio-temporal travel restrictions of travel segments. (5) Each segment is then assigned a precise POI within its cell to create a POI trajectory. Figure 3 shows how a travel diary is translated into a POI trajectory with ASTRA.

3.1 Semantic Extension of EPR

The original EPR model [24] and its extensions do not capture the semantic context of movement. To make synthetic trajectory data more explainable in this regard, we propose an agenda-based semantic extension of density-EPR [19]. This means that we take into account both the spatio-temporal constraints of an agenda's travel segments, and the semantic constraints as imposed by an agenda's non-travel segments and the environment, in which the agent moves.

Spatio-Temporal Travel Constraints: In density-EPR, the target location of exploration is chosen based on the gravity model (see Sect. 2). This allows an agent to pick any destination from a set of candidate locations with a certain probability. This approach, however, is agnostic of the actual travel constraints given by an agenda. In ASTRA, we restrict an agent's movement in the exploration phase to those locations that are reachable in the travel time of the current

Input Data: Travel Diary

Start	End	Activity	Travel Mode
00:00	08:00	sleeping	-
08:00	08:30	eating	-
08:30	09:30	reading	-
09:30	10:00	travel	walk
10:00	13:30	working	-
13:30	13:45	travel	walk
13:45	20:00	education	-
20:00	20:10	travel	drive
20:10	21:00	eating	-
21:00	23:55	sleeping	-

Output Data: POI Trajectory

Start	End	Lat	Lon	POI Label	POI Name	Travel Mode to POI
00:00	09:30	40,743	-73,985	apartment	Prism at Park Avenue South Apartments	-
10:00	13:30	40,746	-73,992	office	Micro Office Solutions	walk
13:45	20:00	40,741	-73,987	university	Princely International University	walk
20:10	23:55	40,737	-73,991	apartment	The Decker Building	drive

Fig. 3. Example of how a travel diary is converted by ASTRA into a POI trajectory.

travel segment considering the specified travel mode. A relaxation of this constraint is allowed if there are no locations reachable that the agent has not yet visited. Note that in the return mode these constraints are not imposed and the agent can pick any of their previously visited locations.

Individual Semantic Constraints: To model the semantic dimension of mobility, we incorporate an agent's daily activities into density-EPR. To do so, we map each of the agenda's non-travel segments to a suitable POI label and restrict the agent's movement to locations that have this label. To find a suitable mapping, potential POI label candidates and the agenda segment's activities are embedded based on a textual description using a sentence transformer approach (SBERT) [21] and compared with a normalized *semantic similarity score* S_{sem}.

Collective Semantic Constraints: The gravity score S_{grv} in density-EPR uses the relevance of a location to describe the influence of a collective force on an agent's movement. We measure the relevance of a cell in the tessellation by its population count, where we additionally consider the semantic relevance of a cell by introducing a normalized *POI label frequency score* S_{frq}. This score aims to favor POIs that are not geographically isolated with regard to their label. For example, if an agent wants to visit a POI with the label *restaurant*, the likelihood to pick a cell in a city's food district where many restaurants are located should be higher than the likelihood to pick a cell that contains only one restaurant.

Scoring of Locations: While the spatio-temporal constraints decide which locations are allowed to be selected by an agent, the final choice of a location is done in proportion to a weighted combination $S = {S_{sem}}^{\alpha} \cdot {S_{frq}}^{\beta} \cdot {S_{grv}}^{\gamma}$, where $\alpha + \beta + \gamma = 1$.

3.2 Data Sources

We now describe the input data required for ASTRA, consisting of travel survey data, census data, and map data. The data sets discussed below are examples that can be replaced by other data sets of the same type.

Travel Survey Data: The basis of ASTRA are mobility diaries of real persons together with a set of their demographic features. Such data can be obtained, for example, from travel survey data sets. In our work we use the Multinational Time Use Study (MTUS) data set [8,9], which is collected at random households in over 20 countries. For each survey respondent, this data set contains sociodemographic features (such as age, gender, ethnicity, or citizen status), their educational and employment background (including employment status, occupation, education level, and income), and details about their social life (like marital status, number of children, or care taking obligations). The respondents' characteristics also include information about their household, such as household size and income, or car or house ownership. As shown in Fig. 3, respondents also provide detailed one-day travel diaries, recording each activity of their day along with its start time, duration, and the travel mode used.

Census and Map Data: To allow for data abstraction, the simulation region is modeled with a spatial *tessellation* that partitions the region into non-overlapping subregions. This can be implemented, for example, using a grid, dividing the region into equally sized square cells. The cell width w of the grid is a user parameter and should be chosen to find a good balance between computational complexity and modeling accuracy. For each cell in the tessellation, census and map data is obtained.

- *Sociodemographic Features*: To create a realistic spatial distribution of the start locations of agents, and to measure the relevance of a cell as a start location for an agent, distributions of sociodemographic features are required. We determine each cell's population numbers by *age* and *gender* groups based on census information obtained from WorldPop[1].
- *Points of Interest*: To allow ASTRA to map activities to real POIs in the simulation area, the POIs of each cell along with their labels are queried from OpenStreetMap (OSM) using OSMnx [5]. We consider only POI labels that are a subcategory of one of the OSM tags *amenity*, *building*, *office*, *shop*, *tourism*, *leisure*, and *sport*, to account for places where people usually spend their time. In each cell, POIs are grouped by their label and counted. This information will serve as the basis for calculating the POI label frequency score S_{frq} as one of the three scores required for the semantic mapping as we described in Sect. 3.1.
- *Travel Times*: For a certain selection of travel modes, such as *car*, *bike* or *walk*, the travel times between pairwise cells are calculated. The calculation is based on the OSM street network which is represented as a directed graph between OSM nodes. Nodes are identified via their respective OSM identifiers and connected via edges, that are weighted with travel speeds for the respective travel modes. Travel times are calculated based on the shortest path between the centroids of the cells.

[1] www.worldpop.org.

3.3 Generating an Agent Population and Their Agendas

After the input data for ASTRA is prepared, a user-defined number of artificial agents n is created and distributed onto the cells of the tessellation in proportion to the actual population size. Each agent is randomly assigned an age and gender following the actual distribution of these features in their cell's geographic area. An agent is further assigned a travel diary and sociodemographic features of a person from the travel diary data set. For this, a person is chosen randomly from all persons with matching gender and (approximate) age. The travel diary is converted into an agenda by applying a non-overlapping sliding window. For each window the longest activity is extracted. Consecutive windows are grouped into travel and non-travel segments according to the nature of their activities.

3.4 Creating POI Trajectories

The goal of the next step is to project each agent's agenda onto the simulation region. ASTRA uses density-EPR with the recency extension described in Sect. 2 and applies the spatio-temporal and semantic extension introduced in Sect. 3.1. This means that in the exploration phase we use the gravity model together with our semantic extension, while in the return phase with a certain likelihood the recency-based approach is chosen.

During exploration, an agenda segment is first mapped to a cell in the tessellation before a specific POI in the chosen cell is derived later on. Candidate cells are obtained according to the spatio-temporal travel time constraints. For each candidate cell, S_{sem} and S_{frq} are calculated. Furthermore, S_{grv} is calculated based on the travel distance between the centroids of the previous and the candidate cell, and by measuring the relevance of a cell by its population size. All three scores are combined into a single score S by weighting the three components according to a user-specific weight assignment (see Sect. 3.1). A POI label and cell are then chosen randomly from the top-k highest scoring cells, where $k \geq 0$ is a user-specific parameter.

To pick a specific POI from the chosen cell, only POIs with a matching label are considered. These POI candidates are evaluated based on a single score. We again use the gravity score as we described in Sect. 2, but calculate it in a different way than before. The calculation of gravity is based on the distance between the centroid of the previous cell's POIs and the POI candidate, and the relevance of a POI candidate is measured by the spatial density of POIs in its surroundings. Following this approach, POIs that are in proximity to other POIs are preferred over those that are spatially isolated. This reflects the increased popularity of POIs in lively areas, such as a store in a shopping center, or a bar in a nightlife district. To realize this calculation, each cell is further divided into subcells using a grid. The relative number of POIs in each subcell in relation to the entire cell then indicates the relevance of the POIs in a subcell.

During the return phase, a POI must be selected that was previously visited by the agent. In order to ensure the greatest possible semantic integrity with the agenda, we ignore the spatio-temporal travel restrictions in this case. ASTRA evaluates each of the POI candidates according to their frequency and recency scores. In addition, ASTRA allows a return to a POI only if its semantic similarity with the segment is greater or equal to a user-defined threshold: $S_{sem} \geq \tau_{sem}$. Otherwise, the return is rejected and an exploration is executed instead.

4 Evaluation

We now evaluate the modeling capabilities of ASTRA[2]. We first assess the semantic mapping quality from agenda segments to POI labels. Then, we compare the spatial, temporal, and semantic characteristics of the generated POI trajectories of ASTRA with a real mobility data set and the traditional density-EPR model, instantiated by the DITRAS framework [20] using default parameters.

4.1 Experimental Setup

We first describe the baseline data set and configurations of ASTRA as used in our experiments.

Baseline Data Set: We compare ASTRA with the Foursquare data set [26], a real mobility data set collected in New York City, USA, between April 2012 and February 2013. It contains 227,428 check-ins (POIs) from 1,083 users who logged their visits to a POI with an app, indicating the POI's label, GPS location, and timestamp. Due to the nature of the app usage, the check-in history of a user can be incomplete and reflect only a part of their daily POIs [28]. Also, because such apps are more likely to be used by young adolescents to connect with their friends, this data set is likely biased towards this demographic group. Even though data sets collected via social media apps usually have these limitations, only they provide the POI information required for comparison with ASTRA.

Simulation Parameters: The simulation region of our experiments consists of the convex hull of all users' POIs in the Foursquare data set. We choose a tessellation cell width of $w = 1500$ m. With ASTRA, we create a population of $n = 10{,}000$ agents along with their POI trajectories. We create the same number of agents and trajectories using DITRAS [20].

Travel Diary Data Basis: We run ASTRA based on an extract of 190,088 MTUS [8,9] travel diaries from the USA between January 2003 and December 2019. This is to reflect travel behavior that is not influenced by the Covid-19 pandemic. The selected data is used by ASTRA as the basis for assigning a real diary to an artificial agent. The travel diaries are divided into agendas with a sliding window. The window size should be chosen appropriately to not miss any travel activities in the preprocessing, but at the same to keep computation time to a minimum. As more than 90 % of non-travel activities in the MTUS extract have a duration of at least 5 min, we choose this value as the window size. The

[2] ASTRA is available open-source at https://github.com/majaschneider/ASTRA.

data set contains travel modes *car*, *bike*, and *walk*, for each of which the travel times between all cells in the simulation area are calculated.

Scoring: Because the level of the semantic similarity score S_{sem} influences the number of POI candidates that an agent can choose from, this might impact the quality of the generated POI trajectories. We therefore test the three semantic similarity score weights of $\alpha = 0.3$, 0.6, and 0.9 and redistribute the remaining weight equally across the two other score components, gravity and POI label frequency score: $\beta = \gamma = \frac{1-\alpha}{2}$. After the combined score S is calculated, we select a POI label and its cell from the top-$k = 5$ highest ranking scores to introduce variability in the data generation process. In the return phase, we test the three minimum similarity score thresholds of $\tau_{sem} = 0.0$, 0.3, and 0.5.

4.2 Evaluation of the Semantic Mapping

The aim of the semantic mapping is to find the most similar POI label for an agenda segment and its activities. Using MTUS and OSM, this involves the mapping of 60 different non-travel activities to 665 unique POI labels.

To calculate the similarity of agenda segments and POI labels, we first embed their textual representation with the sentence transformer SBERT[3]. This model was specifically designed for semantic text comparisons and has been trained on 215 million (question, answer)-pairs. It converts texts into a 384-dimensional vector space and allows semantic nearest neighbor search based on Cosine similarity, which we use as the semantic similarity score.

POIs and activities usually only have a short identifier, or label. Due to SBERT's training data, it can be advantageous to provide more context and convert both identifiers and labels, into a longer textual description so that they resemble a question and an answer. We empirically test which of the following two options achieves better semantic mapping quality: *Label*: Only the identifiers and labels of activities and POIs are used for mapping, or *Desc*: A more comprehensive description of activities and POIs is used for mapping.

To realize the option *Desc*, we use descriptions of POI labels from OSM to obtain an answer-like text. For example, the POI label "Food Court" is converted into "An area with several different restaurant counters and a common dining area. Often found in shopping centers, airports, etc.". Similarly, we manually phrase activity descriptions as questions, in some cases indicating a location. For example, the activity "Food preparation, cooking" is converted to "What is a usual place for personal cooking and food preparation? (Often at apartment, detached house or building)". Optionally, a large language model can be used to help automate this task, with a subsequent manual verification.

In the case that an agenda segment contains multiple activities, several strategies are possible to map it to a POI, varying the impact of certain activities onto the mapping based on their significance. For example, a segment consisting of three activities 6 h *sleeping*, 15 min *eating*, and 15 min *reading*, can probably be

[3] https://huggingface.co/sentence-transformers/multi-qa-MiniLM-L6-cos-v1.

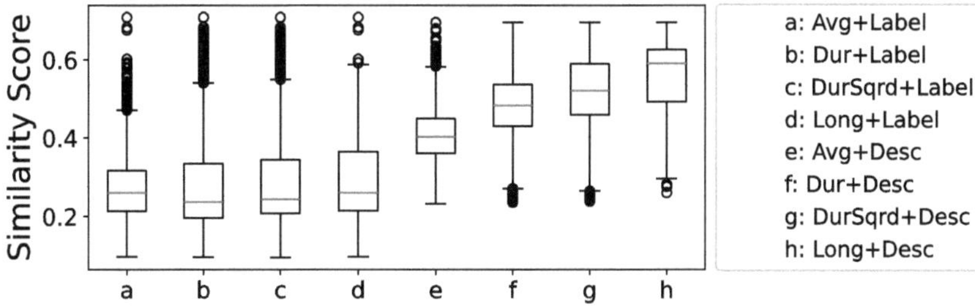

Fig. 4. The semantic similarity scores (based on Cosine similarity) for different mapping approaches between agenda segments and their five respective most similar POIs.

linked to a home location due to the long duration of the sleeping activity. If the duration is ignored, the segment could potentially also be mapped to a restaurant or café. We therefore study different strategies: *Avg*: All activities of an agenda segment are mapped and the scores are averaged for the same POI. *Dur*: All activities are mapped and all single scores are averaged using a weighting according to their duration. *DurSqrd*: This is the same strategy as *Dur*, but the weights are squared in order to test an increased impact of the duration. *Long*: Only the activity with the longest duration is mapped.

To find the best mapping strategy, we use MTUS agenda segments with one of the 400 most frequent activity patterns, covering 75 % of all segments. This subset is chosen to focus only on the most influential activity patterns.

Figure 4 shows the semantic similarity scores of the top five mappings between each of the agenda segments and all POIs, distinguished by the mapping strategies. The figure shows that the similarity score and thus the mapping quality can be increased by using a comprehensive description for POIs and activities. Furthermore, higher scores can be achieved with the mapping strategies *Long* and *DurSqrd*, which assign more weight to longer activities in the embedding.

4.3 Evaluation of Spatio-Temporal and Semantic Properties

To evaluate how well ASTRA captures spatial, temporal, and semantic properties of mobility, we calculate a set of established mobility metrics [13] over the POI trajectories generated with ASTRA, and compare them against the Foursquare data set, and the synthetic POI trajectories created with the DITRAS baseline model. We provide a visual evaluation with comparative plots for each metric in Fig. 5 and calculate the error of fit between synthetic and real data with the Jensen-Shannon divergence (JSD) in Table 1.

JSD is often used to measure the dissimilarity between two distributions, P and Q, and builds on Kullback-Leibler divergence (KLD) [13]. JSD is defined as $JSD(P||Q) = (KLD(P||M) + KLD(Q||M))/2$, where M is the pointwise mean of P and Q and $KLD(P||Q) = \sum_{x \in X} P(x) \log(P(x)/Q(x))$. JSD is normalized and values close to zero indicate high similarity between P and Q.

Table 1. Jensen-Shannon divergence between synthetic and real data per metric. Results for ASTRA are averaged over all nine tested configurations (described in Sect. 4.1). The respective best (lower) result in each column is highlighted in bold.

	Trip Distance	Trips per Hour	Stay Duration	Locations per User	Location Frequency	Visits per Location
ASTRA	**0.240**	0.106	**0.065**	0.224	**0.148**	**0.028**
DITRAS	0.244	**0.006**	0.294	**0.032**	0.237	0.600

Trip Distance: The distance of a trip measures the geographical Euclidean travel distance between two consecutive POI locations in a POI trajectory. We calculate the trip distance for each trip over all POI trajectories and plot the distribution in Fig 5 (a). ASTRA captures the shape of the distribution of real trip distances in the Foursquare data set while slightly overestimating longer trip distances. ASTRA achieves both visually and quantitatively similar results as DITRAS with an average JSD of 0.240 (ASTRA) versus 0.244 (DITRAS). Both approaches do not capture the high ratio of short trip distances that are prevalent in the Foursquare data.

Trips per Hour: The number of trips per hour gives an indication of how realistic the generated data is in terms of its temporal distribution. To calculate this metric, the start times of the agents' visits to a POI are assigned to one hour time windows and counted. We plot the distribution of trips per hour over all POI trajectories over the 24 h of the day in Fig. 5 (b). All three data sets show multiple peaks reflecting the typical human movement behavior to mainly move at specific times of the day [10]. The peaks coincide with the major travel times, for example when people go to work in the morning, go to lunch around noon, and return to their home in the evening. ASTRA is able to reproduce the overall shape of the real distribution with an average JSD of 0.106 while DITRAS can slightly better reproduce all peaks with a JSD of 0.006. For most of the tested configurations of ASTRA, the peaks are present at the typical moving times, however the most relaxed configuration using $\alpha = 0.3$ and $\tau_{sem} = 0.0$ can best reproduce the different peaks. Because ASTRA uses travel diaries that start at midnight, the first POI in each generated POI trajectory has a start time at midnight as well, causing another peak at this time.

Stay Duration: The duration of a stay indicates how much time an individual spends at a POI that they visit. This metric is calculated as the time difference in hours between the start times of consecutive POIs in an agent's POI trajectory. A stay time thus also includes the travel time between two POIs. This is due to the fact that POI trajectories from Foursquare and DITRAS only indicate the start timestamp of a visit to a POI. We plot the distribution of the stay duration over all agents' POI trajectories in Fig. 5 (c). ASTRA can capture the shape of the Foursquare distribution better than DITRAS with a JSD of 0.065 (ASTRA) versus 0.294 (DITRAS). In contrast to the real distribution, ASTRA shows a peak at approximately 8 h. This peak is likely to reflect the time duration that individuals stay at home or at work, whereas this might not be reflected in the Foursquare data set. A possible explanation is that Foursquare users preferably

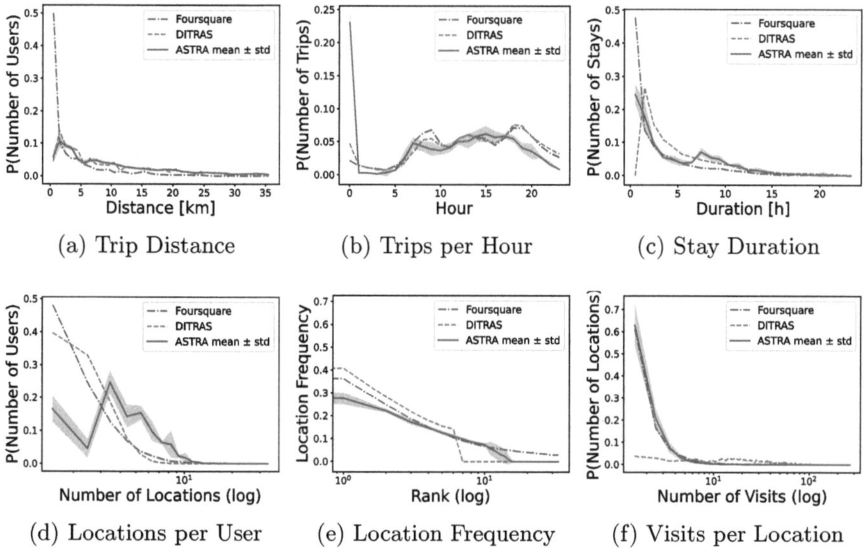

(a) Trip Distance (b) Trips per Hour (c) Stay Duration

(d) Locations per User (e) Location Frequency (f) Visits per Location

Fig. 5. Distributions of spatial, temporal, and semantic metrics.

check in at food places, but not as often check in at home or work [16]. The high ratio of short trip distances in the Foursquare data set (see Fig. 5 (a)) also leads to a high ratio of short or zero durations of stay. DITRAS struggles to reproduce this peak, while ASTRA achieves more similar results.

Locations Per User: The locations per user is the number of distinct POIs, that an agent visits during the period of observation. We plot the distribution of locations per user over all agents in Fig. 5 (d). DITRAS reproduces the distribution of this metric in the Foursquare data set closely (JSD of 0.032), while ASTRA is showing a different pattern (JSD of 0.224) and produces POI trajectories with more POIs than Foursquare. The reason is likely that ASTRA is based on real travel surveys which often include POIs related to a person's home or work. Foursquare users, however, do not check in as often at such a type of POI, reducing the overall number of POIs of a user. Studies of travel diary data and mobile phone records show that about 90 % of human trajectories can be explained by fewer than seven regularly visited POIs [22]. ASTRA reproduces this trend while the majority of agents visits up to ten different POIs.

Location Frequency: The location frequency is the probability of an agent to visit a POI given its rank in the POI trajectory. The rank of a POI is its position in a descending order of visit frequency in the POI trajectory. A low rank thus indicates that a POI was visited often. In Fig. 5 (e) we plot the distribution of the location frequency given the rank over all POI trajectories. Both ASTRA and DITRAS can reproduce the shape of the real distribution approximately with a JSD of 0.148 (ASTRA) and 0.237 (DITRAS), but both approaches underestimate the distribution for higher ranks.

Visits Per Location: The visits per location count the number of visits by all users to a unique POI location, that is a certain GPS coordinate. We plot the distribution of the number of locations given the number of visits in Fig. 5 (f). ASTRA can better reproduce the shape of the real distribution compared to DITRAS leading to a lower JSD of 0.028 versus 0.6 for DITRAS.

5 Discussion

The results of our experiments indicate that ASTRA is a suitable approach to generate realistic synthetic human mobility data that satisfies important spatial, temporal and semantic mobility measures. While the data quality is comparable or better than the DITRAS baseline, ASTRA additionally provides semantically accurate POI information and sociodemographic details about the generated agents. This information is usually not provided by other mobility models but essential for certain questions of (privacy) research.

ASTRA can flexibly be applied to any simulation region where the required input data is available. Census and map data are generally available for most countries, for example from WorldPop and OSM. However, the quality of the map data might vary per region which can limit ASTRA's applicability. Travel surveys can be obtained for many different countries [8]. While studies suggest that human mobility patterns are generally very similar [22], future work should investigate the question whether the semantic and temporal aspects of human travel patterns are similar in different countries. In this case, travel surveys might be interchangeable, which would eliminate the need for a geographically accurate input data set. Because ASTRA uses an embedding approach for the semantic mapping between activities and POI labels, their scope and formatting are not fixed, and therefore different data sources can be used. However, the preprocessing routines might need to be adapted to new source formats.

To create realistic, fair and unbiased data sets, the quality and size of the input data needs to be considered when choosing a simulation region and the respective input data set. In order for ASTRA to reflect realistic travel patterns, the used travel survey data must not contain exceptional events, such as the Covid-19 pandemic, which could distort results. It also needs to be large enough to ensure the synthetic population has diverse sociodemographic features.

Because ASTRA relies on agendas and needs to map agenda segments with specific activities and time range to a suitable POI, the distribution and availability of POIs in the chosen simulation area has an impact on the mapping quality. ASTRA's parameters control how much the semantic accuracy during mapping can be relaxed. Our experiments indicate that these parameters also influence how accurately the temporal patterns are preserved in the synthetic data while the impact on other analyzed metrics seems to be only small. As future work we therefore aim to test ASTRA's applicability to regions with different POI distributions, for example, smaller regions or urban and rural areas.

ASTRA is scalable to the desired number of artificial agents, because travel diaries can be selected multiple times from the travel survey data set if necessary.

This makes the data useful also for other research, for example to study the effects of a growing population on the utilization of a city's transport network.

6 Conclusion

We have presented ASTRA, a novel approach to generate synthetic human mobility data. ASTRA creates POI trajectories that are representative of a society, following an agenda-based approach based on real travel diaries. A major advantage of ASTRA is that its trajectories are explainable, because they retain semantic information about the contained POIs, and they can be linked to the sociodemographic features of their artificial agent. Because the data is not modified by privacy-preserving mechanisms, unlike other synthetic data sets, it is suitable for researching privacy-relevant questions regarding human mobility. ASTRA creates trajectories flexibly in any simulation region and is scalable with regard to the number of created trajectories. Our experiments show that ASTRA can create realistic mobility data that satisfies important metrics of mobility.

Acknowledgments. The authors acknowledge the financial support by the Federal Ministry of Education and Research of Germany and by Sächsische Staatsministerium für Wissenschaft, Kultur und Tourismus in the programme Center of Excellence for AI-research, "Center for Scalable Data Analytics and Artificial Intelligence Dresden/Leipzig", project id: ScaDS.AI. This work was partially funded by Universities Australia and the German Academic Exchange Service (DAAD) grant 57701258.

Disclosure of Interests. The authors have no competing interests to declare.

References

1. Adenaw, L., Bachmeier, Q.: Generating activity-based mobility plans from trip-based models and mobility surveys. Appl. Sci. **12**(17), 8456 (2022)
2. Aharony, N., Pan, W., Ip, C., Khayal, I., Pentland, A.: Social fMRI: investigating and shaping social mechanisms in the real world. Pervasive Mob. Comput. **7**(6), 643–659 (2011)
3. Anda, C., Ordonez Medina, S.A., Axhausen, K.W.: Synthesising digital twin travellers: individual travel demand from aggregated mobile phone data. Transp. Res. Part C: Emerg. Technol. **128**, 103118 (2021)
4. Barbosa, H., de Lima-Neto, F.B., Evsukoff, A., Menezes, R.: The effect of recency to human mobility. EPJ Data Sci. **4**(1), 1–14 (2015). https://doi.org/10.1140/epjds/s13688-015-0059-8
5. Boeing, G.: Modeling and analyzing urban networks and amenities with OSMnx. Working paper (2024). https://geoffboeing.com/publications/osmnx-paper/
6. Brockmann, D., Hufnagel, L., Geisel, T.: The scaling laws of human travel. Nature **439**(7075), 462–465 (2006)
7. Dandekar, A., Bressan, S., Abdessalem, T., Wu, H., Ng, W.S.: Trajectory simulation in communities of commuters. In: 2016 International Conference on Advanced Computer Science and Information Systems (ICACSIS), pp. 39–42 (2016)

8. Fisher, K., et al.: Multinational time use study extract system: version 1.4 [dataset]. Minneapolis, MN: IPUMS (2022)
9. Gershuny, J., Vega-Rapun, M., Lamote, J.: Multinational Time Use Study [dataset] Centre for Time Use Research, UCL IOE, University College London [www.timeuse.org/mtus/] (2020)
10. González, M.C., Hidalgo, C.A., Barabási, A.L.: Understanding individual human mobility patterns. Nature **453**(7196), 779–782 (2008)
11. Gursoy, M.E., Liu, L., Truex, S., Yu, L.: Differentially private and utility-aware publication of trajectory data. IEEE Trans. Mob. Comput. **18**(10), 2315–2329 (2019)
12. Hägerstraand, T.: What about people in regional science? Pap. Reg. Sci. **24**(1), 7–21 (1970)
13. Kapp, A., Hansmeyer, J., Mihaljević, H.: Generative models for synthetic urban mobility data: a systematic literature review. ACM Comput. Surv. **56**(4), 1–37 (2024)
14. Khan, M.T., Hussain, D.J., Tufail, M.: Activity-based tour generation model with peshawar as case study. CONSTRUCTII J. **6**(2), 12–16 (2023)
15. Li, J., Chen, W., Liu, A., Li, Z., Zhao, L.: FTS: a feature-preserving trajectory synthesis model. GeoInformatica **22**(1), 49–70 (2018)
16. Li, Y., Steiner, M., Wang, L., Zhang, Z.L.: Exploring venue popularity in foursquare. In: Proceedings - IEEE INFOCOM, pp. 3357–3362 (2013)
17. Madan, A., Cebrian, M., Moturu, S., Farrahi, K., Pentland, A.S.: Sensing the "Health State" of a community. IEEE Pervasive Comput. **11**(4), 36–45 (2012)
18. Mir, D.J., Isaacman, S., Caceres, R., Martonosi, M., Wright, R.N.: DP-WHERE: differentially private modeling of human mobility. In: 2013 IEEE International Conference on Big Data, pp. 580–588 (2013)
19. Pappalardo, L., Rinzivillo, S., Simini, F.: Human mobility modelling: exploration and preferential return meet the gravity model. Procedia Comput. Sci. **83**, 934–939 (2016)
20. Pappalardo, L., Simini, F.: Data-driven generation of spatio-temporal routines in human mobility. Data Min. Knowl. Disc. **32**(3), 787–829 (2018)
21. Reimers, N., Gurevych, I.: Sentence-BERT: sentence embeddings using Siamese BERT-networks. In: Proceedings of the Conference on Empirical Methods in Natural Language Processing (2019)
22. Schneider, C.M., Belik, V., Couronné, T., Smoreda, Z., González, M.C.: Unravelling daily human mobility motifs. J. R. Soc. Interface **10**(84), 20130246 (2013)
23. Schneider, M., Gehrke, L., Christen, P., Rahm, E.: D-TOUR: detour-based point of interest detection in privacy-sensitive trajectories. In: INFORMATIK 2022, pp. 219–230. Gesellschaft für Informatik, Bonn (2022)
24. Song, C., Koren, T., Wang, P., Barabási, A.L.: Modelling the scaling properties of human mobility. Nat. Phys. **6**(10), 818–823 (2010)
25. Strobel, L., Pruckner, M.: OMOD: an open-source tool for creating disaggregated mobility demand based on OpenStreetMap. Comput. Environ. Urban Syst. **106**, 102029 (2023)
26. Yang, D., Zhang, D., Zheng, V.W., Yu, Z.: Modeling user activity preference by leveraging user spatial temporal characteristics in LBSNs. IEEE Trans. Syst., Man, Cybern.: Syst. **45**(1), 129–142 (2015)
27. Yuan, N.J., Zhang, F., Lian, D., Zheng, K., Yu, S., Xie, X.: We know how you live: exploring the spectrum of urban lifestyles. In: Proceedings of the First ACM Conference on Online Social Networks, pp. 3–14 (2013)

28. Zhang, Z., et al.: On the validity of geosocial mobility traces. In: Proceedings of the Twelfth ACM Workshop on Hot Topics in Networks, pp. 1–7 (2013)
29. Zheng, Q., Hong, X., Liu, J.: An agenda based mobility model. In: 39th Annual Simulation Symposium (ANSS'06), pp. 188–195 (2006)
30. Zhong, Y., Yuan, N.J., Zhong, W., Zhang, F., Xie, X.: You are where you go: inferring demographic attributes from location check-ins. In: WSDM'15, pp. 295–304 (2015)

Data and Machine Learning

Learning a Distance for the Clustering of Patients with Amyotrophic Lateral Sclerosis

Guillaume Tejedor[1(✉)], Veronika Peralta[1], Nicolas Labroche[1], Patrick Marcel[2], Hélène Blasco[3,4], and Hugo Alarcan[3,4]

[1] Laboratoire LIFAT, Université de Tours, Tours, France
{guillaume.tejedor,veronika.peralta,nicolas.labroche}@univ-tours.fr
[2] Laboratoire LIFO, Université d'Orléans, Orléans, France
patrick.marcel@univ-orleans.fr
[3] Service de Biochimie et Biologie Moléculaire, CHRU de Tours, Tours, France
{helene.blasco,hugo.alarcan}@univ-tours.fr
[4] UMR U1253 iBrain, Université de Tours, Inserm, Tours, France

Abstract. Amyotrophic lateral sclerosis (ALS) is a severe disease with a typical survival of 3–5 years after symptom onset. Current treatments offer only limited life extension, and the variability in patient responses highlights the need for personalized care. However, research is hindered by small, heterogeneous cohorts, sparse longitudinal data, and the lack of a clear definition for clinically meaningful patient clusters. Existing clustering methods remain limited in both scope and number. To address this, we propose a clustering approach that groups sequences using a disease progression declarative score. Our approach integrates medical expertise through multiple descriptive variables, investigating several distance measures combining such variables, both by reusing off-the-shelf distances and employing a weak-supervised learning method. We pair these distances with clustering methods and benchmark them against state-of-the-art techniques. The evaluation of our approach on a dataset of 353 ALS patients from the University Hospital of Tours, shows that our method outperforms state-of-the-art methods in survival analysis while achieving comparable silhouette scores. In addition, the learned distances enhance the relevance and interpretability of results for medical experts.

Keywords: Amyotrophic Lateral Sclerosis · Clustering · Weak-supervision

1 Introduction

Amyotrophic lateral sclerosis (ALS) is a fatal neurodegenerative disease of the central nervous system that is difficult to diagnose, especially in its early stages [1].

It primarily affects motor neurons in the brain and spinal cord, which are responsible for controlling voluntary muscle movements. ALS research faces significant challenges due to factors such as small, heterogeneous patient cohorts,

P. K. Chrysanthis et al. (Eds.): ADBIS 2025, LNCS 16043, pp. 201–216, 2026.
https://doi.org/10.1007/978-3-032-05281-0_13

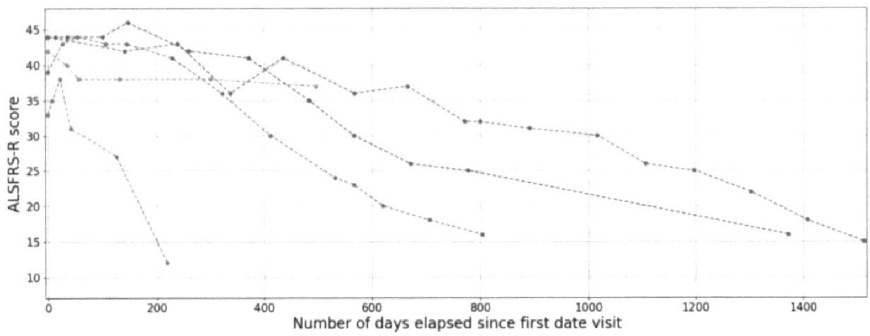

Fig. 1. Example of five ALSFRS-R sequences.

limited longitudinal data, and the lack of clear definitions for clinically meaningful patient clusters (phenotypes) and ground truth. These issues, compounded by data sparsity, make it difficult to develop reliable computational models for diagnosis and prognosis. Datasets are often constrained by costly, invasive procedures and strict data-sharing regulations, requiring substantial resources to maintain high-quality data. As a result, the available samples tend to be scarce and fragmented, which strains machine learning models [2] and weakens their generalization, further complicating efforts to identify distinct patient clusters, develop personalized care options and improve early-stage treatment.

To identify clusters, most existing studies use longitudinal clinical data to represent disease progression [3]. The "ALS Functional Rating Scale-Revised" (ALSFRS-R) remains the standard reference for measuring disease progression [1]. This scale evaluates a patient's physical function using 12 questions, each representing a distinct physical function (e.g., Speech, Salivation) and scored from 0 to 4. The total score ranges from 0 (total physical function loss) to 48 (no functional impairment). It is a revised version of the original ALSFRS score, which had only 10 questions. Figure 1 shows examples of temporal sequences depicting ALSFRS-R progression for 5 patients.

In the literature, some studies assign clusters using threshold rules applied to the ALSFRS-R score at specific time points. This approach is limited because it relies on a single criterion and is highly sensitive to small score variations around the thresholds, which may not reflect meaningful clinical differences. Moreover, these methods do not consider multiple criteria or the temporal progression of the disease, which could improve the robustness of cluster assignment [4].

Without assuming prior identification of specific clusters, we propose an approach for patient clustering that learns a distance measure using a weak-supervised method. In addition, we propose to test different off-the-shelf distances. Theses distances are then paired with clustering methods. Drawing inspiration from clustering of Business Intelligence sequences [5], we develop distances for patients' ALSFRS-R sequences using descriptive variables extracted from these temporal data. The goal is to use these measures in a way that captures patterns of disease progression, and uncover distinct patient clusters, each exhibiting unique forms and rates of progression.

Precisely, the contributions presented in this article include:

1. The identification of a set of descriptive variables for characterizing disease progression within ALSFRS-R sequences.
2. The development of distances based on the extracted variables.
3. An approach to automatically discover patient clusters based on theses distances and off-the-shelf clustering methods.

The remainder of the paper is structured as follows: Sect. 2 presents the current state of research on ALS patient clustering using longitudinal clinical data. Section 3 describes our approach. Section 4 details experiments and results, comparing our approach to state-of-the-art methods. Finally, Sect. 5 provides our conclusions and perspectives.

2 Related Work

In this section, we review the proposals for ALS patient clustering using the ALSFRS-R score and the available datasets.

2.1 Clustering based on progression thresholds

One common approach to cluster patients is to observe ALSFRS-R score at a specific time point and use threshold-based rules to define clusters. For example, Gomeni et al. calculate the percentage of change from first ALSFRS-R score after six months [6]. They define two clusters: one representing patients with slow disease progression (i.e., decline $< 18.6\%$) and fast disease progression (i.e., decline $>= 18.6\%$). The value of 18.6% is determined by computing the 90th percentile of the distribution of ALSFRS-R score at six months.

Grollemund et al. [7] propose a similar approach by defining four clusters directly based on the ALSFRS score one year after the start of follow-up: score ≤ 10, $10 <$ score ≤ 20, $20 <$ score ≤ 30, and score > 30.

Meyer et al. [8] use the D50 model, which outputs the number of months from the start of follow-up where the patient's ALSFRS-R score drops to 24, indicating a halving of functional ability. They define three clusters corresponding to different levels of disease severity: severe (D50 < 20), intermediate ($20 \leq$ D50 < 40), and mild (D50 ≥ 40) [8]. However, using approaches that reduce an entire sequence to a single value limits the ability to identify diverse clusters. Moreover, the heterogeneity of ALS implies that a single criterion may not fully characterize the disease.

2.2 Sequence Comparison Measures

To our knowledge, only Halbersberg et al. propose a comparison measure for evaluating pairs of ALSFRS sequences [4]. Their approach relies on multivariate ALSFRS sequences, where each time point includes 10 dimensions corresponding to ALSFRS subscores (e.g., Speech, Salivation), individually rated on a scale from 0 to 4. To compare pairs of sequences, they use the Dynamic Time Warping (DTW) dissimilarity [9]. They propose the independent warping approach DTW_I, which extends DTW to multi-dimensional sequences. Specifically, the dissimilarity between two multi-dimensional sequences is calculated as the sum of the DTW dissimilarities computed independently for each dimension [10]. Formally, given two patient sequences S and Q, DTW_I is computed as follows:

$$DTW_I(S,Q) = \sum_{i=1}^{10} DTW(S_i, Q_i)$$

where S_i and Q_i denote the ith subscore of sequences S and Q.

After computing DTW_I dissimilarity for each pair of sequences, a dissimilarity matrix is constructed and hierarchical clustering is applied. The number of clusters is set between 4 and 6 and a minimal number of 80 patients in each cluster is sought. The resulting clusters are then used for a classification task that predicts the patient's ALSFRS score at the next visit.

While Halbersberg et al. present promising classification results, they do not evaluate the quality of the clusters using clustering metrics. Furthermore, they do not justify why DTW would be more appropriate than other comparison measures to cluster patients based on ALSFRS sequences. Additionally, DTW was originally developed to align two time series of unequal lengths, under the assumption that the sequences are synchronous with a constant time interval between visits. [9]. However, in most ALS datasets, patient visits occur asynchronously at irregular intervals, making DTW not suitable in such contexts.

2.3 Available Datasets

Few ALS datasets are publicly available online, the most widely used being PRO-ACT, which includes over 10,000 patients and offers a wide variety of clinical and biological parameters. Other datasets from hospitals in Paris (France) and Jena (Germany) are also available but include significantly fewer patients [3].

However, PRO-ACT has several limitations. The patients come from 23 different clinical trials, which leads to inconsistencies in follow-up frequency and unit conventions. Additionally, the dataset only includes information starting from the point of diagnosis, with no data about first follow-up visits. Finally, there is no way to contact the clinicians who followed the patients, which limits the contextual understanding of the data. For these reasons, we consider to work with a dataset from a hospital where we can directly collaborate with physicians who possess expert knowledge of the data and its clinical context.

2.4 Summary

All contributions are summarized in Table 1. The table presents the clustering approaches, number of clusters, used dataset(s) and number of patients per cluster. The first column contains acronyms identifying each paper: GOM for

Table 1. Overview of the datasets used, number of patients after preprocessing, the clustering methods applied, and the number of clusters identified by state-of-the-art techniques.

Id	Approach	Number of clusters	Dataset(s)	Number of patients
GOM	Threshold: % change from baseline at six months [6]	2	PRO-ACT	338
GRO	Threshold: ALSFRS score after one year [7]	4	Trophos	357
			Exonhit	227
			PRO-ACT	2841
			Paris tertiary referral centre	331
MEY	Threshold: D50 value [8]	3	Jena University Hospital	108
HAL	DTW independent [4]	4 to 6	PRO-ACT	3171

Gomeni et al., GRO for Grollemund et al., MEY for Meyer et al., and HAL for Halbersberg et al.. They are also used in experiments in Sect. 4.

For the PRO-ACT dataset, the number of patients may vary depending on the data preparation steps outlined in the respective studies. Finally, we remark that none of the papers use internal metrics to evaluate how well-separated the clusters are. This omission may be limiting, as metrics like silhouette score can offer valuable insights about the intrinsic quality of the clustering and help determine whether the discovered structure is genuinely meaningful.

3 Clustering approach

We propose to represent the disease progression of a patient as a temporal sequence $<s_{t_0} ... s_{t_n}>$, where s_{t_i} denotes the ALSFRS-R score measured within a follow-up visit at relative time t_i (measured in days elapsed from the first visit), with $t_0 = 0$ and t_n corresponding to the last visit time. Due to the decreasing score and increasing time, $s_{t_i} \geq s_{t_{i+1}}$ except in rare cases.

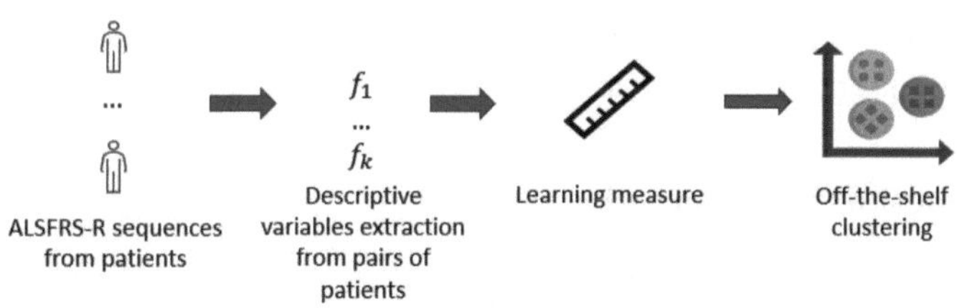

Fig. 2. Framework describing the different steps of the clustering process.

Figure 2 presents our general framework. We take as input the ALSFRS-R sequences of a set of patients. For each pair of sequences, we extract descriptive variables that characterize their differences (detailed in Subsect. 3.1). These

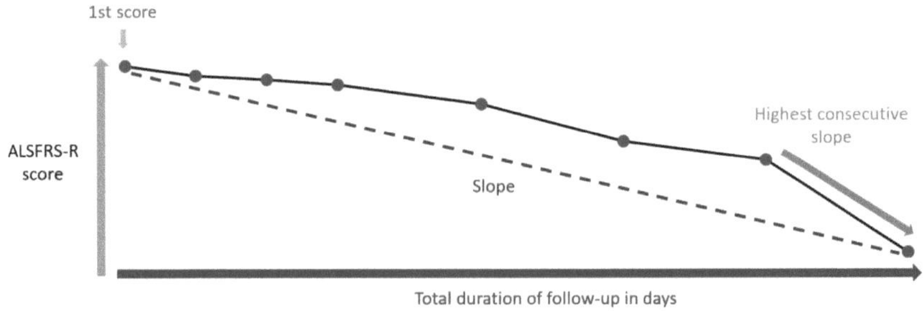

Fig. 3. Extracted information from the ALSFRS-R sequences of each patient.

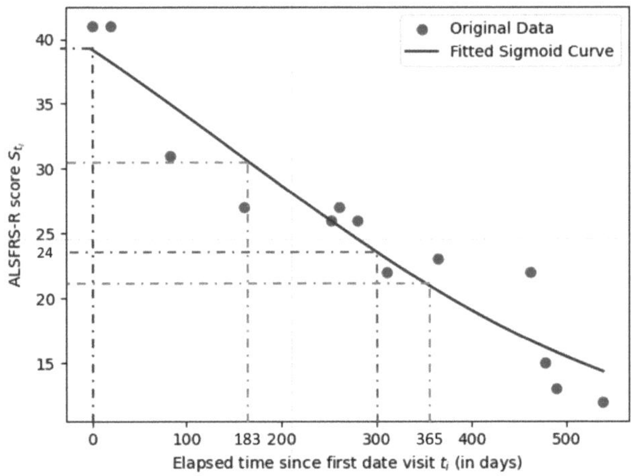

Fig. 4. Extracted information from a fitted sigmoid on an ALSFRS-R sequence. S_{365} is the one-year score, S_0 and S_{183} are the initial and six-month scores used to compute the percentage of decrease, and S_{300} marks when the score reaches 24 (D50) [8].

variables are then used to capture the distance between each pair. We investigate several distances, from classic ones to more advanced methods based on weak-supervised learning (explained in Sect. 3.2). The proposed measures are paired with off-the-shelf clustering methods (i.e., k-means, k-medoids, hierarchical clustering), enabling the identification of patient clusters.

3.1 Extraction of Descriptive Variables

In order to define a set of descriptive variables for each pair of patients, we extract relevant features from ALSFRS-R sequences, as illustrated in Fig. 3 and 4.

Sequence-Based Feature Extraction : Based on feedback from physicians, we extract the following features from each ALSFRS-R sequence: the total follow-up duration; the score at the first visit; the overall slope of the sequence, which is the difference in ALSFRS-R scores between the first and last visits divided by the total follow-up duration, i.e., $\frac{s_{t_n}-s_{t_0}}{t_n}$; and the stiffest slope between two consecutive visits, i.e., $\min_{0\leq i\leq n-1} \frac{s_{t_{i+1}}-s_{t_i}}{t_{i+1}-t_i}$ (see Fig. 3).

Sigmoid-Based Feature Extraction : We extract additional features inspired by related works discussed in Sect. 2. Specifically, we consider the ALSFRS-R score after one year of follow-up, i.e., s_{t_i} at $t_i = 365$; the percentage of decline after six months, i.e., $\frac{s_{t_0}-s_{t_i}}{s_{t_0}}$ at $t_i = 183$; and the D50 value (in days), i.e., t_i where $s_{t_i} = 24$ (see Fig. 4).

Remark that since the sequences are not synchronized, there is not necessarily a visit at (or close to) time 183 and 365, nor a visit observing an ALSFRS-R score equal to 24. To address this, we reuse the state-of-the-art strategy of fitting a sigmoid function [8] to estimate these values, with the sigmoid parameters being learned over multiple runs. The formulation is as follows:

$$sigmoid(x, b, m, a, c) = \frac{b}{1 + e^{m(x-a)}} + c$$

where x represents the number of days elapsed since the start of the follow-up, and b, m, a, and c are the parameters to be learned. This function returns an estimated ALSFRS-R score at time x.

In a preliminary study, we evaluated several sigmoid parameter configurations, ranging from 1 to 5 parameters, and selected this one because it yields the lowest Root Mean Squared Error (RMSE).

From all these features, we compute a set of descriptive variables describing the difference between a pair of ALSFRS-R sequences, namely:

1. DURATION_DIFF: The absolute difference between the total follow-up durations of the two sequences.
2. FIRST_SCORE_DIFF: The absolute difference between the first ALSFRS-R scores of the two sequences.
3. SLOPE_DIFF: The absolute difference between the overall slopes of the two sequences.
4. HIGHEST_CONSECUTIVE_SLOPE_DIFF:
   The absolute difference between the stiffest consecutive slopes of the two sequences.
5. ALS_SCORE_M12_DIFF: The absolute difference between the ALSFRS-R scores after one year of follow-up of the two sequences.
6. PC_CHANGE_M6_DIFF: The absolute difference between percentage of change after six months of follow-up of the two sequences.
7. D50_DIFF: The absolute difference between the D50 values of the two sequences.

Each descriptive variable is normalized between 0 and 1 using the Min-Max Scaler. This method was chosen to ensure all variables share the same scale and magnitude within the [0, 1] range while preserving the data's shape, allowing differences to be interpreted as distances.

Spearman correlation is used to filter out correlated descriptive variables, when correlation coefficient is greater than 0.7.

3.2 Proposed Distances

In this subsection, we investigate several distances to compare two patients P_X and P_Y, and be paired with clustering methods. We consider some standard distances commonly used with such clustering methods. We also present a weak-supervised approach that combines a generative model with a classifier.

In the following, we consider two vectors: $X = (x_1, \dots, x_k)$ and $Y = (y_1, \dots, y_k)$, each x_i (resp. y_i) corresponding to the ith feature computed for the ALSFRS-R sequence of patient P_X (resp. P_Y), with $k = 7$. Our descriptive variable is represented as $|x_i - y_i|$ where $1 <= i <= k$.

Off-the-Shelf Distances: We propose two classic distances: Minkowski and Cosine distances.

- **Minkowski distance** of order p (with p $\geq$ 1) forms a set of metrics within a vector space. It is defined such that:

$$D_p(X, Y) = \left(\sum_{i=1}^{k} |x_i - y_i|^p \right)^{\frac{1}{p}}$$

  We compute the Minkowski distance of order $p = 1$ and $p = 2$, corresponding to Manhattan (MAN) and Euclidean (EUC) distances, respectively.
- **Cosinus distance** (COS) measures one minus the angle between two vectors, the latter computed as the dot product of the vectors divided by the product of their magnitudes. It is defined such that:

$$COS(X, Y) = 1 - \frac{X \cdot Y}{\|X\| \, \|Y\|}$$

Weak-Supervised Distance: We propose the use of a labeling system to decide whether two patients should be grouped together, separated, or left undetermined, and the use of a classifier, trained on such labels, to learn a distance.

To label patient pairs, we propose a voting-based system that relies on weak-supervised labelling functions to assign labels. The design of the labelling functions was set up by frequent discussions with physicians. We devised seven functions, each based on a descriptive variable and relying on its distribution of values. A function assigns label *grouped-together* (T) when the variable value is lower than first quartile, label *separated* (S) when it is greater than third quartile, and label *undetermined* (U) else (in this case the labelling function abstains, representing physicians uncertainty). This is illustrated in Fig. 5.

The final label is determined by a probabilistic voting system across all functions (Fig. 6) [11]. The system models the accuracy and correlations of each labeling function, learning their reliability without ground-truth labels. It then aggregates the outputs into a probability distribution, selecting the label with the highest probability as the final assignment.

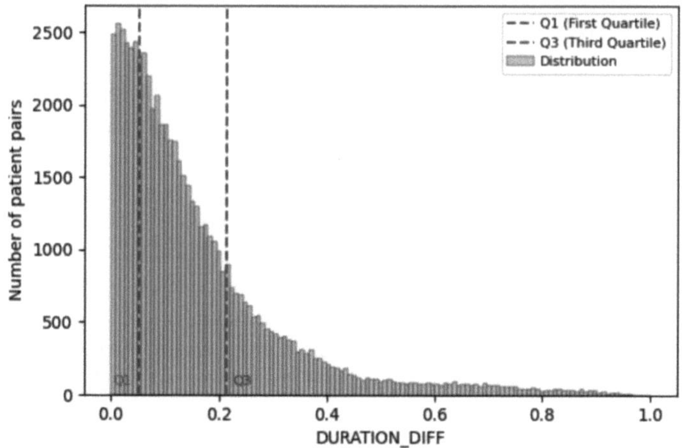

Fig. 5. Distribution of the DURATION_DIFF variable.

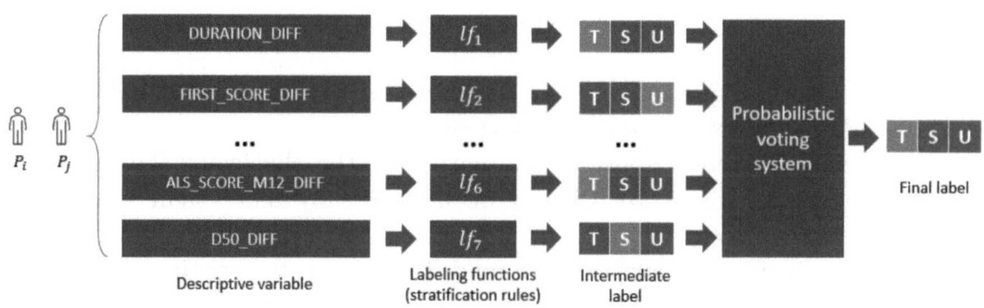

Fig. 6. Example of pair labeling: Each function labels a pair, and the final decision is made by a probabilistic voting system. T, S, and U indicate whether the pair should be grouped, separated, or remain undetermined. The assigned label is shown in green. (Color figure online)

To learn our measure, we use a linear Support Vector Machine (SVM) classifier. It is trained on labeled pairs provided by the voting-based system in order to learn the weights assigned to each proposed labeling function for each descriptive variable. We define our measure named Weak-Supervised Distance (WSD) as follows:

$$WSD(X, Y) = \sum_{i=1}^{k} w_i \cdot |x_i - y_i|$$

where: w_i denotes the weight associated with the ith labeling function, and $|x_i - y_i|$ is the value of the descriptive variable.

The absolute value of w_i reflects the discriminative power of the function (a high value indicates that it is very influential in the decision-making process), while the sign of w_i indicates whether the function favors the grouping of patients or, conversely, their separation. Pairs labeled as U by the voting-based system are not considered in the learning of the weights.

4 Experiments

This section describes our experiments for evaluating the proposed distances, comparing them to state-of-the-art approaches, and analysing cluster quality.

Setting. The experiments are conducted using the data analysis software Konstanz Information Miner (KNIME, version 5.4.2) in combination with the Python programming language (version 3.11.11). The probabilistic voting system is implemented using the generative model Snorkel (version 0.10.0) [11], which provides dedicated Python libraries. The labeling functions are thus defined within this model. To learn parameters of the sigmoid, we use the curve_fit library that use non-linear least squares to fit the ALSFRS-R sequence [12].

Dataset. We work with data from 888 ALS patients diagnosed with motor neuron disorders, who are followed typically every 3 months at the university regional hospital center of Tours. No medical criteria were used to filter patients. The progression is measured using the ALSFRS-R score. We exclude:

1. Patients with fewer than five visits (i.e., $n < 5$) to ensure regular follow-up. This leads to the exclusion of 350 patients.
2. Patients with sequences considered abnormal by the physicians. This includes sequences where the ALSFRS-R score increases by more than two points between two consecutive visits (i.e., $s_{t_{i+1}} - s_{t_i} > 2$), a clinically extremely unlikely scenario, as patients' health tends to deteriorate over time. Therefore, 151 patients are excluded. Additionally, after presenting a detailed visualization of the ALSFRS-R sequences to the physicians, three more patients were excluded due to their sequence being deemed impossible by the experts.
3. Patients whose first recorded ALSFRS-R score arrives more than 30 days after their first visit are excluded, avoiding any bias related to duration. This leads to the exclusion of 19 patients.
4. Finally, since we also compare our approach to the HAL approach, which uses ALSFRS subscores, patients without these available subscores are excluded. This filter excludes 12 patients.

At the end, 353 patients are retained, leading to 62,481 pairs. Spearman correlation test did not reveal any highly correlated descriptive variables (with a correlation coefficient ≥ 0.7). Therefore, we retain all variables described in Sect. 3.1 to develop the distances.

Sigmoid. Using the best-performing sigmoid function, the model achieves a mean error below 2 on the ALSFRS-R score for the majority of patients in the cohort. An error margin comparable to the variability occasionally observed among physicians. The median error is 1.43, indicating that for half of the population, the prediction error remains below this value. This corresponds to a deviation of approximately $\frac{1.43}{48} \approx 3\%$, which is minimal and therefore deemed acceptable for our experiment.

Distance Properties Check. The properties of the WSD and cosine distance measures have been empirically evaluated over the dataset. For both measures, positivity, symmetry, and identity (i.e., the distance from a patient to itself is zero) are satisfied, as expected. Regarding triangle inequality, empirical tests show that violations are rare and very minor in magnitude. Specifically, WSD satisfies the triangle inequality in 98.309% of cases and cosine distance in 71.37%, with WSD showing significantly smaller deviations (i.e., between 1E-06 and 1E-02) when violations occur. It is also worth noting that clustering using non-distances measures is a common and accepted practice in the literature.

Protocol. We compare our method (i.e., clustering and distance), to state-of-the-art methods GOM, GRO, MEY and HAL, described in Sect. 2.

We precompute and store measures in a matrix, which serves as input for various off-the-shelf clustering methods. To enhance cluster detection, we also explore the use of a dimensionality reduction technique on the matrix. Concretely, we use Uniform Manifold Approximation and Projection (UMAP). UMAP achie-ves a compromise between preserving local and global structures in the projected data, and the appropriate choice of its parameters can induce space distortions that are beneficial for cluster identification [13]. The clustering methods are then applied to the resulting low-dimensional representations. To evaluate the impact of dimensionality reduction, we compare clustering performance with and without UMAP. Furthermore, we investigate the impact of using different off-the-shelf clustering methods. The following clustering methods are considered:

- **K-means (KME)**: Applied only to UMAP representations, as it requires vector space coordinates to compute centroids.
- **K-medoids (KMD)**: Applied to both precomputed matrices and UMAP representations.
- **Agglomerative Hierarchical Clustering (AHC)**: Applied to both precomputed matrices and UMAP representations, using the "complete" linkage criterion to favor the formation of compact clusters.

Each of theses possibles combinations distance + clustering method, are called workflows. For each workflow, we vary the number of clusters, testing from 2 to 6. This yields to 100 workflows to assess.

For the state-of-the-art approaches, we use the setting recommended by authors. Approaches using thresholds-based rules, GOM, GRO and MEY, are not paired with all clustering methods nor UMAP and have fixed number of clusters. HAL approach uses AHC, varying between 4 and 6 clusters. Additionally, we investigate whether applying UMAP to their dissimilarity matrix may produce better results. This yields to 9 additional workflows to assess.

Cluster quality. To evaluate the quality of the resulting clusters from different workflows, we propose 3 criteria:

1. **Survival analysis:** We assess whether the identified clusters exhibit distinct survival curves. To compare the survival curves of different clusters, we apply the log-rank test to each pair of curves [14]. The log-rank test evaluates whether the survival curves of two clusters are significantly different. The null hypothesis of this test assumes that there is no difference in survival between the two clusters, i.e., likelihood of survival at any given time is the same for both clusters. A p-value below 0.05 indicates a significant survival difference, suggesting that the two survival curves differ.
The test also provides a Log-Rank Statistic (LRS), calculated as:
$$LRS = \frac{(O - E)^2}{Var(O - E)}$$
where O and E represent the observed and expected number of events (i.e. death of patient) at a particular time point, and Var is the variance that reflects how much variability we expect in the number of events across time. We apply this test to every pair of curves and report the maximum p-value and minimum LRS among all comparisons. P-values range from 0 to 1, where smaller values suggest greater significance. LRS range from 0 to $+\infty$, with higher values indicating better separation between the survival curves, reinforcing the clinical relevance of the findings.
2. **Silhouette score:** Silhouette score is an internal metric that favors compact, well-separated clusters [15]. For each patient, it measures the difference between cohesion (average distance to patients in the same cluster) and separation (average distance to the nearest cluster) [16]. The silhouette score ranges from -1 to +1: a positive score indicates good clustering, with values close to +1 suggesting well-separated and well-defined clusters; a score near 0 indicates overlapping clusters, and a negative score indicates misclassification.
3. **Patient distribution:** We also suggest analyzing how patients are distributed among the clusters by counting the number of patients per cluster, evidencing imbalanced clusters. This is important, as highly uneven clusters with very few patients may indicate outliers and make it challenging to draw meaningful conclusions.

Results. We executed the 109 mentioned workflows on the patients dataset and assessed the described quality criteria. To reduce the analysis of results, we focus on workflows with a silhouette score of 0.5 or higher. This keeps 50 workflows, 41 concerning the proposed measures and 9 from state-of-the-art methods. To align with the log-rank test, we further filter out workflows with non-significant p-values (p-value $\geq$ 0.05), keeping 24 workflows, 21 from our approach, and 3 from state-of-the-art approaches. We remark that all workflows using the HAL approach are not significant.

We present the results in Table 2, which includes the number of clusters, the number of patients in each cluster, the silhouette score, and the log-rank test with its p-value and LRS. The results are sorted in descending order of the LRS. Workflow names follow the format: 'measure'_'UMAP, if applied'_' clustering method'_'number of clusters'. State-of-the-art methods are represented in bold.

We observe that workflows producing five or six clusters are excluded due to insufficient statistical significance in survival distribution. In addition, most of the reported workflows primarily involve UMAP, as those without UMAP result in silhouette scores below 0.5. Workflows concerning the MAN measure are also filtered out.

The best-performing workflow is COS_KMD_2, which achieved the most statistically significant log-rank p-value (6.92E-29), along with a LRS of 124.39. It also achieves the highest silhouette score (0.79) but its high standard deviation ($\pm$ 0.39) indicates instability in cluster cohesion, suggesting that conclusions on silhouette should be drawn with caution.

Other notable workflows include WSD_UMAP_AHC_2 and GOM_2, which yield high and comparable LRS (110.69 and 106.32, respectively) and relatively robust silhouette scores ($\geq$ 0.57). Overall, workflows involving WSD with UMAP followed by AHC, KME, or KMD offer a favorable compromise between cluster compactness and clinical separation, particularly in two and three clusters settings.

Interestingly, COS_AHC_2 attains the highest silhouette score overall (0.93 $\pm$ 0.17), yet exhibits a substantially lower LRS (33.09), suggesting that while the clusters are well separated in the descriptive variables space, they lack meaningful prognostic differentiation. Clinical feedback from collaborating physicians further confirms that the patient groupings produced by this workflow is not clinically relevant or interpretable as there are very few patients in the second cluster. Additionally, configurations involving more than two clusters generally result in diminished LRS, underscoring a decline in clinical clustering power with increased cluster granularity.

In summary, WSD_UMAP_AHC_2, GOM_2 and COS_KMD_2, achieve the most effective balance between intra-cluster consistency and clinical relevance with a high silhouette average. However, the silhouette score for COS-_KMD_2 exhibits a high standard deviation, so we should interpret the results with caution and refrain from drawing definitive conclusions for this workflow. While increasing the number of clusters often dilutes statistical power, most of proposed workflows for three, and four clusters consistently outperform state-of-the-art techniques (i.e., MEY_3 and GRO_4) in terms of LRS, without compromising silhouette quality.

In addition to the proposed metrics, physicians' feedback confirms that for two, three, and four clusters, the first two workflows (i.e., COS_KMD_2, WSD-_UMAP_AHC_2, EUC_UMAP_KMD_3, EUC_UMAP_KME_3, COS_UMAP_KME_4 and COS_UMAP_KMD_4) produce clinically coherent clusters that are generally well-sized and show distinct progression patterns in the survival curves. These findings are promising for the clinical care and the clustering of future patients, as they provide clinically meaningful, well-sized clusters with clearly distinct survival curves.

These results must be interpreted with caution. In particular, although silhouette metric offers useful insights into cluster cohesion, it is inherently influenced by the distribution of distances specific to each metric. As such, compar-

Table 2. Results of different workflows with silhouette and log-rank. State-of-the-art workflows are highlighted in bold.

Combination Workflow name	Number of clusters	Number of patients per cluster	Silhouette	Log-rank p-value (max)	LRS (min)
COS_KMD_2	2	298, 55	0.79 ± 0.39	6.92E-29	124.39
WSD_UMAP_AHC_2	2	236, 117	0.57 ± 0.23	6.91E-26	110.69
GOM_2	2	245, 108	0.61 ± 0.23	2.92E-21	106.32
WSD_UMAP_KME_2	2	181, 172	0.56 ± 0.20	2.22E-20	85.58
WSD_UMAP_KMD_2	2	186, 167	0.56 ± 0.20	5.86E-18	74.57
WSD_AHC_2	2	328, 25	0.56 ± 0.23	2.95E-17	71.38
EUC_UMAP_KME_2	2	183, 170	0.50 ± 0.21	2.59E-15	62.56
EUC_UMAP_AHC_2	2	225, 128	0.50 ± 0.23	6.18E-15	60.84
EUC_UMAP_KMD_2	2	187, 166	0.50 ± 0.21	4.84E-14	56.79
COS_AHC_2	2	348, 5	0.93 ± 0.17	8.80E-09	33.09
EUC_UMAP_KMD_3	3	131, 121, 101	0.50 ± 0.19	4.66E-08	29.85
EUC_UMAP_KME_3	3	132, 123, 98	0.50 ± 0.19	1.10E-07	28.19
WSD_UMAP_AHC_3	3	143, 117, 93	0.61 ± 0.15	3.84E-06	21.35
WSD_UMAP_KME_3	3	142, 118, 93	0.61 ± 0.15	4.79E-06	20.92
WSD_UMAP_KMD_3	3	141, 119, 93	0.61 ± 0.15	5.92E-06	20.51
COS_UMAP_KME_4	4	133, 80, 74, 66	0.59 ± 0.17	3.90E-05	16.92
COS_UMAP_KMD_4	4	133, 78, 74, 68	0.59 ± 0.17	3.90E-05	16.92
MEY_3	3	230, 97, 26	0.59 ± 0.21	5.13E-05	16.40
COS_UMAP_AHC_4	4	141, 85, 66, 61	0.57 ± 0.22	3.08E-04	13.02
WSD_UMAP_AHC_4	4	143, 93, 92, 25	0.55 ± 0.24	6.17E-04	11.72
EUC_UMAP_KMD_4	4	106, 97, 78, 72	0.54 ± 0.18	5.32E-03	7.77
COS_KMD_3	3	218, 116, 19	0.62 ± 0.40	6.10E-03	7.52
GRO_4	4	143, 116, 82, 12	0.54 ± 0.22	2.01E-02	5.41
COS_KMD_4	4	139, 129, 68, 17	0.52 ± 0.38	3.18E-02	4.61

ing silhouette values across different distance measures may confound clustering quality with properties of the distance space itself. Therefore, we place greater emphasis on survival-based metrics like log-rank test, that more directly reflect clinical clustering.

5 Conclusion

We introduce a method to compare ALSFRS-R sequences using both simple distance metrics and an advanced weakly-supervised approach combining a probabilistic voting system with a linear SVM. The resulting distance matrices are paired with an off-the-shelf clustering method.

We propose different workflows by varying the number of clusters, clustering methods, and incorporating UMAP for dimensionality reduction. In total, we evaluated 100 workflows and compared to 9 workflows from state-of-the-art methods. Our approach consistently achieves silhouette scores comparable to state-of-the-art techniques and yields higher log-rank statistics for two, three and four clusters.

These experiments show that several workflows, achieve a strong balance between cluster cohesion and clinical relevance. Physicians' feedback further support that these workflows generate clinically meaningful, well-sized clusters with clearly distinct survival curves. This is a promising result for the clinical management and the clustering of future patients.

Our study shows that high internal clustering metrics like the silhouette score don't always reflect clinical relevance, as survival analysis reveals. While useful for assessing cluster structure, the silhouette score should be combined with clinical metrics to ensure meaningful evaluation.

We identify five key directions for future work. The first one is the integration of the temporal dimension in defining patient clusters. Currently, our method aggregates ALSFRS-R sequences by precomputing descriptive variables, which are then used in clustering methods. However, given the heterogeneous nature of the disease, it is likely that a patient's cluster changes over time. Thus, analyzing these variables across different time windows (e.g., from baseline to 3 months, from 3 to 6 months, etc.) could provide more relevant insights. Incorporating this temporal aspect would enhance the clinical relevance of the patient clusters.

The second key direction is improving the quality of the resulting clusters. Many studies do not assess the quality of their clusters in-depth and often overlook internal metrics, such as the silhouette score, to evaluate cluster separation. Typically, clusters are compared by analyzing statistical differences in clinical variables [3] after clustering. To strengthen the reliability of clustering methods, it is essential to incorporate both internal and external evaluation metrics. A promising direction for external evaluation could be the use of biological data to identify parameters that influence the disease progression.

The third key direction is about the representativeness of the dataset. It is important to ensure that the distribution of descriptive variables is properly verified before normalization. In particular, minimum and maximum values should be checked to detect potential outliers. Additionally, the required number of patients should be assessed to ensure the dataset accurately reflects the target population. Statistical methods, such as Cochran's formula [17], may be considered to estimate an appropriate sample size. Failing to do so could distort the distribution, ultimately leading to a non-representative dataset.

The fourth key direction is to adapt our approach for use in other diseases characterized by progressive and heterogeneous progressions such as Parkinson's, multiple sclerosis, or cancer. To support these diverse disease contexts, we aim to generalize our approach to any sequences with quasi-monotonic behavior, by accommodating various longitudinal clinical or biological parameters.

Finally, we plan to conduct experiments on larger datasets to validate and enhance the robustness of our findings. PRO-ACT appears to be the most suitable option for external validation, as it has been widely used in studies such as GRO, GOM, and HAL. Although it has known limitations, PRO-ACT is widely recognized and accessible, offering better consistency by including more patients than the current dataset. Using PRO-ACT will allow us to compare our results with previous studies, ensuring that our findings are generalizable and reliable despite the imperfections of the data.

Disclosure of Interests. The authors have no competing interests to declare that are relevant to the content of this article.

Availability of Data and Materials. Data and tools used for experiments are available via this link: https://github.com/GuillaumeTejedor/WSD.git.

References

1. Feldman, E.L., et al.: Amyotrophic lateral sclerosis. Lancet **400**(10360), 1363–1380 (2022)
2. Grollemund, V., et al.: Machine learning in amyotrophic lateral sclerosis: achievements, pitfalls, and future directions. Front. Neurosci. **13**, 135 (2019)
3. Tavazzi, E., et al.: Artificial intelligence and statistical methods for stratification and prediction of progression in amyotrophic lateral sclerosis: A systematic review. Artif. Intell. Med. **142**, 102588 (2023)
4. Halbersberg, D., Lerner, B.: Temporal modeling of deterioration patterns and clustering for disease prediction of ALS patients. In: ICMLA, pp. 62–68. IEEE (2019)
5. Drushku, K., et al.: Interest-based recommendations for business intelligence users. Inf. Syst. **86**, 79–93 (2019)
6. Gomeni, R., et al.: Amyotrophic lateral sclerosis disease progression model. Amyotroph Lateral Scler Frontotemporal Degener. **15**(1–2), 119–129 (2014)
7. Grollemund, V., et al.: Manifold learning for amyotrophic lateral sclerosis functional loss assessment: development and validation of a prognosis model. J. Neurol. **268**(3), 825–850 (2021)
8. Meyer, J., et al.: Phosphorylated neurofilament heavy chain (pNfH) concentration in cerebrospinal fluid predicts overall disease aggressiveness (D50) in amyotrophic lateral sclerosis. Front. Neurosci. **19**, 1536818 (2025)
9. Sakoe, H., Chiba, S.: Dynamic programming algorithm optimization for spoken word recognition. IEEE Trans. Acoust. **26**(1), 43–49 (1978)
10. Shokoohi-Yekta, M., et al.: Generalizing DTW to the multi-dimensional case requires an adaptive approach. DMKD **31**, 1–31 (2017)
11. Ratner, A., et al.: Snorkel: rapid training data creation with weak supervision. VLDB Endow. **11**, 269 (2017)
12. Virtanen, P., et al.: SciPy 1.0: fundamental algorithms for scientific computing in Python. Nat. Methods **17**(3), 261–272 (2020)
13. McInnes, L., et al.: UMAP: uniform manifold approximation and projection for dimension reduction. arXiv preprint: arXiv:1802.03426 (2018)
14. Kleinbaum, D.G., et al.: Kaplan–Meier survival curves and the log-rank test. Surv. Anal.: Self-Learn. Text, 55–96 (2012)
15. Rousseeuw, P.J.: Silhouettes: a graphical aid to the interpretation and validation of cluster analysis. J. Comput. Appl. Math. **20**, 53–65 (1987)
16. Shahapure, K.R., Nicholas, C.: Cluster quality analysis using silhouette score. In: DSAA, pp. 747–748. IEEE (2020)
17. Cochran, W.G.: Sampling Techniques. 3rd edn. John Wiley and Sons (1977)

Designing Efficient and Scalable Substructure Discovery Algorithms for Multilayer Networks

Arshdeep Singh, Abhishek Santra, and Sharma Chakravarthy(✉)

IT Laboratory and CSE Department, UT Arlington, Arlington, USA
axs9120@mavs.uta.edu, abhishek.santra@uta.edu,
sharmac@cse.uta.edu

Abstract. To model large, complex data sets – with multiple types of entities and relationships – multilayer networks (or MLNs) have been shown to be more effective and computationally efficient as compared to simple and attributed graphs. Analysis algorithms for MLNs are being developed using the decoupling approach and have been shown to be efficient, scalable, and achieve high accuracy. This paper focuses on substructure discovery in **homogeneous multilayer networks** (or HoMLN, a type of MLN) using a single composition at the end of all iterations for substructure discovery. In this approach, each layer is processed **independently** for several iterations, and then the results from two or more layers are **composed** to identify substructures in the entire MLN. The algorithm is designed and implemented, including the composition part, using one of the distributed processing frameworks to ensure scalability.

After analyzing the tradeoffs between accuracy and efficiency based on composition frequency, we analyze the speedup and response time of the proposed algorithm for the chosen distributed architecture through extensive experimental analysis on large synthetic and real-world data sets with diverse graph characteristics.

Keywords: Homogeneous Multilayer Networks · Substructure Discovery · Decoupling Approach · Composition · Analysis of Complex Data

1 Motivation

Graph data mining approaches have proven to be powerful for analyzing data that can be modeled naturally as a graph. Graph mining provides algorithms to extract and analyze numerous components (including substructures, which are connected subgraphs) that are present in the graph and satisfy some global property. By leveraging graph mining techniques, we can discover these interesting patterns that may be difficult or impossible to identify using other methods.

Substructure discovery is an unsupervised approach to data mining to infer new knowledge and is typically applied to graph representations. Frequent patterns are a common concept in data mining that refers to item sets, subsequences, or substructures that occur frequently in a given data set, meeting or

P. K. Chrysanthis et al. (Eds.): ADBIS 2025, LNCS 16043, pp. 217–233, 2026.
https://doi.org/10.1007/978-3-032-05281-0_14

exceeding a user-defined threshold. By identifying frequent subgraphs, we can gain insights into the underlying structure of the data and identify recurring patterns or motifs.

Graph models have been used to analyze the World Wide Web structure [16], social-media data, bio-informatics data [8], atoms and covalent relationships in chemistry [3]. Graphs use vertices and edges (with labels as needed) where each vertex corresponds to an entity and each edge a relationship between two entities. Applications can be modeled using several graph alternatives: (i) Simple graphs, (ii) Attributed graphs, and (iii) Multilayer networks.

1.1 Multilayer Networks as an Expressive Data Model

Simple graphs (also called a monoplex or a network) have been used for modeling and to derive valuable insights into the functioning of different natural, social, biological, and technological systems in recent decades. However, real-world systems are often interconnected, and they may have multiple interdependencies for which networks can be used but are difficult to understand (structure as well as semantics) or analyze.

Modeling or representing them as simple graphs by using only one type of node and relationship often leads to an oversimplification of reality and information loss. Attributed graphs are better-suited, but currently lack algorithms for analysis. Sociologists have long acknowledged the significance of using distinct types of connectivity among the same group of individuals [26]. Therefore, such systems are better studied through multilayer networks.

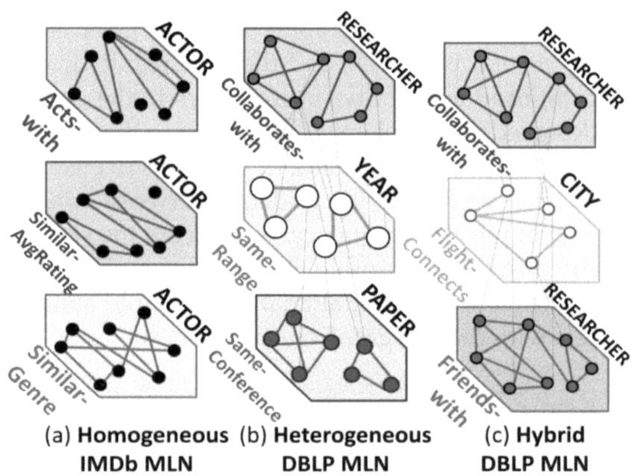

Fig. 1. Multilayer Network Types

In a multilayer network, each layer represents a different type of interaction or relationship between the system's entities. For example, in a social network, one layer may represent *friendship* between individuals, while another layer may represent LinkedIn *connections*. With this representation, it is possible to gain insights on individual layers, subsets as well as the system as a whole.

Multilayer networks can be of several types, including (i) homogeneous (HoMLNs, Fig. 1(a)), (ii) heterogeneous (HeMLNs, Fig. 1(b)), and (iii) hybrid (HyMLNs, Fig. 1(c)), depending on the type of entities in layers and connections between them. Since the focus of this paper is HoMLNs, we start with a formal definition of an MLN that includes HoMLNs.

Definition 1. *A multilayer network $MLN(G, X)$, is defined by two sets of graphs. The set $G = \{G_1, G_2, \ldots, G_n\}$ contains* ***simple*** *graphs (one for each layer), where $G_i(V_i, E_i)$ is defined by a set of vertices V_i and a set of edges E_i. An edge $e(v, u) \in E_i$, connects vertices v and u, where $v, u \in V_i$. The set $X = \{X_{1,2}, X_{1,3}, \ldots, X_{m-1,m}\}$ consists of* ***bipartite*** *graphs*[1]*. Each graph $X_{i,j}(V_i, V_j, L_{i,j})$ is defined by two sets of vertices V_i and V_j and a set of edges (or links) $L_{i,j}$, such that for every link $l(a, b) \in L_{i,j}$, $a \in V_i$ and $b \in V_j$, where V_i (V_j) is the vertex set of graph G_i (G_j). Some of $X_{i,j}$ can be empty.* ***For HoMLNs, inter-layer graphs are implicit as nodes across layers form a common subset having same identity.***

Problem Addressed: Given a homogeneous MLN (HoMLN) with l layers – G_1 (V_1, E_1), G_2 (V_2, E_2), ..., G_l (V_l, E_l), where V_i and E_i are the vertex and edge sets in the ith layer and $V_1 \cap V_2 \cap \ldots \cap V_l$ is not empty – the goal is to discover *interesting substructures* of the HoMLN for any r layers **efficiently** using the decoupling approach by **performing composition only once.** For correctness, r HoMLN layers under consideration are conflated into a simple graph using the Boolean OR operator (as shown in Fig. 1 (a)), and substructures discovered on that simple graph are used as the Ground Truth (GT). We have termed this HoSCA or **Ho**mogeneous **S**ingle **C**omposition **A**lgorithm.

Need for Map/Reduce: We have used the Map/Reduce as a paradigm to exemplify the need and effectiveness of distributed and parallel processing for this problem. Without loss of generality, any other distributed processing paradigm, such as Spark or Pregel can be used in its stead without the need to modify the algorithms presented in this paper. Of course, implementation will change slightly based on the choice of the specific framework. *As our approach supports partitioning of a layer as well as multiple layers for substructure discovery, distributed processing approaches ensure both horizontal and vertical scalability to handle graphs of any size.* **Contributions of this paper are:**

- Map/Reduce **partitioned approach** to substructure discovery in a layer.
- Design and implementation of a **Single Composition Algorithm** for HoMLNs (termed HoSCA) to generate inter-layer substructures using Map/Reduce.
- **Accuracy** tradeoff of the proposed approach with ground truth.
- Extensive experimental analysis of response time and speedup on a large number of **synthetic and real-world** graphs.

Roadmap: Section 2 has related work. MLN analysis alternatives are discussed in Sect. 3. The effect of composition-frequency on the accuracy is analyzed in Sect. 4. Preliminaries are outlined in Sect. 5. Section 6 details the substructure composition algorithm and its Map/Reduce implementation. Experimental evaluation is given in Sect. 7. Section 8 has conclusions.

[1] The number of bipartite graphs can be less or even more than the number of layers as a layer may be connected to multiple layers. That is, m can be $<=$ or even $> n$.

2 Related Work

SUBDUE was the earliest main memory algorithm [9] developed for substructure discovery. It uses an iterative algorithm to systematically generate larger substructures which are evaluated using the Minimum Description Length (MDL [19]). SUBDUE uses the notion of a *beam* to restrict the number of substructures carried to the next iteration.

AGM [2] and FSG [12] are two popular main memory graph mining algorithms that use the *apriori* concept. They generate frequent subgraphs of size $(k+1)$ from frequent subgraphs of size k. FSG identifies repeating substructures in graphs using an apriori approach [12]. Canonical labeling is added to the apriori algorithm. The property that identical graphs have identical canonical labeling has been strategically used to identify frequent substructures.

Disk-based graph mining algorithms [18] were developed to deal with larger graph sizes. HDB-Subdue [17] used the RDBMS for graph representation and SQL to discover substructures. This approach was able to scale to handle graphs with millions of edges. However, self-joins on large relations and the number of joins needed for canonical form generation (column sorting instead of row sorting) limited scaling further.

Many graph mining methods have shown success in cloud-based deployments [25]. Additionally, research has explored patterns in big graphs using the Map/Reduce framework [15]. However, the pattern searching technique in [15] requires a specified pattern to be searched for all its instances in the graph. If we need to find a pattern with the highest compressibility, we cannot supply it beforehand. Splitting/partitioning large graphs into manageable chunks for distributed processing is explored in [27]. Scalable substructure discovery on large simple graphs using Map/Reduce has been developed [10]. A graph can be partitioned in different ways, processed by a distributed and parallel architecture, and results from across partitions combined to obtain correct results.

There are some algorithms developed for MLNs [5,20,21], but not for substructure discovery. *This paper's focus is on MLNs instead of simple graphs, and specifically HoMLNs. Further, the decoupling approach is chosen, which entails the development of a composition function for its efficient and flexible analysis. It also uses extant substructure discovery for each layer.*

3 MLN Analysis Alternatives and Challenges

As we are proposing analysis algorithms on MLNs, we need to understand the alternatives for developing MLN algorithms. These alternatives are shown in Fig. 2. Figure 2(a) shows an MLN conflated into a simple graph by aggregating layers (using Boolean AND/OR operators.) These aggregation approaches, termed type-independent [13] and projection-based [4], ignore type information. Hence, they do not preserve structure and semantics (without elaborate mappings) as they aggregate (or collapse) layers into a simple graph in different ways. As observed in the literature, *without additional mappings*, currently-used aggregation approaches are likely to result in some information loss, distortion of properties, or hide the effect of different entity types and/or different

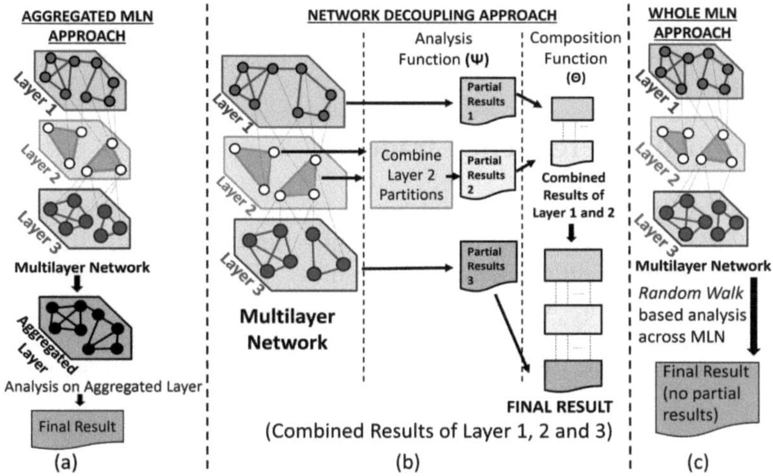

Fig. 2. (a) Lossy, (b) Decoupling, (c) Whole MLN approaches [23].

intra- or inter-layer relationships as elaborated in [11,14]. At the other end of the spectrum, Fig. 2(c) shows the same MLN layers and result computation by traversing the MLN *as is*. This preserves structure and semantics but requires traversing the whole MLN. Hence, the algorithm and bookkeeping are more involved, making it less efficient in most cases.

In contrast, Fig. 2(b) shows yet another alternative proposed in [20,23], the **network decoupling** approach. Due to localized semantics of layers, this approach preserves structure and semantics, making the algorithms simpler. Each layer is processed **independently** (possibly in parallel) during the analysis (Ψ) phase, and a binary composition function (Θ) is applied as shown. This approach has been shown to be flexible, effective, and can be done using Boolean operations for HoMLNs and HeMLNs without losing type information [20–23]. Furthermore, it is more efficient and scalable than the approaches shown in Fig. 2(a) or (c). Finally, the clarity of modeling is retained as well when MLNs are used.

However, the challenges in developing a substructure discovery algorithm are: *i) enumerating all connected subgraphs of any size in a given graph or forest (completeness), ii) identifying duplicates, if any, and removing them (for soundness), iii) counting isomorphic substructures to apply the metric for each distinct substructure and ranking them, and iv) retaining top-beam substructures for the next iteration*. The enumeration is typically done by iteratively increasing the connected subgraph size by 1. Typically, evaluation of each substructure with the desired metric (Frequency or Minimum Description Length) is carried out after each iteration. To contain the search space, a heuristic is applied to carry top-*beam* (*beam* is a parameter) results from each iteration. Current algorithms do this on a simple graph or forest. **How the decoupling approach has been adapted to the HoMLN substructure discovery is the focus of this paper and is elaborated in Section 6.** However, as accuracy is a challenge due to incomplete information, typically, heuristics are used to keep the layer information to a minimum.

For accuracy comparison, the approach shown in Fig. 2(a) is used for ground truth. This paper, however, details an algorithm based on the approach shown in Fig. 2(b).

4 Composition-Frequency Tradeoff

MLN models partition a data set into layers, and the decoupling approach works on each layer independently and composes (using function Θ) *independently generated* results. The substructure discovery algorithm, being iterative, provides multiple opportunities for composition using the decoupling approach. Composition is performed to identify substructures that **span** l **layers** using substructures generated from *each layer individually*. Here, composition can be done after each iteration, or repeated after some fixed number of iterations, or at the end (termed *composition-frequency*.)

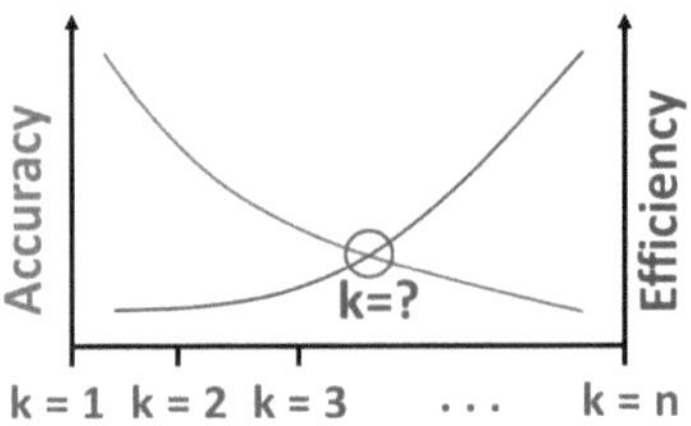

Fig. 3. Composition Trade-off

This trade-off between composition-frequency and accuracy can be hypothesized approximately as shown in Fig. 3. The X-axis shows iterations of the algorithm and opportunities for composition. The two Y-axes show accuracy and efficiency. Our conjecture is that composing after **each iteration** provides complete accuracy but takes more time and resources. Composing after each 'k' iterations ('k' needs to be determined) or at the end (used in this paper) improves efficiency significantly and is likely to reduce accuracy. The circle indicates the sweet spot. Of course, again, layer graph characteristics and other MLN properties (e.g., edge distribution among layers) play a critical role in the value of the sweet spot, and estimating that is a challenge.

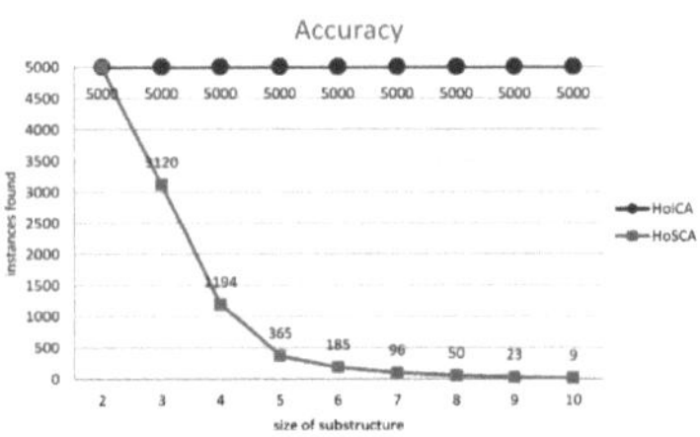

Fig. 4. Accuracy and size of substructure, 50KV100KE

Our experiments also support this conjecture. We have experimented with the two extremes (composition-frequency of 1 and composition at the end), whose results are shown in Fig. 4. For HoSCA, the suspected loss of accuracy is evident, as depicted in Fig. 4, which illustrates the diminishing accuracy with increasing substructure size as compared to HoICA (Homogeneous *Iterative* Composition Algorithm), where composition is done **after each iteration** [24]. For this experiment, we have used a data set of 50K vertices and 100K edges with an embedded substructure of size 10 with a frequency of 5000. The edges were evenly distributed, with a 50/50 split between the layers[2]. Additionally, the beam parameter also plays a role. Accuracy

[2] Accuracy is preserved for k=2 as there is only one iteration. We have also conducted experiments with 70/30 and 90/10 distribution of edges across layers, and as expected, the accuracy **preservation is much better** because when edges in the layers are skewed, less number of inter-layer composed substructures are present.

increases when the beam size increases for HoSCA for the same split. In our composition, the result for size k is obtained by composing the output of size k from each layer from iteration *k-1*. In other words, to generate a size 5 substructure, the size 5 output from each layer (from iteration 4) is used.

The accuracy drops as the size of the substructure increases. This decline is due to the *extended delay* before performing composition and not performing on some substructure sizes and hence not generating some inter-layer substructures. For example, if composition is done at size 5 (i.e., using iteration 4 output), instances from 2 iterations (2 and 3) are missed. However, if composition is delayed until size 10, instances from 9 iterations are missed. For each iteration where composition is not applied, increasingly more information is lost. This leads to an increased loss of inter-layer substructure instances. Consequently, the missed instances accumulate, and by iteration 9, there is minimal information left to regenerate all inter-layer substructures.

5 Preliminaries and Terminology

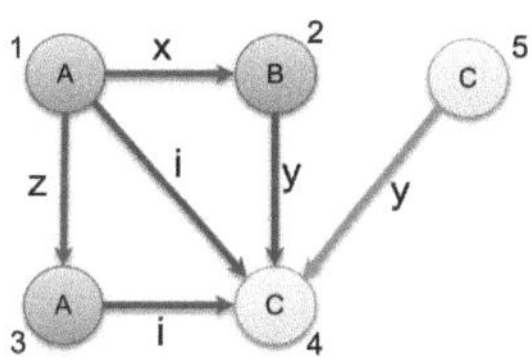

Fig. 5. Input Graph

Table 1. Edge List

Edge List
(x,1,A,2,B)
(z,1,A,3,A)
(i,1,A,4,C)
(y,2,B,4,C)
(i,3,A,4,C)
(y,5,c,4,c)

Table 2. Adjacency List

Vertex ID	Adjacency List
1	(x,1,A,2,B), (z,1,A,3,A), (i,1,A,4,C)
2	(x,1,A,2,B), (y,2,B,4,C)
3	(z,1,A,3,A), (i,3,A,4,C)
4	(i,1,A,4,C), (y,2,B,4,C), (i,3,A,4,C), (y,5,c,4,c)
5	(y,5,c,4,c)

Table 3. Adjacency Lists for Range Partitions of Fig. 5

Vertex ID	**Adjacency List for Partition p1**
1	(x,1,A,2,B), (z,1,A,3,A), (i,1,A,4,C)
2	(x,1,A,2,B), (y,2,B,4,C)
3	(z,1,A,3,A), (i,3,A,4,C)
Vertex ID	**Adjacency List for Partition p2**
4	(i,1,A,4,C), (y,2,B,4,C), (i,3,A,4,C), (y,5,c,4,c)
5	(y,5,c,4,c)

A graph (layer, inter-layer, or a partition) is represented as a list of unordered edges. Each edge is completely represented by a 5-element tuple $< E_l, V_{sid}, V_{sl}, V_{did}, V_{dl} >$ where, E_l is edge label, V_{sid} is source vertex ID, V_{sl} is source vertex label, V_{did} is destination vertex ID and V_{dl} is destination vertex label. In this context, it is important to note that the vertex IDs (V_{sid} and V_{did}) are guaranteed to be unique. However, vertex labels (V_{sl} and V_{dl}) and the edge label (E_l) need not be unique. Table 1 shows the edge list representation for the input graph (or a layer) shown in Fig. 5.

This 5-element tuple edge representation (that includes direction) is used to represent a k-edge substructure (a connected graph with k edges and *at most* $k+1$ nodes) as a collection of k 1-edge substructures. Our algorithm takes an input graph represented as a text file with a 1-edge substructure per line.

Adjacency List: Adjacency list is a representation where each vertex is associated with the edges that are incident (both in and out) on it, using a list (of 1-edge substructures.) Table 2 shows the edge list and adjacency list of the graph shown in Fig. 5.

Graph Partitioning: An MLN layer L_i can be partitioned into p partitions (L_1^i, L_2^i,..., L_p^i) for distributed processing. We use *range-based partitioning* [10] to create the partitions using the vertex IDs. Each graph partition[3] is a range of node IDs and the size of each partition need not be the same. There can be missing vertex IDs in a given range. As the ranges are disjoint, nodes in the adjacency list partitions are also disjoint. Each vertex ID in the range and its adjacency list corresponds to a single adjacency list partition. If neighboring nodes are in two partitions, the edge connecting them will be in the adjacency list of *both partitions*. As a result, during expansion, the same substructure can belong to many graph partitions. As adjacency list partitions are indexed on vertex IDs, each substructure is expanded only once in each partition.

Figure 5 also shows 2 range partitions of the graph as partition $p1$ with vertex IDs from 1 to 3, and partition $p2$ with vertex IDs 4 to 5. Both partitions are connected by red edges, which appear in different adjacency list partitions. The adjacency list partitions are displayed in Table 3. The edges that occur in both adjacency list partitions are $< i, 1, A, 4, C >$, $< y, 2, B, 4, C >$, and $< i, 3, A, 4, C >$.

Independent Expansion of Substructures: Substructure discovery uses independent, unconstrained expansion of k-edge substructures into $(k + 1)$-edge substructures starting from an edge (k = 1 or a 1-edge substructure.) Each expansion of each vertex with an edge becomes a separate substructure. This independent unconstrained expansion ensures the generation of all possible substructures in a principled manner, thus ensuring completeness of the substructure generation process. However, this independent expansion also results in the generation of duplicate substructure instances, which must be removed to prevent incorrect counting of isomorphic substructures (soundness).

To remove duplicate substructure instances, we use lexicographic ordering to convert a substructure instance into its **canonical representation**. First, we order the edges on the edge label. In case edges have the same edge labels, we order them based on the source vertex label. If source vertex labels are also the same, then we further order them based on the destination vertex label. If edge labels and vertex labels are all identical, then we order them based on the source vertex IDs. Finally, if source vertex IDs are also the same, then we order them based on the destination vertex IDs. Using this lexicographic order for **each 1-edge**, a substructure can be uniquely represented, which is called a canonical k-edge instance. If two k-edge substructure instances are duplicates, they must have the same ordering of their vertex IDs and therefore, will have the same canonical k-edge instance.

[3] Graph partition and partition are used interchangeably. Adjacency List partition is an adjacency list for a specific graph partition.

The canonical representation of instances is helpful in identifying and removing duplicate instances. However, counting the frequency of isomorphic substructures requires a canonical form that does not rely on absolute vertex IDs. Therefore, we use a canonical form of the substructure that converts vertex IDs to relative vertex IDs.

Graph isomorphism refers to the concept of two graphs being structurally identical, with a one-to-one mapping of their vertices, preserving edge connections and edge labels, and edge orientations. Exact matching of substructures is needed to compute frequent substructures accurately. Isomorphs have the same vertex labels and edge labels but different vertex IDs (as they are different instances). To identify them, we need to create a canonical substructure from the canonical instance. For this vertex IDs are changed from absolute to relative ordering. Any two exact substructures will have the same relative ordering of vertex IDs [10]. This is used to identify and compute the frequency of a substructure.

Metrics for Ranking Substructures: MDL is an information-theoretic, domain-independent metric [9] that has been demonstrated to emphasize the significance of a substructure in terms of its ability (global property) to compress a complete graph or forest. Although it is defined, originally, in terms of bits used for graph representation, we use the number of nodes and edges for that purpose. The general formula for MDL (DL(G)/(DL(S) + DL(G|S))) represents the description length of the substructure S being evaluated, DL(G|S) represents the description length of the graph G when compressed by replacing each instance of the substructure S as a node, and DL(G) (or DL(S)) represents the description length of the original graph (or the substructure S.) The substructure of the graph achieves the highest compression when the MDL value is the highest. Both the frequency of the subgraph and its connectivity have an impact on compression.

6 Composition Algorithm and Map/Reduce Architecture

Existing algorithms are designed for simple and attributed graphs and do not work natively on MLNs. Decoupling approach and composition are used beneficially to process layers independently, but use composition for generating the missing (or substructures spanning layers) to get correct results for the MLN. For verification, as indicated earlier, the Boolean *OR* aggregation of the layers is used to find substructures and compared with the algorithm proposed.

Below we present the **S**ingle **C**omposition **A**lgorithm for **Ho**mogeneous MLNs (**HoSCA**), which performs composition *once at the end of all iterations.* Before that, the substructure discovery iterative process is applied for all layers independently, producing layer-wise outputs of size k from each layer (or when the detected substructures do not change). Upon the completion of *all* layer-wise iterations, their outputs are used for composition. Its purpose is to identify substructures that span two or more layers.

Inter-layer substructures generated by composition undergo duplicate elimination, and then are ranked using the desired metric and evaluated along with the intra-layer substructures. *The key advantage of this approach lies in the fact that the composition is only performed once, rather than after each iteration.* This leads to an efficient substructure discovery, although it does involve some trade-off in accuracy due to information loss as explained in Sect. 4.

6.1 HoSCA: HoMLN Single Composition Algorithm

Figure 6 provides a comprehensive overview of HoSCA approach for substructure discovery in a Homogeneous Multilayer Network.

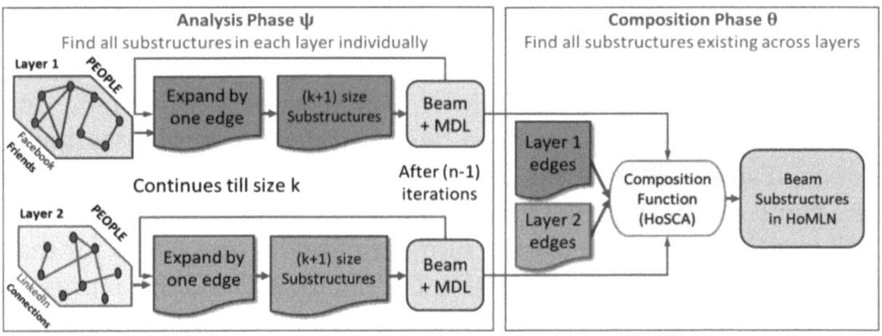

Fig. 6. Overview of the Decoupling Approach for Single Composition Using Two Layers.

The input to the algorithm is all intra-layer substructure instances **tagged with layer number**, each of size *k*. The initial composition step involves generating connected subgraphs of sizes ranging from 1 to *k-1* from each of the participating intra-layer substructures. This process is accomplished by iteratively breaking down each substructure instance into different-sized substructure instances whose algorithm is not shown due to space constraints. The resulting output includes all possible connected subgraphs of the specified size range. This approach allows us to circumvent the generation of disconnected subgraphs, avoiding incurring connectivity. This algorithm is not shown due to space constraints giving preference to the composition algorithm.

The next phase of the algorithm involves combining the generated instances (of different sizes) based on their vertex IDs by applying the *Combine* function for each vertex ID (shown in Algorithm 1.) Only substructures with tags from different layers are composed. This recursive function examines all possible combinations of the connected subgraphs originating from different layers that combine to a k-edge inter-layer substructure. The composition process requires the grouping of subgraphs generated in the previous step based on vertex IDs. This grouping ensures that the composition algorithm can efficiently identify and combine subgraphs that will generate inter-layer composed substructure instances. Thus, the setup of input parameters for the *Combine* function becomes

Algorithm 1. Composing inter-layer substructures of size k (Recursively)

```
Require: Tagged layer-wise substructure instances of size k − 1;
accumulator: Initially an empty string, accumulates edges in each recursive call;
combinations: Set to store composed substructures; size: Size of substructures to generate
Ensure: Composed substructures added to combinations set
Initialize combinations ← ∅
for each vertexid in currentSet do
    Generate Subgraphs List for vertex id
    accumulator ← empty string
    Call combine (Subgraphs, accumulator, combinations, size) {size = k }
end for
Recursive Combine Function
islast ← Boolean(Subgraphs contains only one set)
currentSet ← first set in Subgraphs
for each substructure in currentSet do
    newSubstructure = accumulator + substructure
    if islast is true then
        if size of newSubstructure = size then
            Add to combinations set
        end if
    else
        newSubgraphs = Subgraphs − currentSet
        combine (newSubgraphs, newSubstructure, combinations, size)
    end if
end for
```

essential. We utilize the Map/Reduce paradigm to bring subgraphs that share a common vertex ID together.

As the input subgraphs are connected, the accumulator string, which generates and stores the inter-layer substructure instances, will consistently maintain a valid connected substructure. Furthermore, since the sets in the *Subgraphs* contain substructures from different layers, we can be certain that the substructure generated is an inter-layer substructure, meaning it includes edges from different layers.

Composition, as is done independently on each vertex ID, also generates duplicates. Duplicate elimination and canonical substructure conversion are the same as has been described earlier.

6.2 Implementation Using Map/Reduce

Our input graph is represented as a sequence of unordered edges (or 1-edge substructure). We have discussed the representation in detail in Sect. 5. To recap, each edge is represented by a 5 element tuple ⟨ *edge label, source vertex ID, source vertex label, destination vertex ID, destination vertex label* ⟩. A substructure instance is a sequence of canonical edge instances. A substructure is a sequence of canonical edge instances represented using a relative vertex ID to identify exact subgraphs. Both the range partitioning and substructure discovery are done inside the MapReduce framework.

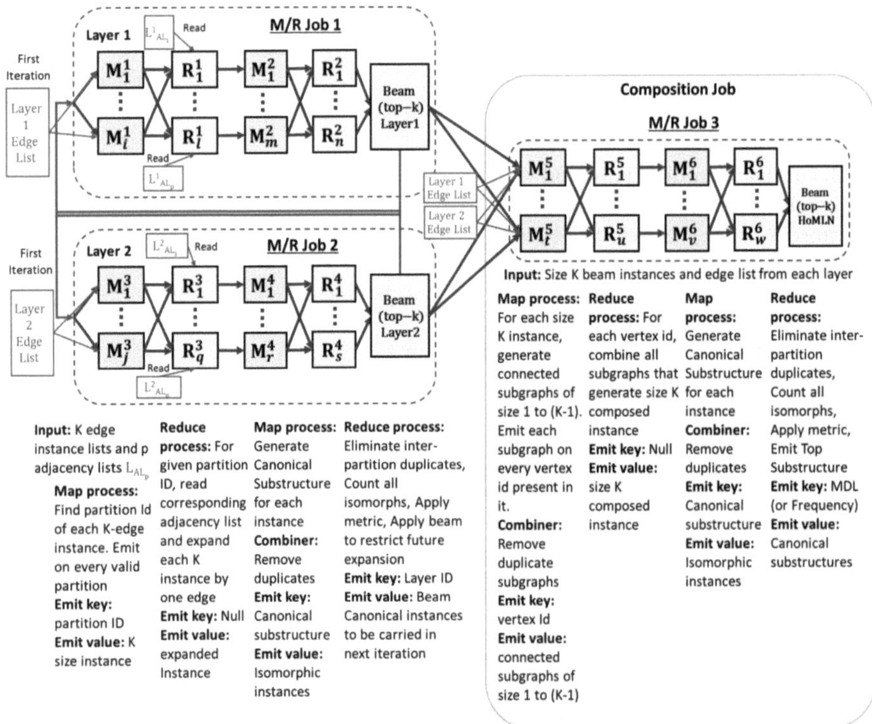

Fig. 7. HoSCA Architecture using MapReduce.

The single composition algorithm for substructure discovery in HoMLNs (HoSCA) is divided into two distinct phases of the decoupling approach - Analysis phase and Composition phase. Figure 7 shows the overall flow of HoSCA using the Map/Reduce paradigm. Each Map and Reduce step is described in Figure 7 in detail.

7 Experimental Evaluation

Experiments were conducted on the Expanse cluster at SDSC (San Diego Supercomputer Center) whose details are shown in Table 4.

Table 4. Expanse System Details

Component	Configuration
Node Count	728
Cores/Node	128 built on 2 processors (64 cores each)
Processor	AMD EPYC 7742
Memory	256 GB DDR4 DRAM
Storage	1TB Intel P4510 NVMe PCIe SSD

Data Sets and Graphs: In the context of our substructure discovery algorithm, when we model a data set as a HoMLN, each layer is considered as an independent graph. Consequently, for synthetic HoMLNs we transform a large simple graph into an MLN by partitioning it into layers. This approach provides the flexibility to assess the algorithm's performance across MLNs of varying sizes while maintaining control over other graph characteristics, such as the edge distribution among layers. Furthermore, for empirical validation, we can embed substructures with known frequencies to ensure consistent identification of substructures and their frequency, whether the data set is processed as a single graph or as an MLN.

Table 5. Data Set Description

Data Set	Used For	Beam Value	M/R configs (per layer)
Synthetic	Accuracy, Response Time	4, 8, 12	2M/2R, 4M/4R, 8M/8R
Synthetic Large	Response Time, Scalability	4	4M/4R, 8M/8R, 16M/16R, 32M/32R
Amazon	Response Time, Scalability	4	8M/8R, 16M/16R, 32M/32R, 64M/64R
LiveJournal	Response Time, Scalability	4	8M/8R, 16M/16R, 32M/32R, 64M/64R

We have conducted experiments on real-world and synthetic data sets to evaluate the effectiveness, response time, and speedup concerning different configurations of mappers and reducers, and scalability of our approach. Table 5 outlines the data sets used. Additionally, real-world data sets from Amazon [6,7] and LiveJournal [1] were also used for our experiments. For these data sets, diverse two-layer HoMLNs were generated with three different *randomly chosen* edge distributions (50/50, 70/30, and 90/10) for testing as detailed in Table 5.

Table 6. Data Sets with Edge Distributions

Base Graph #nodes, #edges	Edge Dist.	#L1 edges	#L2 edges
50KV, 100KE	50/50	49797	50203
	70/30	69862	30138
	90/10	89899	10101
100KV, 500KE	50/50	249953	250047
	70/30	349722	150278
	90/10	449691	50309
400KV, 1ME	50/50	500006	499994
	70/30	700077	299923
	90/10	899994	1000006
1MV, 4ME	50/50	1998995	2001005
	70/30	2799577	1200423
	90/10	3599822	400178
Amazon .74MV, 2.6ME	50/50	1305431	1305160
	70/30	1827821	782770
	90/10	2349423	261168
LiveJournal 4.9MV, 69ME	50/50	34489570	34504203
	70/30	48295818	20697955
	90/10	62096307	6897466

7.1 Response Time and Scalability: Synthetic and Real-World Graphs

To validate speedup using distributed processing, we varied the number of partitions of the same graph, along with increasing the number of mappers and reducers to maximize parallelization. The objective was to observe how the response time varies when resources are increased. To analyze the speedup achieved on synthetic data sets, various data set sizes were explored as shown in Table 6. Regardless of the layer size, each layer was partitioned into the same number of partitions. We experimented with 8, 16, 32, and 64 partitions, maintaining an equal number of mappers and reducers. We saw a similar trend in each data set.

For this discussion, we have focused on two data sets – the largest synthetic (1MV4ME) and another real-world (Amazon). Results for all data sets are shown in Fig. 9. For these two data sets, the absolute time taken differs based on the edge distribution. In general, a higher speedup is seen for the synthetic data set as seen in Fig. 8 (a) as compared to the real-world data set as shown in Fig. 8 (b). Also, 50/50 edge distribution shows maximum speedup as expected, as layers are not skewed and resources are utilized better. However, the speedup achieved was not linear; doubling the mappers and reducers does not halve the time taken. *As the number of partitions increases, the overhead of writing more files and shuffling comes into effect.*

In HoSCA, each layer is processed in a separate MapReduce job. In case of skewed layer distributions, where a smaller layer finishes processing earlier than the larger one, its processors remain idle while waiting for the larger job to complete. Consequently, *skewed distributions lead to increased processing time compared to uniform distributions, where load distribution among processors is more balanced.* To mitigate this, alternative partitioning strategies based on layer size could be explored for HoSCA.

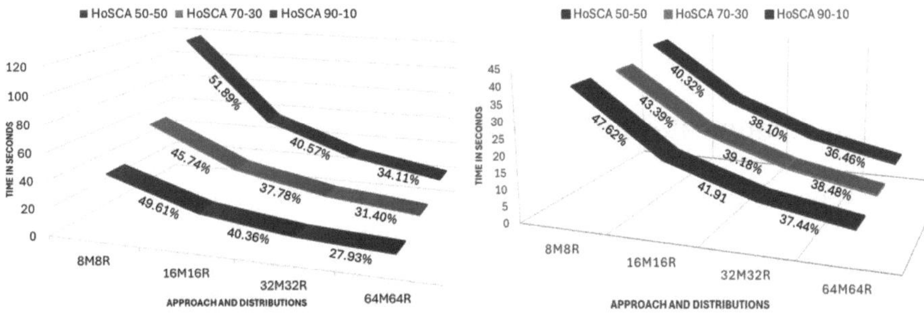

(a) Speedup: 1MV4ME Data Set (b) Speedup: Amazon Data Set

Fig. 8. Speedup Comparison by Varying Map and Reduce Configurations.

7.2 Response Time for All Experiments

In addition to synthetic graphs, two real-world data sets - Amazon and LiveJournal - were used with sizes of 0.74MV2.6ME and 4.9MV69ME, respectively. These data sets were partitioned into 8, 16, 32, and 64 partitions, and experiments were performed using the same number of mappers and reducers for each. Figure 9 shows all the experiments that were performed for all the data sets shown in Table 6.

To contrast scalability, we have shown the response time for 100KV500KE data set using Subdue on the same hardware. There is no way to achieve scalability using main-memory algorithms, such as Subdue, which is the motivation for horizontal and vertical partitioning to scale to arbitrary sizes. Also, the speedup is similar to the one shown for 2 data sets, establishing the scalability of the composition as well.

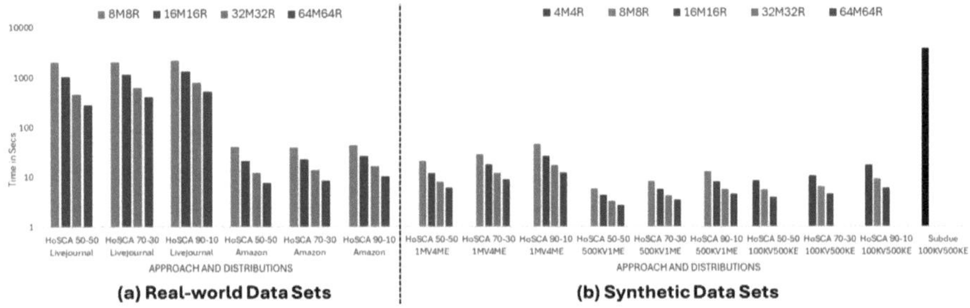

Fig. 9. Comprehensive SpeedUp Experiment Results.

8 Conclusions and Future Work

This paper has described a scalable algorithm for substructure discovery for HoMLNs, using a decoupling-based strategy along with a possible accuracy-efficiency tradeoff. A composition algorithm was presented along with its implementation using Map/Reduce to show scalability. Future work includes finding the sweet spot based on layer characteristics and edge distributions.

Acknowledgment. This work was supported by NSF award CCF-1955798.

References

1. Live-journal. http://snap.stanford.edu/data/com-LiveJournal.html
2. Agrawal, R., Srikant, R., et al.: Fast algorithms for mining association rules. In: Proceedings of the 20th International Conference on Very Large Data Bases, VLDB, vol. 1215, pp. 487–499 (1994)
3. Balaban, A.T.: Applications of graph theory in chemistry. J. Chem. Inf. Comput. Sci. **25**(3), 334–343 (1985)
4. Berenstein, A., Magarinos, M.P., Chernomoretz, A., Aguero, F.: A multilayer network approach for guiding drug repositioning in neglected diseases. PLoS **10**(1), 1–33 (2016)
5. Boden, B., Günnemann, S., Hoffmann, H., Seidl, T.: Mining coherent subgraphs in multi-layer graphs with edge labels. In: Proceedings of the 18th ACM SIGKDD International Conference on Knowledge Discovery and Data Mining, pp. 1258–1266 (2012)
6. Boldi, P., Rosa, M., Santini, M., Vigna, S.: Layered label propagation: a multiresolution coordinate-free ordering for compressing social networks. In: Proceedings of the 20th international conference on World Wide Web, pp. 587–596. ACM Press
7. Boldi, P., Vigna, S.: The WebGraph framework I: compression techniques. In: Proceedings of the 13th WWW Conference, pp. 595–601 (2004)
8. de Bruijn, N.G.: A combinatorial problem. Proc. Akad. Wet. Amsterdam **49**, 758–764 (1946)
9. Cook, D.J., Holder, L.B.: Substructure discovery using minimum description length and background knowledge. J. Artif. Intell. Res. **1**, 231–255 (1994)

10. Das, S., Chakravarthy, S.: Partition and conquer: map/reduce way of substructure discovery. In: International Conference on Big Data Analytics and Knowledge Discovery, pp. 365–378. Springer (2015)
11. De Domenico, M., Solé-Ribalta, A., Gómez, S., Arenas, A.: Navigability of interconnected networks under random failures. In: Proceedings of the National Academy of Sciences (2014)
12. Deshpande, M., Kuramochi, M., Karypis, G.: Frequent Sub-Structure-Based Approaches for Classifying Chemical Compounds. In: IEEE International Conference on Data Mining, pp. 35–42 (2003). https://doi.org/10.1109/ICDM.2003.1250900
13. Domenico, M.D., Nicosia, V., Arenas, A., Latora, V.: Layer aggregation and reducibility of multilayer interconnected networks. CoRR **abs/1405.0425** (2014)
14. Kivelä, M., et al.: Multilayer networks. J. Complex Netw. **2**(3), 203–271 (2014)
15. Liu, Y., Jiang, X., Chen, H., Ma, J., Zhang, X.: MapReduce-Based pattern finding algorithm applied in motif detection for prescription compatibility network. In: Advanced Parallel Programming Technologies, pp. 341–355 (2009)
16. Coimbra, M.E., Francisco, A.P., Veiga, L.: An analysis of the graph processing landscape. J. Big Data **8**(1), 1–41 (2021). https://doi.org/10.1186/s40537-021-00443-9
17. Padmanabhan, S., Chakravarthy, S.: HDB-Subdue: a scalable approach to graph mining. In: DaWaK, pp. 325–338 (2009)
18. Pei, J., et al.: PrefixSpan,: mining sequential patterns efficiently by prefix-projected pattern growth. In: ICDE, pp. 215–224 (2001). https://doi.org/10.1109/ICDE.2001.914830
19. Rissanen, J.: A universal prior for integers and estimation by minimum description length. Ann. Stat. **11**(2), 416–431 (1983)
20. Santra, A., Bhowmick, S., Chakravarthy, S.: Efficient community re-creation in multilayer networks using Boolean operations. In: International Conf. on Computational Science, ICCS 2017 (2017)
21. Santra, A., Bhowmick, S., Chakravarthy, S.: Hubify: efficient estimation of central entities across multiplex layer compositions. In: IEEE International Conference on Data Mining Workshops, ICDM Workshops 2017 (2017)
22. Santra, A., Irany, F.A., Madduri, K., Chakravarthy, S., Bhowmick, S.: Efficient community detection in multilayer networks using Boolean compositions. Front. Big Data **6** (2023)
23. Santra, A., Komar, K.S., Bhowmick, S., Chakravarthy, S.: From base data to knowledge discovery - a life cycle approach - using multilayer networks. Data Knowl. Eng. **141**, 102058 (2022)
24. Singh, A., Santra, A., Chakravarthy, S.: Scalable substructure discovery algorithm for homogeneous multilayer networks. arXiv preprint arXiv:2504.19328 (2025)
25. Suri, S., Vassilvitskii, S.: Counting triangles and the curse of the last reducer. In: World Wide Web Conference Series. pp. 607–614 (2011). https://doi.org/10.1145/1963405.1963491
26. Wasserman, S., Faust, K.: Social network analysis: methods and applications (1994)
27. Yang, S., Yan, X., Zong, B., Khan, A.: Towards effective partition management for large graphs. In: SIGMOD Conference, pp. 517–528 (2012)

Bitemporal Property Graphs: Dealing with Both Valid and Transaction Time

Luca Anselma[1(✉)], Marco Ballerini[1], Paolo Giordano[1], Erica Raina[2,3], and Paolo Terenziani[2,3]

[1] Dipartimento di Informatica, Università Degli Studi di Torino, Turin, Italy
luca.anselma@unito.it, {marco.ballerini, paolo.giordano494}@edu.unito.it

[2] DISIT, University of Eastern Piedmont, Alessandria, Italy
{erica.raina,paolo.terenziani}@uniupo.it

[3] Integrated Laboratory of AI and Medical Informatics DAIRI + DISIT, Alessandria, Italy

Abstract. The need to deal with both *valid* and *transaction* time in temporal databases has been identified already in the 1980s, and bitemporal databases are fully consolidated in the relational context. However, only few graph databases deal with both times. In this paper, we extend the recent graph database AeonG to consider both transaction and valid time. We propose an extended data model, and demonstrate its expressiveness showing how it copes with a benchmark example in the literature. We extend the query language, and experimentally prove that, despite the increased expressiveness, our framework adds only a limited overhead to AeonG.

Keywords: Temporal graph databases · Valid Time · Transaction Time · Query Language · Experimental Evaluation

1 Introduction

Already in 1986, in their seminal paper [1], Snodgrass and Ahn clearly pointed out the necessity of distinguishing between ***valid time***, i.e., *the time when the information is valid* in the modelled mini-world, and ***transaction time***, i.e., *the time the information was stored in the database*. Since then, such a distinction has pervaded the area of temporal DB. For instance, in the recent Encyclopedia of Database Systems [2], the following definitions of valid and transaction time are reported.

"The valid time of a fact is the time when the fact is true in the modelled reality. Any subset of the time domain may be associated with a fact." [3].

"A database fact is stored in a database at some point in time, and after it is stored, it remains current, or part of the current database state, until it is logically deleted. The transaction time of a database fact is the time when the fact is current in the database. As a consequence, the transaction time of a fact is generally not a time instant, but rather has duration. The transaction time of a fact cannot extend into the future. Also,

The original version of the chapter has been revised. A correction to this chapter can be found at https://doi.org/10.1007/978-3-032-05281-0_21

P. K. Chrysanthis et al. (Eds.): ADBIS 2025, LNCS 16043, pp. 234–249, 2026.
https://doi.org/10.1007/978-3-032-05281-0_15

as it is impossible to change the past, meaning that (past) transaction times cannot be changed." [4].

Both valid and transaction time are included in TSQL2 [5], a milestone in the area of temporal relational DBs, ATSQL [6], and SQL standard (i.e., system-versioned tables for transaction time and application-time period tables for valid time in SQL:2023 [7]).

Already the seminal paper [1] reports several motivations for considering both valid and transaction time, such as support to historical queries about the past (e.g., for trend analysis), need to support retroactive and proactive changes, need to support error corrections while preserving the possibility to access past versions of the Database (e.g., for legal purposes). As an introductory running example, we adopt one of the examples provided in [1] slightly modified to include additional aspects. In Sect. 4 we will provide an additional example to consider also other aspects of graph databases.

Example 1.

A. Merrie joins as an instructor. Inserted in June 1973, valid from September 1973
B. Merrie is promoted to full professor. Inserted in December 1973, valid from December 1973
C. Correction: Merrie was promoted to assistant professor. Corrected in February 1974, valid from December 1973
D. Merrie is promoted to full professor. Inserted in December 1978, valid from June 1978.

Example A-D above shows that valid time (VT) and transaction time (TT) are orthogonal. In particular:

- B shows an example in which TT and VT are the same
- A shows an example of proactive change: a piece of information is inserted into the DB before it starts to hold in the modelled world
- D shows an example of retroactive change: a piece of information is inserted into the DB after it has already started to hold in the modelled world.

Example C shows a correction. Keeping track of the past states of the DB may be important. For instance, between December 1973 and February 1974 the DB contained the information that Merrie was a Full Professor. If such information is not retained (as a past version of the DB), administrative stuff may occur in legal problems, since they paid Merrie as a Full Professor, while she was not (in the real world).

The approach in [8] provides a very interesting framework to cope with temporal Graph DB. In particular, AeonG is a graph database with a built-in mono-temporal support. AeonG relies upon two physically isolated storage systems, one based on RocksDB [9], a persistent key-value store for fast storage environments, and one on Memgraph [10], a graph DB that provides resiliency, low latency and multi-tenancy. AeonG, for the sake of efficiency, proposes the "delta + anchor": it periodically creates a complete version (namely anchor) of a graph object and maintains every change (namely delta) between two adjacent anchors of the same object. The deltas from the most recent anchor are stored in main memory through Memgraph, while anchors and past deltas are stored in secondary memory in RocksDB. AeonG has been shown to be extremely fast, with only a 9.74% performance degradation for supporting temporal features.

A major limitation of such an approach is that it only copes with one type of time. Indeed, though the paper does not clearly specify that, we infer that it is transaction time from the fact that the starting time associated with each vertex, edge and property starts at the exact time when they are inserted in the DB (and ends when they are modified or deleted). An additional indication in favour of the "transaction time interpretation" is the fact that the start and end times cannot be inserted by users, but are automatically managed by the DBMS (which is exactly the behaviour adopted for transaction time; see, e.g., TSQL2 [5]).

The goal of our work is to extend the approach by [8] to also deal with valid time. In Sect. 2, we extend AeonG's definition of Temporal Property Graph to consider both valid and transaction time, and exemplify the expressiveness of our data model considering Example 1 above. In Sect. 3, we briefly describe our extended query language. Since Example 1 is in the tradition of temporal relational DBs literature (and a single node suffices to cope with it), in Sect. 4 we propose an additional example, to further stress and clarify the impact of our approach in the Temporal Graph context. In Sect. 5 we describe our prototypical implementation, and experimentally demonstrate that, despite the increased expressiveness, our approach only adds a limited overhead to AeonG's one. Finally, Sect. 6 contains conclusions, related and future work.

2 Bitemporal Property Graphs

In this Section, we propose our data model. We define Bitemporal Property Graphs, describe the temporal constraints we impose, and show their expressiveness through its application to Example 1 above.

2.1 Data Model

The starting point of our approach is the definition of the domains of time. As in BCDM (Bitemporal Conceptual Data Model) [11] (a unifying data model, isolating the "core" semantics underlying many temporal relational approaches, including TSQL2 [5]) we consider both *valid time* and *transaction time*. Both time domains are totally ordered and isomorphic to the subsets of the domain of natural numbers.

Our definition of BiTemporal Property Graphs is an extension of the definition in [8], to include also transaction time.

Definition 1 (BiTemporal Property Graph). Let $\mathcal{N}$ and $\mathcal{E}$ denote sets of vertex ids and edge ids, and $\mathcal{L}$, $\mathcal{K}$ and $\mathcal{V}$ be countable sets of labels, property names and property values. A BiTemporal Property Graph is a tuple $G = \langle \Omega^{vt}, \Omega^{tt}, N, E, \rho, \lambda, \sigma, \tau \rangle$ where:

- Ω^{vt} is a temporal domain, which is a finite set of consecutive time points, that is, $\Omega^{\mathrm{vt}} = \{i \in T^{\mathrm{v}} \mid a \leq i \leq b\} \cup \{\infty\}$ for some $a, b \in T^{\mathrm{v}}$ such that $a \leq b$. T^{v} represents the universe of valid-time points;
- Ω^{tt} is a temporal domain, which is a finite set of consecutive time points, that is, $\Omega^{\mathrm{tt}} = \{i \in T^{\mathrm{t}} \mid a \leq i \leq b\} \cup \{\infty\}$ for some $a, b \in T^{\mathrm{t}}$ such that $a \leq b$. T^{t} represents the universe of transaction-time points;
- N is a finite subset of $\mathcal{N}$, whose elements are referred to as the vertices of G;
- E is a finite subset of $\mathcal{E}$, whose elements are referred to as the edges of G and $N \cap E = \emptyset$;

- ρ: $E \rightarrow (N \times N)$ is a total function mapping each edge to its source and destination vertices;
- λ: $(N \cup E) \rightarrow 2^{\mathcal{L}}$ is a total function mapping vertices and edges to finite sets of labels (including the empty set);
- $\sigma^{vt,tt}$: $(N \cup E) \times \Omega^{vt} \times \Omega^{tt} \rightarrow \{true, false\}$ is a total function that maps a vertex or an edge and a bitemporal point to a Boolean variable, indicating whether this vertex or edge holds at that bitemporal point. Notably, a vertex or an edge holds in a bitemporal point $\langle c^{vt}, c^{tt} \rangle \in \Omega^{vt} \times \Omega^{tt}$ if it holds in the modelled world at $c \in \Omega^{vt}$, and it exists (is logically present) in the DB at $c^t \in \Omega^{tt}$.
- $\tau : (N \cup E) \times \mathcal{K} \times \Omega^{vt} \times \Omega^{vt} \times \Omega^{tt} \times \Omega^{tt} \rightarrow \mathcal{V}$ is a partial function that maps a vertex or an edge and a property key, a time period ω^{vt} for the valid time and a period ω^{tt} for the transaction time to a value.

Notably, though not explicitly introduced in the definition of Temporal Property Graph, the approach in [8] also adopts a special value, "∞", to denote the fact that the given entity (node or event) holds in the DB until it will possibly be changed/removed.

Terminology. In the following and above, we call ***bitemporal point*** each pair of time points $\langle c^v, c^t \rangle \in \Omega^{vt} \times \Omega^{tt}$.

Note. We assume that intervals are closed to the left and open to the right.

Constraints. In our bitemporal property graph model, we impose two constraints to enforce that the graph at any time point corresponds to a valid property graph. At any ***transaction*** time point $c^t \in \Omega^{tt}$:

(1) An edge exists in the DB only if both source and destination vertices exist in the DB at c^t. Formally, if $e \in E$ and $\exists c^v \in \Omega^{vt}$ $\sigma(e, c^v, c^t) = true$ and $\rho(e) = (v1, v2)$, then $\exists c^{v'} \in \Omega^{vt}$ $\sigma(v1, c^{v'}, c^t) = true$ and $\exists c^{v''} \in \Omega^{vt}$ $\sigma(v2, c^{v''}, c^t) = true$;
(2) A property can only take on a value during the bitemporal time period when the corresponding vertex or edge exists. Formally, if $\tau(o, k, c^{vs}, c^{ve}, c^{ts}, c^{te}) = val$, where $o \in (N \cup E)$, $k \in \mathcal{K}$, $val \in \mathcal{V}$, then $\forall c^t \in [c^{ts}, c^{te})$ and $\forall c^v \in [c^{vs}, c^{ve})$ we have $\sigma(o, c^v, c^t) = true$.

Note. As a remarkable side-effect, the deletion of a node implies the deletion of all its entering and exiting edges. Notably, no piece of information is physically deleted from the bitemporal graph. The deletion of an entity e with transaction time $[c^{ts}, +\infty)$ at a transaction time $c^t \in \Omega^{tt}$ is achieved by setting its transaction time to $[c^{ts}, c^t)$.

Intuitively, temporal constraints in [8] impose that properties and relationships of any vertex are "synchronous" with the vertex (i.e., they may hold only during the time of the vertex). Such constraints are reasonable when considering transaction time: if a vertex v is inserted in the Graph at time s and is deleted at e (i.e., its transaction time is [s,e)), it cannot have properties/relationships in the graph out of such a period.

However, such a constraint would be too restrictive when applied to valid time, since many phenomena in the "real" world would not be captured by the graph DB.

Example 2. Suppose, e.g., that one wants to model family trees, considering a property (or a relationship) Great-Grandfather. Obviously, the great-grandfather's life is not necessarily contained in the life of the great-grandson.

Example 3. As a further example, suppose that one wants to model in a graph DB the notion of *causation*, modelling it through an edge connecting two vertices, representing cause and effect. Temporal constraints between causes and effects are quite various, and have been widely studied (see, e.g., [12]), but in general effects start after the beginning of their causes (or even after their end, as in "one-shot" causation). An example concerning causation is proposed in Sect. 4.

2.2 An Example: The Evolution of Merrie's Rank

As an example, we consider Example 1 in the introduction. We assume the granularity of months both for valid and transaction time, and we take January 1973 as the origin of time (so "1" is the month of January 1973, "2" is February 1973 and so on).

Independently of the temporal issues, there are (at least) two different ways to model the example: (i) considering a node for employees, and representing ranks as an attribute, or (ii) considering a node for employees, one for ranks, and an edge to connect them. In the following, for the sake of brevity, we chose the first alternative, but also the second can be modelled in our approach.

A. **Merrie joins as an instructor. Inserted in June 1973, valid from September 1973**

For the temporal attributes, we represent them as a pair [Valid time start, Valid time end) followed by a pair [Transaction time start, Transaction time end).

V0: Employee $[6, \infty)[6, \infty)$
V01: Name: Merrie $[6, \infty)[6, \infty)$
V02: Rank: Instructor $[9, \infty)[6, \infty)$

The time point 6 (both in valid time and transaction time) denotes June 1973. The node employee, with attribute *Name* with value "*Merrie*" is inserted (transaction time) at 6, and starts to hold at 6 (valid time). However, the attribute *Rank* assumes its value at time 9.

B. **Merrie is promoted to full professor. Inserted in December 1973, valid from December 1973**

V0: Employee $[6, \infty)[6, \infty)$
V01: Name: Merrie $[6, \infty)[6, \infty)$
V02: Rank: Instructor $[9, \infty)[6, 12)$
V03: Rank: Instructor $[9, 12)[12, \infty)$
V04: Rank: Full Professor $[12, \infty)[12, \infty)$

From transaction time 6 to 12 (excluded) the DB contained the information that Merrie was an instructor (from 9 on). Such information is retained. However, starting from transaction time 12, the DB has the information that Merrie was an Instructor from 9 to 12 (excluded), and is a Full Professor since 12.

C. **Correction: Merrie was promoted to assistant professor. Corrected in February 1974, valid from December 1973**

V0: Employee $[6, \infty)[6, \infty)$
V01: Name: Merrie $[6, \infty)[6, \infty)$
V02: Rank: Instructor $[9, \infty)[6, 12)$
V03: Rank: Instructor $[9, 12)[12, \infty)$

V04: Rank: Full Professor [12, ∞)[12, 14)
V05: Rank: Assoc Professor [12, ∞)[14, ∞)

Until time 14, the DB contains the information described at step B above. At time 14, the information that Merrie is a Full Professor is logically deleted (by setting to 14 the upper bound of transaction time), while it is added (lower bound of transaction time) the information that, starting from month 12 (valid time [12, ∞)), Merrie is an Associate Professor.

D. **Merrie is promoted to full professor. Inserted in December 1978, valid from June 1978**

V0: Employee [6, ∞)[6, ∞)
V01: Name: Merrie [6, ∞)[6, ∞)
V02: Rank: Instructor [9, ∞)[6, 12)
V03: Rank: Instructor [9, 12)[12, ∞)
V04: Rank: Full Professor [12, ∞)[12, 14)
V05: Rank: Assoc Professor [12, ∞)[14, 72)
V06: Rank: Assoc Professor [12, 66)[72, ∞)
V07: Rank: Full Professor [66, ∞)[72, ∞)

Until time 72, the DB contains the information described at step C above. At (transaction) time 72, we add the information that Merrie ended to be Associate Professor at (valid) time 66, and since that time she is a Full Professor.

Comment (how to read). At each snapshot (A, A + B, A + B + C, A + B + C + D) of the evolution of the DB, the "current status" can be obtained by considering only those properties whose right bound of the transaction time is "∞".

For example, in the snapshot A + B + C the "current situation" was:
V0: Employee [6, ∞)[6, ∞)
V01: Name: Merrie [6, ∞)[6, ∞)
V03: Rank: Instructor [9, 12)[12, ∞)
V05: Rank: Assoc Professor [12, ∞)[14, ∞)

More generally, to see the DB situation at a given Transaction Time t, one simply has to consider only those properties whose transaction time period contains t. For instance, at transaction time 13 (January 1974), it was believed that Merrie had been an Instructor from September to December 1973, and she became a Full Professor, i.e.,
V0: Employee [6, ∞)[6, ∞)
V01: Name: Merrie [6, ∞)[6, ∞)
V03: Rank: Instructor [9, 12)[12, ∞)
V04: Rank: Full Professor [12, ∞)[12, ∞)

3 Manipulation and Query Language

AeonG incorporates, as a manipulation and query language, a temporal-enhanced Cypher, which extends the standard syntax defined in OpenCypher [13] to support temporal queries. In short, AeonG introduces two temporal extensions in the MATCH clause:

- `FOR TT AS OF` *t*, which retrieves all graph objects legal at time *t* (called *time-point queries*),
- `FOR TT FROM` *t*1`TO` *t*2, which locates all graph objects consistently legal within the time range from *t*1 to *t*2 (called *time-slice queries*).

Such limited extensions suffice, since AeonG only considers transaction time which, by definition, must be automatically managed by the DBMS (it represents the DB insertion and deletion/update time). We retain such extensions in our approach, but, in order to deal also with *valid time*, several extensions are needed:

i. extend the MATCH construct to query also the valid time,
ii. extend the *manipulation* operators to provide the possibility to specify the valid time,
iii. extend the *pattern* definition in order to be able to specify valid times for vertices, edges and properties.

Notably, while in (i) valid time is a time point or a time period, in (ii) and (iii) valid time consists of a -possibly unitary- set of non-overlapping periods (see Definition 1). In this paper, we focus only on aspects (i) and (ii), while (iii) is left as future work.

3.1 Extensions to the Query Language

In this paper, we extend the AeonG framework specifically with the purpose of expressing time-point and time-slice queries considering valid time, or transaction time, or both. We thus have, besides the cases above (in our approach, "VT" stands for valid time), also the cases:

- `FOR VT AS OF` *t*, which retrieves all graph objects holding in the modelled world at time *t* (time-point query),
- `FOR VT FROM` *t*1 `TO` *t*2, which locates all graph objects holding in the modelled world within the time range from *t*1 to *t*2 (time-slice query).

Example 4. What was Merrie's rank in January 1974 considering the info in the DB in January 1974?

```
MATCH (n:Employee {Name: Merrie}) FOR TT AS OF 13 FOR VT AS OF 13
RETURN n.
```

- Full Professor $[12, +\infty)[12, 14)$

Example 5. What was Merrie's rank in January 1974 considering the info in the DB from January 1974 to March 1974?

```
MATCH (n:Employee {Name: Merrie}) FOR TT FROM 13 TO 15 FOR VT AS OF
13 RETURN n.
```

- Full Professor $[12, \infty)[12, 14)$
- Assoc Professor $[12, +\infty)[14, 72)$

Example 6. What was Merrie's rank from October 1973 to January 1974 considering the info in the DB from January 1974 to March 1974?

```
MATCH (n:Employee {Name: Merrie}) FOR TT FROM 13 TO 15 FOR VT FROM
10 TO 13 RETURN n.
```

- Instructor [9, 12)[12, $+\infty$)
- Full Professor [12, $+\infty$)[12, 14)
- Assoc Professor [12, $+\infty$)[14, 72)

3.2 Manipulation Operators

To illustrate our extensions to manipulation operation, we re-consider Example 1. Results obtained by the application of the operations below are the ones already presented in Sect. 2.

A. **Merrie joins as an instructor. Inserted in June 1973, valid from September 1973**

At transaction time 6, `CREATE (:Employee {Name: Merrie}) FOR VT FROM 6 TO "+INF"`.

At transaction time 6, `MATCH (n:Employee {Name: Merrie}) FOR VT FROM 9 TO "+INF" SET n.Rank= 'Instructor'`.

B. **Merrie is promoted to full professor. Inserted in December 1973, valid from December 1973**

At transaction time 12, `MATCH (n:Employee {Name: Merrie}) FOR VT FROM 9 TO 12 SET n.Rank='Instructor'`.

At transaction time 12, `MATCH (n:Employee {Name: Merrie}) FOR VT FROM 12 TO "+INF" SET n.Rank= 'Full Professor'`.

C. **Correction: Merrie was promoted to assistant professor. Corrected in February 1974, valid from December 1973**

At transaction time 14, `MATCH (n:Employee { Name: Merrie}) FOR VT FROM 12 TO "+INF" SET n.Rank= 'Assoc Professor'`.

D. **Merrie is promoted to full professor. Inserted in December 1978, valid from June 1978**

At transaction time 72, `MATCH (n:Employee { Name : Merrie}) FOR VT FROM 12 TO 66 SET n.Rank = 'Assoc Professor'`.

At transaction time 72, `MATCH (n:Employee { Name : Merrie}) FOR VT FROM 66 TO "+INF" SET n.Rank = 'Full Professor'`.

4 A Further Example: Nodes and Edges

Example 1 above is one of the most famous examples in the tradition of temporal relational DBs, to show the need of coping with both valid time and transaction time. However, Example 1 can be "naturally" modelled by a one-node graph, so that one may wonder whether the same concepts are also important in the context of graphs. In the following we propose a small additional example consisting of three nodes and two edges. Though "minimal", such an example allows us to point out important concepts concerning the need to manage both valid and transaction times also in graphs.

Our example is in the domain of medical informatics since, though our bitemporal graph DB approach is completely domain-independent, we intend to apply our approach within such a context. Indeed, our group has an almost thirty-year tradition in providing AI and temporal DB supports to clinical decision support and medical educational systems, within the long-term Guideline Representation, Acquisition and Execution (GLARE) project [14–19].

Example 7. Let us model a DB of therapies, which are administered to patients and produce effects, consisting of the following facts:

7a. Anna (born on 1980–03-03) followed the Corticosteroid therapy from 2023–01-10 to 2023–01-20 (included). The therapy caused an elevated intraocular pressure from 2023–02-01 to 2023–02-15 (notice that the effect starts after the end of the therapy). Such information was stored in the DB on 2023–03-01.

7b. On 2023–04-01 (i.e., after 2023–03-01), an error is noticed: the effect was not elevated intraocular pressure but mild retinal edema, starting on 2023–02-03 and ending on 2023–02-17. Such an update was applied to the DB.

7c. Anna died on 2023–10-01.

7d. On 2023–11-01, the queries Q1 e Q2 were asked:

Q1 What was the effect of the Corticosteroid therapy on Anna, given the info in the DB on 2023-03-15?

Q2 What was the effect of the Corticosteroid therapy on Anna, given the info in the DB on 2023-10-15?

We can support the preceding facts with the following query statements (using dates as time points).

7a. At transaction time 2023–03-01,

- `CREATE (:Patient {Name:Anna}) FOR VT FROM "1980-03-03" TO "+INF"`
- `CREATE(:Therapy {Type:Corticosteroid})FOR VT FROM "2023-01-10" TO "2023-01-20" CREATE(:Effect {Type:elevated intraocular pressure}) FOR VT FROM "2023-02-01" TO "2023-02-15"`
- `MATCH (t:Therapy {Type:Corticosteroid}), (p:Patient {Name:Anna}) CREATE (t)-[:Applied_to]->(p) FOR VT FROM "2023-01-10" TO "2023-01-20"`
- `MATCH (t:Therapy {Type:Corticosteroid}), (e:Effect {Type:elevated intraocular pressure}) CREATE (t)-[:Has_Effect]->(e) FOR VT FROM "2023-02-01" TO "2023-02-15"`

The transaction time of all the nodes and edges is [2023-03-01, +INF). On the other hand, we have different VTs. Notably, the VT of the effect "elevated intraocular pressure" (the effect of the therapy) is [2023-02-01, 2023-02-16) and is later than the VT of the corticosteroid therapy. This aspect cannot be captured in all those models that (like AeonG) dictate that edges can only exist at the time of the nodes they connect.

7b. At transaction time 2023–04-01,

- `CREATE (:Effect {Type:mild retinal edema}) FOR VT FROM "2023-02-01" TO "2023-02-15"`
- `MATCH (:Therapy {Type:Corticosteroid})-[a:Has_Effect]->(:Effect {Type:elevated intraocular pressure}) DELETE a`
- `MATCH (t:Therapy {Type:Corticosteroid}), (e:Effect {Type:mild retinal edema}) CREATE (t)-[:Has_Effect]->(e) FOR VT FROM "2023-02-01" TO "2023-02-15"`

The update is obtained by deleting the Has_Effect edge from Corticosteroid to Elevated intraocular pressure (i.e., its TT becomes [2023-03-01, 2023-04-01)) and inserting a new Has_Effect edge from Corticosteroid therapy to Mild retinal edema effect with TT [2023-04-01,+INF).

7c. At transaction time 2023–10-01,

- `MATCH (p:Patient {Name:Anna}) DELETE p`
- `CREATE (:Patient {Name:Anna}) FOR VT FROM "1980-03-03" TO "2023-10-02"`

Notably, the death of Anna leads to an update of her VT from [1980-03-03, +INF] to [1980-03-03, 2023-10-02). Such new info is inserted with TT [2023-10-01, +INF).

Q1: What was the effect of the Corticosteroid therapy on Anna, given the info in the DB at 2023–03-15?

- ```
  MATCH (e:Effect ) <- [:Has_Effect] - (t:Therapy
    {Type: Corticosteroid}) - [a:Applied_to] - >
    (p:Patient {Name:Anna}) FOR TT AS OF "2023-03-15"
    RETURN e
    >>elevatedintraocularpressure[2023-02-01,2023-02-16)
  ```
  ```

Q2: What was the effect of therapy Corticosteroid on Anna, given the info in the DB at 2023–10-15?

- `MATCH (e:Effect) < - [:Has_Effect] - (t:Therapy {Type: Corticosteroid}) - [a:Applied_to] - > (p:Patient {Name:Anna}) FOR TT AS OF "2023-10-15" RETURN e`
  `>> mild retinal edema [2023-02-01, 2023-02-16)`

5 Implementation and Experimental Evaluation

In this section, we sketch our implementation and describe some experimental results.

5.1 Implementation

In this section, we briefly introduce our extension of AeonG, called AeonG-VT, along with the important changes we made to AeonG's system architecture to cope with valid time.

To store multiple valid time intervals, we added to each graph object a newly created object called *VTStore*. This object stores, for each graph object property (and for the object itself), its valid time intervals using vectors of a custom-made class called *VTDateTime*.

We added three new attributes to the structure of delta:

- *applied_vt* (VTDateTime start, VTDateTime end): The valid time in the query,
- *new_value* (propertyValue): The new value of that specific property,
- *vt* (VTDateTime start, VTDateTime end): The applied valid time for the delta and, for extension, the value of valid time when a rollback is required.

Regarding the historical storage, we changed the structure of the key, adding the valid time. With this small change, we can store the valid time in the physical disk.

5.2 Experiments

We performed the experiments on a physical device with 11th Gen Intel® Core™ i7-11370H @ 3.30GHz × 8, 15.4 GiB RAM on Ubuntu 20.04.6 LTS. The garbage collector's interval for both AeonG and AeonG-VT was set to 60 s.

We devised two main experiments.

1. The first experiment compares the performance of a set of queries without a specific valid time, on both AeonG and AeonG-VT, with the same previously described dataset in the database.
2. The second experiment considers AeonG-VT only and evaluates its performance on valid time queries. We performed each query three times and report the average time.

5.3 Comparing AeonG-VT and AeonG

Following AeonG's paper, we have experimentally evaluated the performance of the approach considering the temporal-enhanced workload of T-mgBench, which is based on data derived from the Pokec real-world Slovak social network [20]. As the initial dataset, T-mgBench consists of 10,000 vertices and 121,716 edges. Following AeonG's approach, we added to the initial dataset a randomly generated set of updates, specifically 80,000 updates: MATCH-SET, MATCH-DELETE, and MATCH-CREATE (not considering VT). We generated the update dataset with the *create_temporal_query.py* script to compare AeonG-VT and AeonG in a realistic environment and under the same conditions. This dataset is a mix of 80% updates, 10% non-temporal creates, and 10% deletes.

In this experiment, we run the four types of queries used in AeonG's evaluation [8] (Table 1).

Table 1. Transaction-time queries in AeonG and AeonG-VT.

Query	Statement
Q1.1	`MATCH (n:User) FOR TT AS OF t RETURN n`
Q1.2	`MATCH (n:User) FOR TT FROM t1 TO t2 RETURN n`
Q1.3	`MATCH (n:User)-[r]->(m) FOR TT AS OF t RETURN m, n, r`
Q1.4	`MATCH (n:User)-[r]->(m) FOR TT FROM t1 TO t2 RETURN m, n, r`

After creating the database, we divided the current transaction time into ten intervals. With those intervals, we replaced the parameters in the queries, and we evaluated the performance in AeonG and AeonG-VT. In Fig. 1 we show the results for queries of type Q1.4 (other results are analogous).

The experiments clearly show that, considering TT only, despite the increased expressiveness of our approach, the overhead we added to AeonG is almost negligible.

5.4 Evaluating VT Queries in AeonG-VT

For evaluating VT queries, we created a different workload enriching the T-mgBench dataset. Our new dataset is based on the plain T-mgBench dataset and is generated by a script, to add to the vertices their valid time. This script takes two *.cypher* files: the chosen cypher dataset (in our case T-mgBench small) and the updated dataset already generated with the "create_temporal_query.py" script. The script adds to each vertex a single valid-time interval, evenly selected from a set of ten non-overlapping intervals.

The following types of VT queries are asked on the newly-created dataset (Table 2).

In Fig. 2, we show the time needed to answer queries of type Q2.2, considering different degrees of selectivity (from 10% to 100%) of the queries. Notably, the query

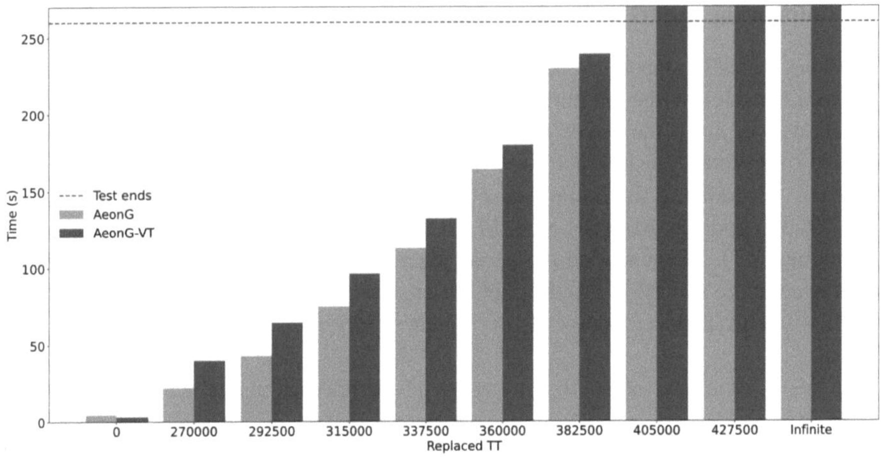

Fig. 1. Comparison of Q1.4 queries between AeonG and AeonG-VT.

Table 2. Valid time queries in AeonG-VT

Query	Statement
Q2.1	`MATCH (n:User) FOR VT AS OF t RETURN n`
Q2.2	`MATCH (n:User) FOR VT FROM t1 TO t2 RETURN n`
Q2.3	`MATCH (n:User)-[r]->(m) FOR VT AS OF t RETURN m, n, r`
Q2.4	`MATCH (n:User)-[r]->(m) FOR VT FROM t1 TO t2 RETURN m, n, r`

time increases linearly with the number of selected vertices. In all the cases, all the vertices in the dataset are selected, in less than 1 s.

In Fig. 3 we show times for queries of type Q2.4, which consider vertex + edge + vertex triples. In all the cases (i.e., all the possible triples in the graph are selected), the answer time is less than 15 s.

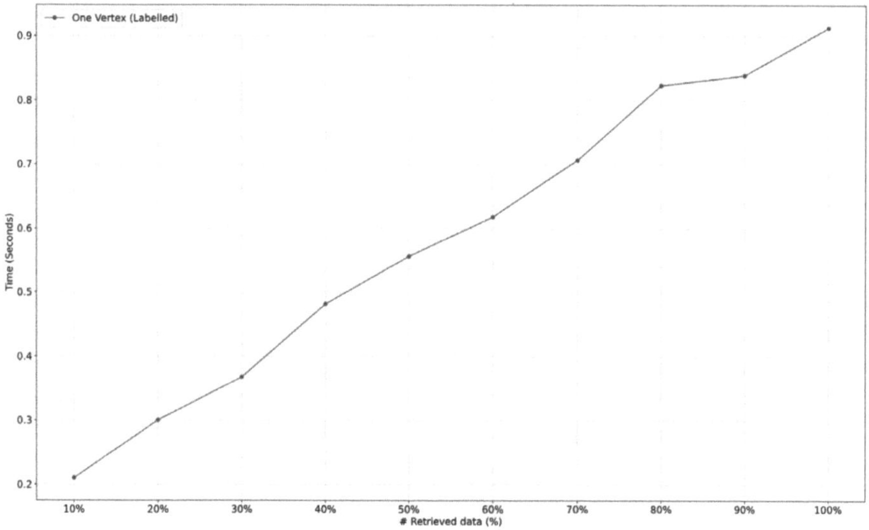

Fig. 2. Q2.2 queries performance evaluation.

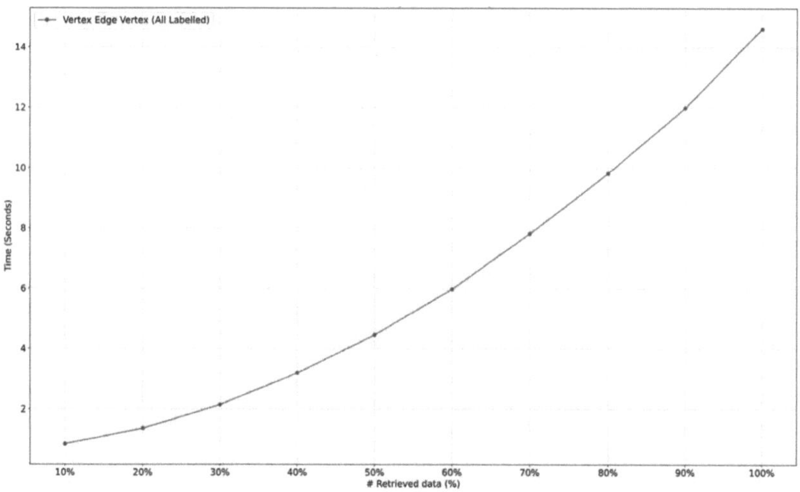

Fig. 3. Q2.4 queries performance evaluation.

6 Conclusions

Even though the treatment of both *valid* and *transaction* time has a long tradition in relational DBs, in the area of Graph DBs such a topic has only be partially explored. Indeed, to the best of our knowledge, only a few approaches considered time in the Graph DB context, and none of them supporting two temporal dimensions (i.e., ***both*** transaction ***and*** valid time). This is a relevant drawback of the area, given the wide relevance of bitemporal data, as ascertained by 30 years of relational and object-oriented temporal

DB literature. The approach in [21] deals with one temporal dimension (indeed, since an explicit definition is not provided, it is not totally clear whether the approach in [21] deals with transaction time or with valid time). However, the approach does not define a ***bitemporal*** model. The implementation is built on top of Neo4J and manages time through an external layer. Furthermore, it does not provide an exhaustive evaluation of the system, as the authors only evaluate the impact of the factors that influence the system's performance. A very promising approach to temporal DB Graph has been recently proposed by AeonG [8], considering only one temporal dimension.

In this paper we extend AeonG's data model, and demonstrate the expressiveness of our extended model showing how it copes with a benchmark example in the literature. We extend the query language of AeonG, and experimentally prove that, despite the increased expressiveness, our framework adds only a very limited overhead to AeonG.

As future work, we aim to extend the query language to specify (1) different valid times for vertices, edges and properties in the same query, and (2) qualitative temporal constraints such as, e.g., Allen's relations along the lines suggested by [21].

Acknowledgments. Erica Raina is a PhD student enrolled in the National PhD program in Artificial Intelligence, XXXIX cycle, course on Health and life sciences, organized by Università Campus Bio-Medico di Roma.

Disclosure of Interests. The authors have no competing interests to declare that are relevant to the content of this article.

References

1. Snodgrass, L.A.: Temporal databases. Computer **19**, 35–42 (1986). https://doi.org/10.1109/mc.1986.1663327
2. Liu, L., Özsu, M.T. (eds.): Encyclopedia of Database Systems. Springer New York, New York (2018). https://doi.org/10.1007/978-1-4614-8265-9
3. Jensen, C.S., Snodgrass, R.T.: Valid Time. In: Liu, L., Özsu, M.T. (eds.) Encyclopedia of Database Systems. pp. 4359–4360. Springer, New York (2018). https://doi.org/10.1007/978-1-4614-8265-9_1066
4. Jensen, C.S., Snodgrass, R.T.: Transaction time. In: Liu, L., Özsu, M.T. (eds.) Encyclopedia of Database Systems. pp. 4200–4201. Springer, New York (2018). https://doi.org/10.1007/978-1-4614-8265-9_1064
5. Snodgrass, R.T. (ed.): The TSQL2 Temporal Query Language. Springer US, Boston (1995). https://doi.org/10.1007/978-1-4615-2289-8
6. Böhlen, M.H., Jensen, C.S., Snodgrass, R.T.: Temporal statement modifiers. ACM Trans. Database Syst. **25**, 407–456 (2000). https://doi.org/10.1145/377674.377665
7. ISO/IEC 9075-16:2023 Information technology — database languages — SQL — part 16: Property graph queries (SQL/PGQ). International Organization for Standardization, Geneva (2023)
8. Hou, J., et al.: AeonG: an efficient built-in temporal support in graph databases. Proc. VLDB Endow. **17**, 1515–1527 (2024). https://doi.org/10.14778/3648160.3648187
9. RocksDB | A persistent key-value store. http://rocksdb.org/. Accessed 06 May 2025
10. Memgraph. https://memgraph.com/. Accessed 06 May 2025
11. Jensen, C.S., Snodgrass, R.T.: Semantics of time-varying information. Inf. Syst. **21**, 311–352 (1996). https://doi.org/10.1016/0306-4379(96)00017-8

12. Terenziani, P.: Towards a causal ontology coping with the temporal constraints between causes and effects. Int. J. Hum Comput Stud. **43**, 847–863 (1995). https://doi.org/10.1006/ijhc.1995.1078
13. Francis, N., et al.: Cypher: an evolving query language for property graphs. In: Proceedings of the 2018 International Conference on Management of Data, pp. 1433–1445. Association for Computing Machinery, New York (2018). https://doi.org/10.1145/3183713.3190657
14. Terenziani, P., Molino, G., Torchio, M.: A modular approach for representing and executing clinical guidelines. Artif. Intell. Med. **23**, 249–276 (2001)
15. Anselma, L., Terenziani, P., Montani, S., Bottrighi, A.: Towards a comprehensive treatment of repetitions, periodicity and temporal constraints in clinical guidelines. Artif. Intell. Med. **38**, 171–195 (2006). https://doi.org/10.1016/j.artmed.2006.03.007
16. Terenziani, P., German, E., Shahar, Y.: The temporal aspects of clinical guidelines. Stud Health Technol Inform. **139**, 81–100 (2008)
17. Piovesan, L., Anselma, L., Terenziani, P.: Temporal detection of guideline interactions. In: Presented at the HEALTHINF 2015 - 8th International Conference on Health Informatics, Proceedings; Part of 8th International Joint Conference on Biomedical Engineering Systems and Technologies, BIOSTEC 2015 (2015). https://doi.org/10.5220/0005186300400050
18. Spiotta, M., Terenziani, P., Dupré, D.T.: Temporal conformance analysis and explanation of clinical guidelines execution: an answer set programming approach. IEEE Trans. Knowl. Data Eng. **29**, 2567–2580 (2017). https://doi.org/10.1109/TKDE.2017.2734084
19. Bottrighi, A., et al.: A symbolic AI approach to medical training. J. Med. Syst. **49** (2025). https://doi.org/10.1007/s10916-024-02139-y
20. Pokec.sk - chat, kde je vždy najviac ľudí. https://pokec.azet.sk/. Accessed 06 May 2025
21. Debrouvier, A., Parodi, E., Perazzo, M., Soliani, V., Vaisman, A.: A model and query language for temporal graph databases. VLDB J. **30**, 825–858 (2021). https://doi.org/10.1007/s00778-021-00675-4

Tutorials

Networks of Knowledge: Graph Analytics for Bridging Human and Data Sciences

ADBIS-2025 Tutorial - Exploring Intellectual Influence and Epistemic Violence

Genoveva Vargas-Solar[1(✉)] and Alejandra Josiowicz[2]

[1] CNRS, Univ Lyon, INSA Lyon, UCBL, LIRIS, UMR5205, 69221 Lyon, France
genoveva.vargas-solar@cnrs.fr

[2] Instituto de Letras, Universidade Estadual do Rio de Janeiro (ILE-UERJ) and Fundação Carlos Chagas Filho de Amparo à Pesquisa do Estado do Rio de Janeiro (FAPERJ), Rio de Janeiro, Brazil
alejandra.josiowicz@uerj.br

Abstract. This paper presents the scientific foundation of the ADBIS 2025 tutorial on graph analytics for estimating reciprocity in intellectual communities using property graphs. By integrating Natural Language Processing (NLP), graph analytics, and embeddings, we analyse culturally rich datasets—such as Wikipedia entries and literary corpora from Latin America and Eastern Europe—to uncover epistemic violence, highlight marginalized intellectual contributions, and advance decolonial approaches to information systems.

Keywords: Graph querying · graph analytics · Wikipedia mining · decolonial computing

1 Introduction

In a world increasingly shaped by data science and algorithmic systems, questions of cultural memory, knowledge representation, and epistemic justice demand urgent attention. As data infrastructures encode and circulate dominant narratives, they also risk perpetuating historical exclusions—silencing voices, distorting contributions, and reinforcing power asymmetries. This paper reports on the ADBIS 2025 Tutorial on *"Networks of Knowledge: Graph Analytics for Bridging Human and Data Sciences"*. The tutorial calls on the data and computer science communities to confront such challenges by embedding critical, humanistic inquiry into technical practice. Who appears—and who disappears—in our datasets, queries, and models? How are knowledge systems constructed, weighted, and made legible in the digital sphere?

ADBIS-2025 Tutorial - Exploring intellectual influence and epistemic violence—Supported by FAPERJ and the AP program of the CNRS.

P. K. Chrysanthis et al. (Eds.): ADBIS 2025, LNCS 16043, pp. 253–263, 2026.
https://doi.org/10.1007/978-3-032-05281-0_16

Focusing on intellectual networks from Latin America and Eastern Europe, we examine how computational methods can surface patterns of epistemic violence and support decolonial approaches to digital scholarship. Using multilingual, culturally embedded datasets such as Wikipedia biographies and literary corpora, we repurpose natural language processing (NLP), graph analytics, and embeddings not merely as tools of classification and efficiency, but as instruments for recognition, reparation, and radical imagination.

Technically, we present a framework for modelling, querying, and analyzing cultural knowledge using property graphs derived from semi-structured data in Wikipedia biographies. Nodes represent intellectual agents, concepts, or institutions; edges encode relationships such as influence, thematic affinity, or institutional linkage. Graph metrics— including centrality, reciprocity, visibility ratios, and absence scores— are used to define indices that estimate how knowledge is distributed, whose contributions are emphasized, who is considered central in intellectual networks and who is silenced or marginalized.

The tutorial furthers shows how integrating embedding-based models (e.g., BERT) can capture semantic proximity between entities, enabling deeper reasoning over linguistic and contextual nuance. The experiments introduced in the tutorial, implemented with tools like NetworkX and spaCy, demonstrate how these representations can reveal structural disparities and support exploratory analysis of cultural datasets.

Accordingly, the remainder of the paper is organised as follows. Section 2 provides a synthesis of digital humanities and graph exploration works which are the background of this study. Section 3 introduces graph driven cultural modelling as an approach for representing cultural knowledge and exploring it. Section 5 concludes the paper and discusses future work.

2 Related Work

Graph-based methods have become central to both technical infrastructures and critical cultural inquiry. In the context of digital humanities, these approaches enable the modelling of knowledge systems, the interrogation of epistemic inequality, and the surfacing of silenced or marginalized perspectives. This section reviews foundational literature in graph querying and representation, recent innovations in humanities-focused graph analytics, and the growing body of tutorials and workshops that blend computational and cultural methods.

Property Graphs Design and Querying. Graph-based models have emerged as powerful tools for representing, querying, and analysing complex semantic relationships in both technical and humanistic contexts. Foundational surveys and frameworks for graph data management and querying, such as those by Angles and Gutierrez [2], Bonifati et al. [5], and Rodriguez and Neubauer [17], provide a solid basis for modelling knowledge networks at scale. In parallel, embedding-based language models like BERT [7] allow for semantic enrichment of textual data, complementing structural graph metrics and enhancing context-aware analytics.

Graph Analytics in Humanities. Recent advances in data science and AI have foregrounded the role of computational methods in shaping cultural memory and knowledge systems. In the humanities, the area of graph analytics has supported investigations of social and narrative networks in literature [10], cultural history [18], and distant reading methodologies [15]. Scholars such as Aeschbach et al. [1] and Jannidis [13] highlight the interpretive potential of graphs to uncover latent power relations, intertextuality, and gendered absences in literary and historical corpora. These approaches intersect with critical scholarship on epistemic injustice and technocolonial infrastructures [3, 16, 19], demonstrating how computational tools can both reinforce and resist dominant hierarchies. Frameworks like data feminism [9] and postcolonial computing [12] further call for reflexive, inclusive methodologies in algorithmic design and analysis.

Furthermore, network science in the Humanities, specifically social network analysis, have contributed to the study of scientific dynamics—specifically, the ways in which scientists cite, reference, and engage in forms of collaboration and co-authorship- pointing to exclusion mechanisms, hierarchies, and inequalities in the circulation of knowledge [8]. The study of citation dynamics in scientific production enables the exploration of cognitive networks through data visualization techniques involving large datasets [6]. This makes it possible to map, through the authors cited, named, and referenced, different ways of knowing, thinking, and mobilizing sociocultural meanings.

Tutorials on Graph Analytics in Humanities. Over the past three years, several tutorials have addressed the intersection of graph technologies and the humanities, reflecting growing interdisciplinary momentum. The 2024 JOWO tutorial on Exploring Multiple Knowledge Graphs in Digital Humanities focused on cultural heritage graph integration, while the 2021 DHQ tutorial on Visualizing Historical Networks Using Palladio emphasized accessible network analysis for non-technical scholars. At ESWC 2023, semantic technologies were showcased for cultural heritage applications using linked open data workflows. In parallel, technically focused sessions like the Learning on Graphs tutorials (2022-2023) delved into GNNs and advanced graph learning, but with limited engagement with humanistic questions. Other initiatives, such as SemLab's workshops on linked data art archives, highlight preservation and interoperability concerns.

Discussion and Rationale. The intersection of graph analytics, digital humanities, and critical data studies enables new approaches to knowledge representation, marginality, and power. Technical systems—databases, graphs, algorithms—are not neutral; they reflect epistemic biases. Foundational works on property graphs [2, 5, 17], interpretive cultural network analysis [10, 13, 18], and critical frameworks like data feminism and postcolonial computing [9, 12, 16] show how computation can both challenge and reproduce exclusion. Graph methods offer scalable ways to encode relationships like influence or absence in multilingual, fragmented corpora. Embedding models such as BERT [7] add semantic nuance, but risk reinforcing high-resource biases.

Our tutorial centres critical, multilingual graph experimentation to reveal epistemic visibility, absence, and reciprocity. In contexts marked by global digital inequality, we introduce and discuss pipelines that quantify epistemic violence and support participatory, sovereignty-aware knowledge infrastructures. Ultimately, bridging analytics with justice demands tools that reflect plural histories, redistribute visibility, and remain accountable to represented communities.

3 Leveraging Graphs for Cultural Modelling

Introduction to Cultural Analytics. Cultural analytics in digital humanities refers to the computational study of cultural patterns, texts, and artifacts at scale. It enables the detection of structural, semantic, and historical patterns that might remain opaque through traditional humanistic methods. Rooted in the work of scholars like Melanie Walsh [20], cultural analytics harnesses tools from statistics, graph theory, and machine learning to explore how meaning, influence, and representation circulate across time, space, and social categories.

In this tutorial, we propose a graph-based approach to cultural analytics that operationalizes the notion of intellectual reciprocity: the mutual acknowledgment, citation, and influence among thinkers, especially across gendered, racialized, and geopolitical divides. We focus on Wikipedia as a primary data source, due to its global reach, semi-structured format, and crowdsourced dynamics of inclusion and omission. Wikipedia serves as a unique cultural and epistemic mirror:

- Globally influential, shaping search engine results, curricula, and public understanding;
- Multilingual, allowing for comparative cultural and linguistic representations;
- Structured, with links, categories, and infoboxes that can be parsed into graph data;
- Contested, reflecting real-world asymmetries in recognition and marginalization.

However, Wikipedia is not neutral, editorial biases, notability standards, and language asymmetries often lead to epistemic violence, particularly against intellectuals from the Majority World, women, and marginalized communities [4,11,14].

Graph-Based Cultural Analytics Pipeline. To analyse intellectual reciprocity using Wikipedia, we show how to model intellectual communities using property graphs where:

- Nodes represent intellectuals, institutions, publications, and concepts;
- Edges capture relationships such as influence, co-citation, thematic similarity, or mentorship.

Nodes represent entities of interest, intellectuals, texts, institutions, or concepts-while edges denote semantically meaningful relationships such as citation, co-authorship, correspondence, ideological influence, or thematic proximity. Property maps annotate these elements with descriptors (e.g., `gender: female`, `region: Latin America`, `period: 1960-1980`) to support comparative cultural analytics. Using graph processing libraries like networkX, we compute:

- Reciprocity indices: measuring the balance of recognition between actors;
- Centrality and absence scores: to highlight under-recognized figures;
- Cultural flows: tracing directional influence across regions (e.g., Latin America –> Europe vs. Europe –> Latin America);
- Community detection: uncovering epistemic clusters shaped by political, linguistic, or disciplinary affinities.

The tutorial demonstrates how to integrate these metrics into composite indices that estimate intellectual reciprocity, reveal epistemic violence, and identify gaps in information. We complement structural analytics with embedding models (e.g., BERT-based vector representations of biographies and articles) to capture semantic proximity and latent influence not reflected in hyperlinks.

The approach introduced in the tutorial contributes to answering the question: *What is cultural analytics in digital humanities?* By demonstrating that cultural analytics is not just about counting, clustering, or classifying cultural data, but about interpreting power, absence, and recognition, we extend the practice to include:

- A decolonial analytics of knowledge graphs that challenges canonical hierarchies;
- A framework for detecting asymmetries in mutual recognition across gender, race, and geography;
- A methodology for reparative querying, where absent or underlinked figures can be algorithmically surfaced for editorial and archival attention.

This graph-based cultural analytics methodology, grounded in critical digital humanities and technical graph querying, transforms Wikipedia from a static data source into a dynamic site for epistemic inquiry, enabling scholars to map, measure, and intervene in the politics of knowledge circulation.

Graph-Based Methodology for Modelling Intellectual Communities. To model intellectual communities in digital humanities, we adopt a graph-theoretic approach that conceptualizes knowledge production and circulation as a network of entities and relationships. In this framework, we employ *property graphs* that allow to represent heterogeneous relationships with rich metadata, such as temporal context, gender, affiliation, or epistemic domain.

Graph Construction. Graphs are built from structured and semi-structured sources such as Wikipedia, Wikidata, and bibliographic corpora. Entity resolution links concepts across datasets, while edges are instantiated using semantic rules (e.g., author A cites author B). Graphs can also integrate multilingual, multimodal, or community-generated knowledge. Table 1 shows an example of the steps of the pipeline that implements the construction of an intellectual graph mesh from her Wikipedia page and the people cited in the page.

Table 1. Pipeline for Building Gabriela Mistral's Intellectual Graph

Step	Description
1. Seed Node	Extract Gabriela Mistral's attributes from her Wikipedia infobox (e.g., gender, field, region) and store in `gabriela_nodes.csv`.
2. Outgoing Links	Parse all outgoing Wikipedia links; keep only those referring to people; extract name mentions from text → `gabriela_outgoing_raw.csv`.
3. Enrich Attributes	Scrape infoboxes of referenced people to collect attributes and generate `people_attributes.csv`.
4. Recursive Graphs	Repeat steps 2-3 for each person cited in Gabriela Mistral's page to build their own ego graphs.
5. Incoming Links	Identify pages where Gabriela Mistral is cited; build `gabriela_incoming.csv` using previously generated subgraphs.
6. Edge Integration	Merge outgoing and incoming links with directionality and context into a unified `edges.csv`.
7. Graph Assembly	Build the full graph using NetworkX or Neo4j; use edge color to indicate direction (green: out, purple: in, gold: reciprocal).
8. Mesh Navigation	Enable interactive visualization; clicking a node displays the merged mesh with Gabriela's graph.
9. Graph Analysis	Compute intersection metrics (shared links, mutual citations, centrality) to compare ego networks.

Operations on Graphs. Two broad categories of operations are used: navigational and analytical.

- **Navigational operations** retrieve subgraphs or paths that match a given pattern or satisfy certain constraints. Given a predicate ϕ, a navigational query extracts paths $P = (v_1, e_1, v_2, \ldots, v_k)$ such that $\phi(P)$ holds. For example: "all female authors citing a Latin American scholar within two hops." These are executed using traversal algorithms (e.g., BFS, DFS) or declarative graph query languages like Cypher or GQL.
- **Analytical operations** compute global or local properties over the graph:

- *Degree Centrality*: $d(v) = |\{e \in E : e = (u, v)\}|$
- *Betweenness Centrality*:

$$BC(v) = \sum_{s \neq v \neq t} \frac{\sigma_{st}(v)}{\sigma_{st}}$$

  where σ_{st} is the total number of shortest paths from s to t and $\sigma_{st}(v)$ is the number passing through v.
- *Reciprocity*:

$$R = \frac{|\{(u, v) \in E \wedge (v, u) \in E\}|}{|E|}$$

  useful to assess mutual recognition in knowledge exchange.
- *Community detection* via modularity optimization or label propagation.
- *Clustering coefficients* and *graph density* to assess cohesion.

We also compute custom metrics such as:

- **Visibility Ratio:** compares centrality scores across gender or region.
- **Absence Score:** quantifies structurally missing links (e.g., absent citations between marginalized scholars).

Critical Framing. The graph is not merely a data structure but a model of epistemic space—encoding not only who is connected, but how and under what cultural assumptions. Comparing graphs derived from Latin American and Eastern European intellectuals, concepts and themes enables the identification of systemic disparities in representation, reciprocity, and influence. This methodology integrates formal graph definitions with critical humanistic inquiry. It equips researchers with tools to measure, visualize, and interrogate knowledge hierarchies, while embedding inquiries about epistemic justice into the analytical pipeline.

4 Experiment: Graph-Based Modelling of Intellectual Communities

To demonstrate the methodology proposed for modelling cultural knowledge networks, we designed an experiment centred on constructing and analysing property graphs of Latin American and Eastern European intellectuals. This experiment focuses on graph operations, structural metrics, and epistemic visibility indicators using open data, particularly from curated CSV files and Wikipedia-derived metadata.

Dataset Description. The experiment uses two core resources:

- `intellectuals_dictatorships.csv`: A curated dataset listing intellectuals active during and after dictatorial regimes in Latin America and Eastern Europe, including metadata such as name, gender, field, region, influence links, and biographical notes.
- `intellectuals_network_enriched_styled.gexf`: A styled and enriched property graph exported for Gephi-based exploration.

The main objectives of the experiment are:

1. Construct a property graph representing intellectual relationships.
2. Compute and visualize structural graph metrics (degree centrality, betweenness, communities).
3. Enrich nodes with metadata (gender, field, region).
4. Export the graph for visual inspection and dissemination.

Graph Construction. The property graph is built using the `networkx` Python library. Each intellectual is modelled as a node, with directed edges denoting influence, collaboration, translation, or any other relationship that is worth mentioning as a linked reference in the Wikipedia article.

Listing 1.1. Constructing an Influence Graph from CSV Data

```
import pandas as pd
import networkx as nx

df = pd.read_csv("intellectuals_dictatorships.csv")
G = nx.DiGraph()

# Add nodes
for _, row in df.iterrows():
    G.add_node(row['name'], gender=row['gender'],
               field=row['field'], region=row['region'])

# Add edges
for _, row in df.iterrows():
    if pd.notnull(row['influenced_by']):
        influencers = row['influenced_by'].split(';')
        for influencer in influencers:
            G.add_edge(influencer.strip(), row['name'])
```

Graph Metrics and Community Detection. We calculate classical graph metrics and apply community detection to assess the structure and epistemic reciprocity within and across regions.

Listing 1.2. Computing Centrality Measures and Detecting Communities

```
import community as community_louvain

centrality = nx.degree_centrality(G)
betweenness = nx.betweenness_centrality(G)

# Louvain community detection (converted to undirected)
undirected = G.to_undirected()
partition = community_louvain.best_partition(undirected)

nx.set_node_attributes(G, centrality,
    'degree_centrality')
nx.set_node_attributes(G, betweenness, 'betweenness')
nx.set_node_attributes(G, partition, 'community')
```

The enriched graph is exported to GEXF format, supporting styled and interactive exploration in Gephi.

Use in Pedagogy and Public History. The experiment is intended as part of a hands-on datathon where participants query and analyse multilingual Wikipedia entries, build graphs using provided templates, and apply critical analysis based on epistemic visibility and marginality. Metrics such as *reciprocity ratio*, *gender visibility index*, and *absence score* are computed and discussed collaboratively. Through the analysis of network topology and community structure, we:

- Identify under-recognized contributors in each region.
- Visualize cross-regional epistemic bridges and silences.
- Evaluate the representation of gender, race, and geography.

This experiment illustrates how graph analytics, situated in critical frameworks, can support decolonial digital humanities practices by surfacing epistemic violence and proposing indicators for measuring and assessing intellectual recognition and/or marginalization.

5 Conclusions and Future Work

This tutorial demonstrates the value of combining graph theory, digital humanities, and critical data studies to model and interrogate cultural and intellectual knowledge. Through the construction of multilingual property graphs and the computation of epistemic visibility metrics, such as degree centrality, absence scores, and representation ratios, we illustrated how graph analytics can expose epistemic violence and highlight structural inequalities across regions and languages. Participants developed practical skills in querying Wikipedia's multilingual API, extracting relational data, and using analytical and visual tools to understand influence, marginality, and reciprocity within intellectual communities. Comparative analyses of Latin American and Eastern European networks revealed not only quantitative disparities but also culturally specific silences shaped by historical and geopolitical contexts.

The datathon setting fostered collaborative inquiry, encouraging participants to reflect critically on the ways data structures, algorithms, and source biases can either reinforce or challenge hegemonic knowledge systems. Nevertheless, several open questions remain. How can we formalize the notion of epistemic violence in a culturally sensitive yet computationally tractable way? What are the best practices for co-creating such graphs with affected communities, ensuring accountability, consent, and contextual integrity?

Future work should explore the integration of Indigenous and feminist graphs, multilingual embeddings, and hybrid participatory pipelines that make epistemic reparation actionable in data science workflows. Further research is also needed on scalable, sovereignty-aware infrastructures that allow cultural datasets to remain locally governed while contributing to global analytical ecosystems.

References

1. Aeschbach, L., Mariani, J., Kaplun, A., Kaplan, F.: Graphs in the humanities: a survey. Digit. Scholarship Hum. **35**(1), 100–118 (2020)
2. Angles, R., Gutierrez, C.: A survey of graph database models. ACM Comput. Surv. **40**(1), 1–39 (2008)
3. Benjamin, R.: Race After Technology: Abolitionist Tools for the New Jim Code. Polity Press (2019)
4. Beytía, P., Wagner, C.: Visibility layers: a framework for systematising the gender gap in wikipedia content. Internet Policy Rev. **11**(1) (2022). https://doi.org/10.14763/2022.1.1621
5. Bonifati, A., Martens, W., Timm, T.: Querying graphs. ACM Comput. Surv. **50**(5), 1–40 (2018)
6. Brasil, A., Jr., Carvalho, L.: Por dentro das ciências humanas: um mapeamento semântico da área via base scielo-brasil (2002–2019). Revista de Humanidades Digitales **5**, 149–183 (2020)
7. Devlin, J., Chang, M.W., Lee, K., Toutanova, K.: Bert: pre-training of deep bidirectional transformers for language understanding. In: Proceedings of NAACL-HLT (2019)
8. Dworkin, J., Zurn, P., Bassett, D.S.: (in)citing action to realize an equitable future. Neuron **106**(6), 890–894 (2020). https://doi.org/10.1016/j.neuron.2020.05.011
9. D'Ignazio, C., Klein, L.F.: Data Feminism. MIT Press (2020)
10. Elson, D.K., Dames, N., McKeown, K.R.: Extracting social networks from literary fiction. In: Proceedings of the 48th Annual Meeting of the Association for Computational Linguistics, pp. 138–147 (2010)
11. Ferran-Ferrer, N., Boté-Vericad, J.J., Minguillón, J.: Wikipedia gender gap: a scoping review. Profesional de la Información **32**(6) (2023). https://doi.org/10.3145/epi.2023.nov.17
12. Irani, L., Vertesi, J., Dourish, P., Philip, K., Grinter, R.E.: Postcolonial computing: a lens on design and development. In: Proceedings of the SIGCHI Conference on Human Factors in Computing Systems, pp. 1311–1320 (2010)
13. Jannidis, F., Kohle, H.: Analysis of literary history through social network analysis. Int. J. Digit. Art History (2) (2016)

14. Langrock, I., González-Bailón, S.: The gender divide in wikipedia: quantifying and assessing the impact of two feminist interventions. J. Commun. **72**(3), 297–321 (2022). https://doi.org/10.1093/joc/jqac004
15. Moretti, F.: Network theory, plot analysis. In: Distant Reading, pp. 211–238. Verso Books (2013)
16. Ricaurte, P.: Technocoloniality: digital infrastructures and epistemic injustice in latin america. AI & Soc. **37**(1), 255–267 (2022)
17. Rodriguez, M.A., Neubauer, P.: Constructions from dots and lines: graph analytics for complex network analysis. Proc. VLDB Endowment **3**(2), 1055–1060 (2010)
18. Schich, M., et al.: A network framework of cultural history. In: Science, vol. 345, pp. 558–562. AAAS (2014)
19. Siles, I.: Epistemic inequalities and data coloniality in the digital age. AI & Soc. (2023)
20. Walsh, M.: Introduction to cultural analytics (2020). https://melaniewalsh.github.io/Intro-Cultural-Analytics/welcome.html. Accessed 13 July 2025

Data Warehousing: The Industrial Perspective

Henrietta Dombrovskaya(✉)

DRW Holdings, Chicago, IL 60661, USA
hdombrovska@drwholdings.com

Abstract. Like with many other concepts, the theoretical ideas regarding how data warehouses should be organized are rarely followed in real-life systems. The notion of data warehouses as storage for historical, rarely or never changing records used for summary reports, slicing and dicing, and trend analysis do not reflect the most common business needs. In this tutorial, we will closely examine existing business demands and discuss how to support them.

Keywords: data warehouse · OLAP · temporal data · columnar storage · hybrid architecture

1 Tutorial Overview

1.1 Objective

In this tutorial, we are going to discuss the data warehousing industry needs and whether these needs are covered by existing commercial offerings. We will outline some known challenges and discuss how new research could address them. We want to convince attendees that data warehouses are more complicated than they are usually thought of and are therefore interesting as a source of research problems.

1.2 Topic Relevance

As one can hardly find an application that does not use- a database, it is hardly possible to come across an organization that does not need a data warehouse. All businesses must produce financial reports, analyze marketing trends and plan for the future.

For all these tasks, data warehouses are essential. Multiple commercial offerings are available, but without understanding which of the characteristics are important, it's difficult to make a call. This tutorial provides the guidelines on how to evaluate market offerings against the actual business needs, identify the gaps between the two and outline the areas where new scientific research could help to move the industry forward.

P. K. Chrysanthis et al. (Eds.): ADBIS 2025, LNCS 16043, pp. 264–271, 2026.
https://doi.org/10.1007/978-3-032-05281-0_17

2 Target Audience

This tutorial is intended for database researchers, post-docs and students who would benefit from being exposed to the real-world challenges of data warehousing. The topic of data warehousing is not popular these days, and most researchers do not even consider exploring it believing that everything regarding data warehousing is conceptually set in stone, and nothing except for some minor algorithm tweaks can be added. We intend to demonstrate that there is a few real-life problems related to data warehousing that await to be addressed.

3 Recent Tutorials Overview

The topic of data warehousing has rarely been chosen as a tutorial subject in the past ten years, except for the recently revitalized interest in connection with the use of AI. One of the examples of this trend is [1]. This tutorial focuses is on the benefits of AI for database maintenance and database support for AI and does not concern any data warehousing-specific challenges.

Other tutorials, like [2], focus on demonstrating the usage of the existing data warehousing and reporting tools. While it is important to provide academic audience with hands-on examples, tutorials of that kind do not touch the underlying principles of building and maintenance of data warehouses.

In contrast, a 2012 tutorial by Golab and Johnson [3] addresses many of the above-mentioned challenges, including an often-overlooked need for real-time reporting. The authors of this tutorial explore the concept of Data Stream Warehousing, including the business need for real-time vs. long-term reporting. While the Data Streams address many of the challenges of the latter one, this approach maintains some limitations, including its insert-only nature and limited ability to correlate incoming data with the historical records.

A wide variety of tutorials on commercial data warehousing systems ([4, 5] and [6]) are focused exclusively on specific products and rarely touch on data warehousing principles (with [4] being a rare exception).

4 Tutorial Outline

For over thirty years, both database researchers and practitioners have maintained a clear distinction between OLTP and OLAP systems, between application databases and data warehouses.

However, this ideal picture does not match the reality of industrial practices. Often, businesses built de-facto data warehouses as parts of their real-time applications without utilizing the proper techniques. At the same time, many data warehousing and reporting commercial tools exploit the idea that business needs specialized software for efficient reporting, thus locking companies into product dependencies.

4.1 Data Warehouses Evolution

First databases didn't have any special features to be classified as OLTP or OLAP: they simply provided the ways to store and retrieve data. The term *data warehouse* (DW) was already in use in these early days (even the punch cards that contained information about the US Census in 1948 were referred to as a "data warehouse"). Back then, however, it referenced not specific system characteristics but rather the fact that the data from different external sources was collected and stored in one place, where it could be processed and restructured for reporting and analytical purposes.

Although Database Management Systems as specialized software came to life in the late 1960s, and relational databases appeared in the 1980s, the specialization of databases based on their functionality was not yet on the picture. Until the 1990s, the hardware limitations (especially the RAM size) didn't allow efficient processing of large data sets. Databases were still mostly used for storing and retrieving data, and data analysis was performed outside databases.

As more powerful hardware became available and relational databases became mainstream, both researchers and practitioners started to recognize different database usage patterns. In 1993, the terms OLTP and OLAP were coined by Ted Codd [7]. Shortly after, in 1996, the Star Model was introduced by Ralph Kimball [8].

The late 1990s-early 2000s mark a spike of interest in data warehousing research and development. The topic became extremely popular, and a substantial number of research papers and books on data warehouse architecture, models, tools and management principles was published. The Chaudhuri and Dayal 1997 paper [9] is a great example. Surprisingly, this paper appears to be mostly relevant to today's readers, which, on the one hand, attests to its high quality, but on the other hand, presents evidence of the fact that not much of the research was conducted afterward.

In subsequent years, we see an increasing number of commercial products ([10–12]) and reporting tools[13] focused on data analytics, business intelligence, etc. While the number of commercial offerings is steadily increasing, we do not see a lot of research activity in this area. The books published after the early 2000s, like [14], essentially summarize the principles and best practices outlined in the earlier works.

4.2 A Textbook Data Warehouse

There are some well-established views regarding the principles and techniques on how OLAP systems should look like and what we can expect of them. These ideas date back to the 1990s, and although we can't exactly label them as wrong, they do not cover a substantial portion of real-life use cases.

Data Warehouses are Static. Although there is no formal definition of what makes data historical (as an opposite of current), the conventional assumption is that a data warehouse stored yesterdays and earlier data (T + 1 data in financial terminology), i.e., the transactions which are already settled and not going to be changed. That assumption reaffirms the views that data warehouses are *insert-only* and *write ones/read many times*, thus dismissing the need to consider specific update algorithms.

Specific Data Model Requirements. Another commonly held opinion is that once loaded, the data should be transformed from its original format to a “more suitable” format for data warehousing. That includes broadly understood denormalization, converting data to STAR schema, the usage of columnar storage, etc. Combined with the presumption of the DW static nature, the assumption is that this transformation has to be done once. The implicit expectation is that the DW users would most often retrieve the data from one gigantic fact table, which often becomes indistinguishable from the reporting table.

There is No Need to Optimize ETL and Reports. The ETL (Extract-Transform-Load) stands for the process of loading data into a data warehouse. While a fast response time for OLTP systems is usually considered critical, the situation with OLAP systems is different. Users are generally more relaxed regarding the ETL times, maintaining that “there is a lot to be loaded” and “if the load is finished by morning, there are no issues.” Indeed, the milliseconds and even seconds difference in the loading time is not critical. Besides, the assumption is that after the transformation the data is structured perfectly to satisfy all reporting needs, so data extraction will be almost instantaneous. In other words, the “online” part of the OLAP means that users can retrieve the results of the reports that were run in advance.

Specific Set of Operations. Since the data is already transformed (and transformation might include complex operations, including multiple joins), the assumption is that there will be no joins between fact tables. The only operations a DW should support are, summarization, group functions calculation, and “slicing and dicing”.

The Need for Specialized DBMSs. Since data warehouses are viewed as databases with fundamentally different characteristics compared to OLTP, including different data models and operations, a conventional assumption is that specialized DBMS (RedShift, Snowflake, BigQuery, etc.) will work best. This statement is overwhelmingly supported by companies advertising, as well as by some researchers who experimentally show that columns provide 20–30% speedup on scans and over-emphasize scalability.

Special Hardware Needs. When a company starts to plan for a data warehouse and spec out the hardware needs, it is almost always assumed (without any analysis) that a database for a data warehouse should of size XL. It’s automatically assumed that it will need more disk space, more RAM and more cores than an OLTP system.

4.3 Why Does It not Work?

A textbook data warehouse described in the previous section has multiple limitations and does not cover a significant portion, if not most of practical needs.

First, the concept of a data warehouse as a historical data storage is constantly being challenged. More and more often, we observe the need to analyze marketing trends or to adjust the course (marketing strategy, pricing, specific offerings) based on the live feedback. That means that the business demands having “last minute data” available in the same space as “yesterday’s data” and “last year’s data” and being able to produce reports that incorporate all of them.

This demand leads to the necessity of multiple updates of the stored data, and multiple partial recalculations of the aggregated values, along with constant adjusting of the marketing models. As one of the consequences, data warehouse maintenance loses the luxury of "ETL windows"- we can't pause the operations to allow time for new data loads, everything has to happen simultaneously.

Secondly, "keeping the history" acquires a different meaning. The concept of data cleansing before loading to the data warehouse lost its popularity not only because this process is difficult, but also because the notion of "what is correct" might change when we obtain additional information. For example, when a credit card vendor posts a transaction completed in a different currency, the exchange rate for a pending transaction may differ from the exchange rate of the settled transaction. It is impossible to "cleanse" this data before loading, and we can't wait and not post a transaction until it is settled. These demands call for the necessity of storing the history of changes and snapshots of data, as it was asserted at different moments of time, since different decisions could be made based on different statuses of data.

Thirdly, real data does not fit into the STAR model, nor can it be represented by a combination of facts and dimensions. Trends analysis calls for a more sophisticated approach, mostly for correlated queries, like "How did the price for coal changed when the price for gas went up?" or "how did the sales spikes correlate with new sales points openings."

4.4 Industrial Offerings

There are multiple data warehousing solutions that are marketed as specialized databases for data warehouses. Each of them claims to be the best tool for data analysis, being able to manage petabytes of data with ease.

Proprietary data warehouses, like Teradata Vantage and Oracle Autonomous Data Warehouse, combined with top-notch hardware, are the best-performing offerings. The only downside is their price, which is prohibitive for smaller companies.

Other data warehousing solutions include Snowflake, Amazon Redshift, Google BigQuery, ClickHouse and more specialized offerings, like Timescale. All mentioned systems provide an option of Cloud hosting with some (ClickHouse and Timescale) supporting on-premises solutions as well. It's not easy to determine whether these systems utilize some fundamentally different storage and query processing techniques that can't be replicated in general-purpose databases.

All Cloud offerings emphasize the "unlimited" resources, and the recommended Cloud instance size is massive. That does not surprise potential customers as, as we noted earlier, the assumption is that any data warehouse needs a large database. For on-premises solutions, the hardware requirements are even more demanding. For example, ClickHouse recommendations call for a 1:50 ratio between RAM and disk space, and a 4:1 ratio between vCPU and RAM. The calculation example from the ClickHouse website [10] suggests having 11TB total RAM for 540 TB of disk space and a total of 2700 of vCPU (distributed to 40 nodes). With that resource allocation, almost any query will run fast regardless of the search engine specifics.

Most data warehousing solutions support columnar storage and data compression, and some (Timescale) support time series.

4.5 Warehousing Techniques

In this section, we will cover several techniques that are widely used both by specialized systems and by practitioners when they need some data warehousing solution but do not want to be locked into any proprietary solutions and want to be able to use their "regular database features."

Tables vs. Materialized views. Since early 1990s, materialized views are strongly associated with data warehouses, since they embody the concept of "write once/read many times". An impression that they can be refreshed completely automatically makes them attractive.

The major disadvantage of the materialized view is that partial refreshes are still challenging. This might result in long waits, inconsistencies or even data loss. In practice, using regular tables which are incrementally updated provides significantly better outcomes,. The refresh algorithm would include the following steps:

1. Identifying new data in the source system, including inserting, modifying, and removing existing data elements
2. Inserting/updating/deleting from a data warehouse table

Columnar Storage vs. Storing Rows. Advantages of columnar storage include space saving and very quick response time on specific queries. Basically, if we think about a textbook DW where we mostly retrieve a small (often just one) column value from a fact table, the execution time may decrease dramatically if we store data in the columnar format. However, the opposite is also true: some queries may be painfully slow. Examples include "wide" fact tables and a large number of joins in one query. In addition, updates are prohibitively slow.

Time Series vs. Temporal Tables. Time series is a trigger word, and whenever somebody mentions they need to store data related to each day/hour/etc., an immediate reaction is "We need a time series database!" Time series databases (TSDBs) are specialized databases designed to store, query, and analyze data that changes over time. Examples of such data include monitoring data, financial data, sensor reading, etc.

Commercial and Open-Source TSDBs, like Timescale or Influx, have built-in mechanisms for fast data ingestion, efficient compression, and specialized analytical capabilities for time-series data. Also, many database developers built their own time-series-based solutions within general-purpose databases, like PostgreSQL.

The non-obvious problem with the time series solutions is that they assert time as discrete (the system receives data at specific moments of time). However, in reality, time is a continuum, and the values should have some meaning even between measurements. Thus, time series solutions are most advantageous when the point in time is predictable, for example, when a monitoring system checks the system status every minute. However, if you need to fill in the gaps (for example, one of the sensors didn't respond in time), a query searching for the "last value before a certain point in time" might be extremely slow.

From that perspective, using time ranges (any value stays the same until changed) appears to be more practical, but many practitioners treat time rangers as "big unknows," and there is a reluctance in switching a paradigm. In the systems that support time ranges, like PostgreSQL, this approach yields significant performance improvements.

Partitioning and Sharding. Sharding is another buzz word, meaning that anytime somebody talks about data warehouse, the concept of sharding comes up as a cure to all worlds' problems. It is believed that if we can keep data physically in many separate places, ideally "closer to the consumer," it will take less time to extract data and deliver it to the customer. This might be true in some cases, but without careful planning, the performance might suffer dramatically.

Building Analytical Models. Marketing analytics use data warehouses when building predictive models, for which they need to load the historical data. The most common problem arises in the process is the inability to adjust the models incrementally. We are not aware of any good solutions to that problem; thus, we often observe Python scrips reading all records from a DW table (starting from January 2015 to today's date).

4.6 Research Challenges

There are several research topics that could yield a significant improvement in the data warehousing practices. These topics include implementing incremental materialized views refreshing, including incremental aggregates and rollup updates, as well as incremental building of analytical models.

Other potential research areas involve working with multi-versioned data (multiple time dimensions) specifically supporting temporal aggregates.

5 Summary

This tutorial discusses the history of data warehouse, the data warehousing industry needs and whether these needs are covered by existing commercial offerings. The need in data warehousing is greater than we think: practitioners often do not realize that they need a data warehouse as a part of their OLTP system.

Widely available specialized systems promise perfect solutions to all data management needs, but the users often find themselves locked into one-system dependency. It is often difficult to evaluate the variety of commercial offerings, and the knowledge of universal algorithms helps making an informed decision.

A textbook data warehouse does not address many a big portion of the industrial needs: data warehouses are more complex than they are usually thought of and are therefore interesting as a source of research problems.

References

1. Li, G., Zhou, X., Cao, L.: AI meets database: AI4DB and DB4AI. In: Proceedings of the 2021 International Conference on Management of Data (SIGMOD '21), pp. 2859–2866. ACM, New York (2021). https://doi.org/10.1145/3448016.3457542
2. Guimaraes, M.: Data warehouse and data mining with SQL server. J. Comput. Sci. Coll. **34**(1), 233–234 (2018). https://doi.org/10.5555/3280489.3280528

3. Golab, L., Johnson, T.: Data stream warehousing. In: Proceedings of the 2013 ACM SIGMOD International Conference on Management of Data (SIGMOD '13), pp. 949–952. ACM, New York (2013). https://doi.org/10.1145/2463676.2465337
4. Snowflake: https://www.snowflake.com/en/resources/learn/training/
5. Amazon Redshift: https://aws.amazon.com/redshift/
6. Grafana: https://grafana.com/tutorials/
7. Codd, E.F., Codd, S.B., Salley, C.T.: Providing OLAP (On-Line Analytical Processing) to user analyst: an IT mandate. Available from Arbor Software's web site http://www.arborsoft.com/OLAP.html
8. Ralf Kimball, T: The Data Warehouse Toolkit (1996)
9. Chaudhuri, S., Dayal, U.: An overview of data warehousing and OLAP technology. SIGMOD Rec. **26**(1), 65–74 (1997). https://doi.org/10.1145/248603.248616
10. ClickHouse. https://clickhouse.com/docs/guides/sizing-and-hardware-recommendations
11. Google BigQuery. https://cloud.google.com/bigquery
12. Timescale: https://www.timescale.com/
13. Tableau: https://www.tableau.com/
14. Jarke, M., Lenzerini, M., Vassiliou, Y.: Fundamentals of Data Warehouses. Springer-Verlag (2003). https://doi.org/10.1007/978-3-662-05153-5

Vector Representations of Multi-modal Data

Toni Taipalus[1(✉)] and Jiaheng Lu[2]

[1] Tampere University, Tampere, Finland
toni.taipalus@tuni.fi
[2] University of Helsinki, Helsinki, Finland
jiaheng.lu@helsinki.fi

Abstract. Multi-modal data processing is about exploring the interactions between various types of data to produce a more comprehensive or accurate understanding of a phenomenon such as health, emotions, or circumstances. Vectors as data representation methods have emerged as an important component in modern data management, driven by the growing importance for the need to computationally describe multi-modal data such as texts, images and video in various domains. In this tutorial, we provide a fundamental introduction on vector representations of multi-modal data, which includes intra-modal representation and inter-modal representation. The goal of our tutorial is to provide a centralized and condensed introduction regarding theories and applications of multi-modal data vectorization technologies for both database researchers and practitioners. We also discuss how to use vector database management systems for the management of multi-modal data.

Keywords: multi-modality · vector databases · vectorization · database · data management

1 Introduction

Multi-modality and vector databases have both gained substantial momentum in both research and practice [13,21]. Multi-modality means several modalities. These modalities can include verbal (such as spoken language), visual (such as images or videos), auditory (such as sounds, noises or music), and spatial (such as physical space or arrangement), and efficiently utilizing multi-modality can achieve, for example, more accurate recommender systems [23] and richer medical diagnoses [1]. Vectorization of data, on the other hand, has the potential for more efficient storage, as well as representing data objects of different nature with similar terms, i.e., vectors. By combining the advances in the fields of both multi-modality data and vector databases, managing multi-modality data can be fundamentally more efficient, and reach new use-cases in various multidisciplinary domains.

Tutorial Overview: In this tutorial, we will discuss five modals of data: *text*, *image*, *audio*, *video* and *time-series* data. For each modal, we will present their

P. K. Chrysanthis et al. (Eds.): ADBIS 2025, LNCS 16043, pp. 272–279, 2026.
https://doi.org/10.1007/978-3-032-05281-0_18

respective vector representation method, and review cases for the combination of multiple modal data, including text+image, text+audio, image+audio, and text+image+audio. For each combination of modals, we will review the algorithms for vector alignment and representation fusion. In this tutorial, we will also show a fundamental introduction to applications of using multi-modality data with vector databases. Our tutorial discusses the basics behind vector database management systems and the process and unification of vectorization, as well as open problems and future opportunities of the intersection of multi-modality and vector databases.

The learning outcomes of this tutorial are to *(i)* understand the need behind vectors and multi-modality data management in today's landscape, *(ii)* understand how different data objects can be vectorized and why vectorization is useful for multi-modal data, *(iii)* learn different methods to unify and align vector representations for multi-modal data, and *(iv)* know of the current challenges and opportunities in the intersection of multi-modality and vector databases.

Intended Audience: This tutorial is designed for a broad audience, including academic researchers, students, industrial developers, and practitioners, who seek to understand vector representations for multi-modal data, explore how the convergence of vector databases and multi-modality enhances data management, and examine current challenges and future opportunities in inter-modality vectorization. A foundational understanding of databases and machine learning concepts is required to follow the tutorial effectively.

Related Tutorials: We are aware of tutorials which are tangential to this work. A SIGMOD'24 tutorial [15] gives a general survey on vector database management systems and vectorization techniques. A RecSys'21 tutorial [23] shows how multi-modality can relieve the challenge of sparsity of vectors in recommender systems. Finally, a 2020 *Information Fusion* tutorial [25] discusses multi-modality emotion recognition using a medical dataset.

Contributions: To the best of our knowledge, this is the first tutorial to discuss the methods of vectorization and use-cases of multi-modal data processing through vector databases. While previous tutorials have described both vector database management systems and multi-modality separately, this intersection has received no scientific tutorials, especially from a general perspective. The topic of vector databases and multi-modality covered in this tutorial may help industry professionals, researchers, educators, and students in understanding how vector databases can be used in multi-modality use-cases now and in the future, and what challenges this intersection currently faces.

2 Intra-modal Vector Representations

Vectors can be used to represent various types of data, including text, audio, video, images and time-series data. By vectorizing the features of data objects such as images, various features are given numerical representations. Vectors created from various data objects can be relatively efficiently stored and compared

[21]. It is highly context-dependent on which features of data objects are vectorized, as vectorizing all features is rarely feasible or even possible. A vectorized data object is often called a *feature vector* or a *vector embedding.*

Text: Feature vectors have been used to enhance search accuracy by understanding synonyms and contextual relevance [16], in classifying documents into predefined categories for spam detection and sentiment analysis [5], for identifying and classifying entities within texts using contextual embeddings, for improving machine translation quality by representing words and phrases in a continuous vector space, and for grouping similar documents together using their vector representations for more accurate information retrieval [18].

Image: Vector representations generated by convolutional neural networks (CNNs) have been utilized in finding similar images in a database and for classifying images into categories by using the CNN features as inputs to a classifier [11]. Furthermore, vectors can be used in identifying objects within images using embeddings to represent features, and for generating descriptive captions for images by combining image embeddings with language models [6].

Audio: Vectors can be used in speech recognition by simply converting speech to text using audio embeddings such as Speech2Vec [2]. Vectors can also be used in speaker identification based on their unique voice characteristics captured in embeddings [9], song recognition, sound classification [10], and music recommendations.

Video: Converting a video into a vector representation involves extracting meaningful features from selected frames and encoding them as vectors. This process is useful for various tasks such as video classification and understanding. For example VideoBERT [19] passes each frame through a pre-trained CNN to obtain a feature vector from a fully connected or global pooling layer. The final result is a single vector representing the video.

Time-Series: Time series data refers to a sequence of data points collected at successive points in time, typically at regular intervals. Converting time series data into a vector representation involves various feature extraction techniques including statistical methods, wavelets, and embedding-based approaches [8].

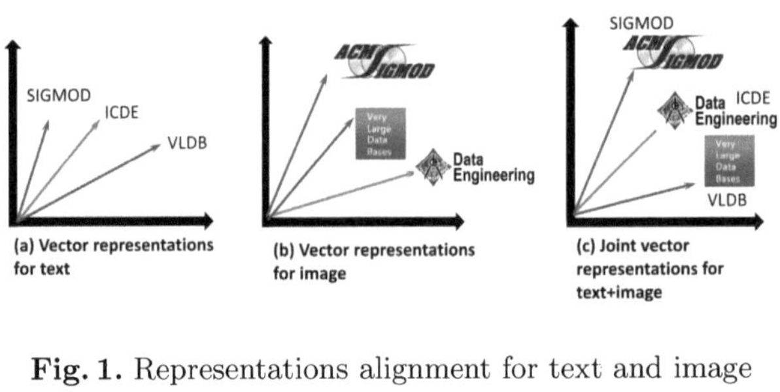

Fig. 1. Representations alignment for text and image

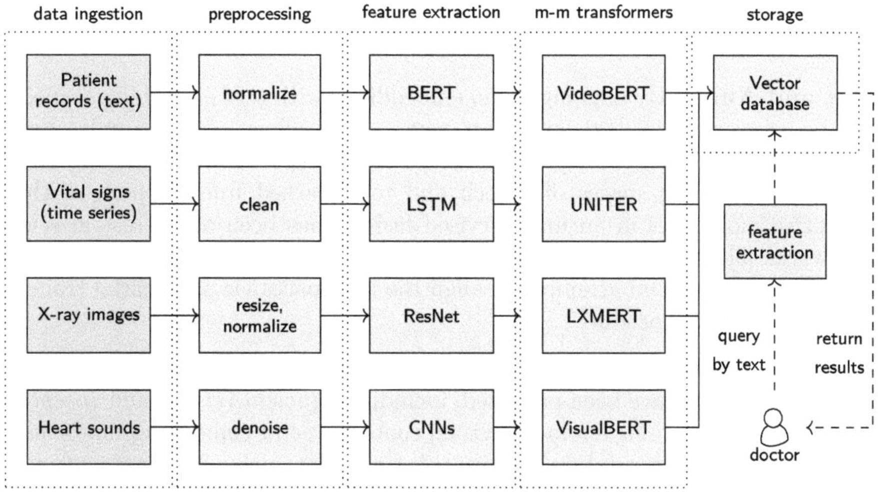

Fig. 2. A general process of storing multi-modality data into a common vector space, and into a vector database management system with an example of healthcare data; illustrated software, libraries, and techniques such as BERT and UNITER are examples; multi-modality transformers (abbreviated *m-m transformers* above) are used to ensure that multi-modality data objects are stored into a common vector space, which can subsequently be queried by a doctor using only textual input

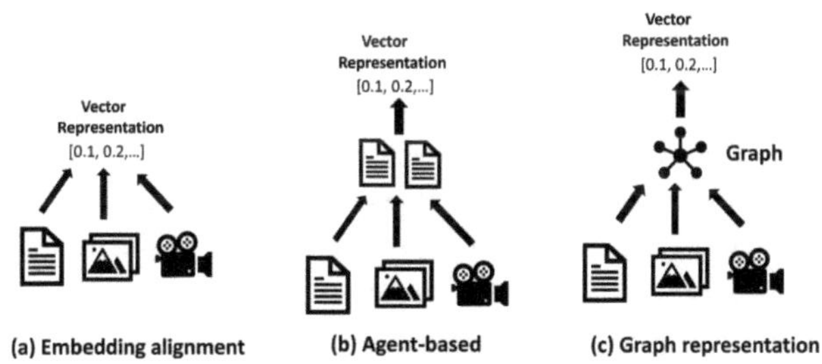

Fig. 3. Various approaches for unifying inter-modality vector representations

3 Inter-modal Vector Representations

In inter-modality contexts, vector representations can be used in tandem with multiple modals. In this section, we describe the cross-modal alignment between the embedding spaces of multiple modalities (such as text+image, cf. Fig. 1) learned from corpora of their respective modalities.

Text and Image: Image captioning inherently combines visual data with textual data. Visual features extracted from images and the sequential nature of text generation can be integrated into a unified model. Furthermore, this enables cross-modal retrieval, as querying by text can retrieve images and vice versa, as both text and images are embedded to a common embedding space [4]. Additionally, this approach enables answering questions about images when the cross-modal embeddings are passed to a language model.

Text and Audio: By aligning audio embeddings with text embeddings, speech can be converted to text and vice versa. This allows, for example, automated summarization of long speeches such as podcasts. Cross-modal alignment between embedding spaces of speech and text learned from corpora of their respective modalities in an unsupervised fashion has been of interest in scientific works [3]. The proposed framework learns the individual speech and text embedding spaces, and attempts to align the two spaces via adversarial training and a refinement procedure.

Image and Audio: For audio-image vector representations, several CNN-based feature extractors have been proposed, including *EfficientNet* [22] and *Inception ResNet* [20]. By aligning audio and visual content, vector embedding can be used in audiovisual synchronization, for example for synchronizing lip movements with speech. Multi-modality through image and audio can also be used to enhance emotion recognition by considering both facial or bodily expressions as well as audio cues [25].

Text, Image and Audio: The combination of text, image, and audio data into a common embedding space enables multi-modal search engines which can retrieve results across several types of data objects. Similar embeddings can also be used in multi-modal recommender systems [23]. This allows content recommendations which are not based on data objects of similar type. That is, the recommender system may suggest, e.g., text documents based on videos watched, and these recommendations are not limited to metadata, but to analyzed (and vectorized) content. Additionally, inter-modality with feature vectors can serves as a basis for both inputs and outputs of digital assistants. Fig. 2 illustrates a general example use-case and example techniques for storing multi-modal data objects into a vector database, using a common vector space.

4 Methods for Unifying Inter-modality Vectors

In this section, we introduce basic approaches on how to unify different vector representations for multi-modal data to integrate and reconcile vectors from different modalities with a representation that can capture the essence of several modalities simultaneously. Figure 3 illustrates various approaches to produce a unified vector representation.

Embedding Alignment: Models are trained with the aim of learning a joint embedding space where representations from different modalities are embedded such that similar instances across modalities are closer together in the space, and the distance between dissimilar instances is maximized. This is typically achieved using a contrastive learning objective or a triplet loss function [24]. For example, Fig. 1 illustrates the vector alignment for text and image data in a joint space.

Agent-Based Alignment: Another method of dealing with multi-modality data is simply converting all data objects to a single modality, e.g., textual descriptions, which are then vectorized. This approach circumvents the challenge of vectorizing multi-modality data objects directly into the same vector space, as the vectorization process is only applied to text. We call this an agent-based alignment, rooted in the presumption that different models (or *agents*) are responsible for creating feature vectors for different data objects. **-to-text* models such as CLIP [17] with GPT for image-to-text, BART [12] for text-to-text (e.g., summarizing and translation), and *Whisper* for audio-to-text have already shown substantial results in their respective tasks.

Fusion Techniques: These techniques include early fusion and late fusion [26]. Early fusion involves concatenating or stacking vectors from different modalities into a single vector representation before inputting them into a neural network, which outputs the final vector to be stored in a vector database. These vectors can be high-dimensional, making it more difficult for the model to learn efficiently. In contrast, late fusion combines the outputs from models trained separately on each modality, which can involve averaging or weighting predictions from the individual models to generate a final representation. The challenges in late fusion are typically related to relatively low coupling between the modalities.

Graph Representation: Construct a graph where nodes represent instances and edges capture relationships or similarities between instances across different modalities [14]. Graph neural networks can then be applied to learn a unified representation by aggregating information from the graph.

5 Vector Databases for Multi-modal Data

After relevant data objects have been vectorized, these feature vectors can be stored for subsequent use by, e.g., customer-facing applications. These applications, such as the healthcare-related systems described in Fig. 2, may either use libraries, a dedicated vector DBMS, or a more general-purpose DBMS to index and query the vectors. Each of these approaches relieve the need to implement, e.g., vector indexing techniques such as *product quantization* or *hierarchical navigable small world*, and have distinct use-cases.

Vector management-dedicated software libraries such as FAISS [7] typically offer different ways of indexing vectors and querying them with similarity search. Typically, this approach is often computationally fast, yet provides little more for effective data management [21]. Dedicated vector DBMSs such as Milvus or Chroma, on the other hand, have been built from the ground up to serve the needs of vector data management. This relatively novel approach typically supports several vector index types, and strives for interoperability between the DBMS and other system components. Finally, more general DBMSs such as PostgreSQL, SQLite, MongoDB, Cassandra, and Redis also support vector data management, some through third party extensions. This approach potentially adds data management features such as concurrency control, access control,

ACID-compliance, back-up and recovery, and an expressive query language to support vector data management. For example, with PostgreSQL's `pgvector` extension, a software developer may query vectors using SQL, as well as utilize vector metadata in queries. For example, a query with a query vector can be complemented with more traditional SQL expressions.

References

1. Cheng, D., Liu, M.: CNNs based multi-modality classification for ad diagnosis. In: 2017 10th International Congress on Image and Signal Processing, BioMedical Engineering and Informatics (CISP-BMEI), pp. 1–5 (2017). https://doi.org/10.1109/CISP-BMEI.2017.8302281
2. Chung, Y., Glass, J.R.: Speech2vec: A sequence-to-sequence framework for learning word embeddings from speech. In: Yegnanarayana, B. (ed.) 19th Annual Conference of the International Speech Communication Association. pp. 811–815. ISCA (2018https://doi.org/10.21437/INTERSPEECH.2018-2341
3. Chung, Y.A., Weng, W.H., Tong, S., Glass, J.: Towards unsupervised speech-to-text translation. In: ICASSP 2019-2019 IEEE International Conference on Acoustics, Speech and Signal Processing (ICASSP), pp. 7170–7174. IEEE (2019)
4. Gong, Y., Cosma, G.: Improving visual-semantic embeddings by learning semantically-enhanced hard negatives for cross-modal information retrieval. Pattern Recogn. **137**, 109272 (2023). https://doi.org/10.1016/j.patcog.2022.109272
5. Guo, S., Yao, N.: Document vector extension for documents classification. IEEE Trans. Knowl. Data Eng. **33**(8), 3062–3074 (2021). https://doi.org/10.1109/TKDE.2019.2961343
6. Herdade, S., Kappeler, A., Boakye, K., Soares, J.: Image captioning: transforming objects into words. Adv. Neural Inf. Processing Syst. **32** (2019)
7. Johnson, J., Douze, M., Jegou, H.: Billion-scale similarity search with GPUs. IEEE Trans. Big Data **7**(3), 535–547 (2021). https://doi.org/10.1109/tbdata.2019.2921572
8. Kazemi, S.M., et al.: Time2vec: Learning a vector representation of time (2019). https://arxiv.org/abs/1907.05321
9. Kinnunen, T., Karpov, E., Franti, P.: Real-time speaker identification and verification. IEEE Trans. Audio Speech Lang. Process. **14**(1), 277–288 (2006). https://doi.org/10.1109/TSA.2005.853206
10. Ko, K., Park, S., Ko, H.: Convolutional feature vectors and support vector machine for animal sound classification. In: 2018 40th Annual International Conference of the IEEE Engineering in Medicine and Biology Society (EMBC), pp. 376–379 (2018). https://doi.org/10.1109/EMBC.2018.8512408
11. Kriegel, H.P., Brecheisen, S., Kröger, P., Pfeifle, M., Schubert, M.: Using sets of feature vectors for similarity search on voxelized CAD objects. In: Proceedings of the 2003 ACM SIGMOD International Conference on Management of Data, pp. 587–598. SIGMOD '03, ACM (2003). https://doi.org/10.1145/872757.872828
12. Lewis, M.: Bart: Denoising sequence-to-sequence pre-training for natural language generation, translation, and comprehension (2019). https://arxiv.org/abs/1910.13461
13. Lu, J., Holubová, I.: Multi-model databases: A new journey to handle the variety of data. ACM Comput. Surv. **52**(3) (2019). https://doi.org/10.1145/3323214

14. Lu, Y., Zhao, W., Sun, N., Wang, J.: Enhancing multimodal knowledge graph representation learning through triple contrastive learning. In: Proceedings of the Thirty-Third International Joint Conference on Artificial Intelligence, pp. 5963–5971. ijcai.org (2024)
15. Pan, J.J., Wang, J., Li, G.: Vector database management techniques and systems. In: Companion of the 2024 International Conference on Management of Data, pp. 597–604. ACM (2024)
16. Perrin, P., Petry, F.E.: Extraction and representation of contextual information for knowledge discovery in texts. Inf. Sci. **151**, 125–152 (2003)
17. Radford, A., et al.: Learning transferable visual models from natural language supervision. In: International conference on machine learning, pp. 8748–8763. PMLR (2021)
18. Singh, S.P., Kumar, A., Darbari, H., Singh, L., Rastogi, A., Jain, S.: Machine translation using deep learning: An overview. In: 2017 International Conference on Computer, Communications and Electronics (Comptelix), pp. 162–167 (2017). https://doi.org/10.1109/COMPTELIX.2017.8003957
19. Sun, C., Myers, A., Vondrick, C., Murphy, K., Schmid, C.: VideoBERT: A joint model for video and language representation learning. In: Proceedings of the IEEE/CVF International Conference on Computer Vision, pp. 7464–7473 (2019)
20. Szegedy, C., Ioffe, S., Vanhoucke, V., Alemi, A.A.: Inception-v4, Inception-ResNet and the impact of residual connections on learning. In: Singh, S., Markovitch, S. (eds.) Proceedings of the 31st AAAI Conference on Artificial Intelligence, pp. 4278–4284. AAAI Press (2017). https://doi.org/10.1609/AAAI.V31I1.11231
21. Taipalus, T.: Vector database management systems: Fundamental concepts, use-cases, and current challenges. Cogn. Syst. Res. **85**, 101216 (2024). https://doi.org/10.1016/j.cogsys.2024.101216
22. Tan, M., Le, Q.V.: Efficientnet: Rethinking model scaling for convolutional neural networks. In: Chaudhuri, K., Salakhutdinov, R. (eds.) Proceedings of the 36th International Conference on Machine Learning. vol. 97. PMLR (2019)
23. Truong, Q.T., Salah, A., Lauw, H.: Multi-modal recommender systems: Hands-on exploration. In: Proceedings of the 15th ACM Conference on Recommender Systems, pp. 834–837. RecSys '21, ACM (2021). https://doi.org/10.1145/3460231.3473324
24. Yuan, X., et al.: Multimodal contrastive training for visual representation learning. In: IEEE Conference on Computer Vision and Pattern Recognition, pp. 6995–7004. Computer Vision Foundation / IEEE (2021)
25. Zhang, J., Yin, Z., Chen, P., Nichele, S.: Emotion recognition using multi-modal data and machine learning techniques: A tutorial and review. Inf. Fusion **59**, 103–126 (2020). https://doi.org/10.1016/j.inffus.2020.01.011
26. Zhang, Y., Sidibé, D., Morel, O., Mériaudeau, F.: Deep multimodal fusion for semantic image segmentation: A survey. Image Vision Comput. **105**, 104042 (2021). https://doi.org/10.1016/j.imavis.2020.104042

Utilizing Quantum Computing to Improve the Quality of Data

Valter Uotila[1(✉)], Soror Sahri[2], and Sven Groppe[3]

[1] University of Helsinki and Aalto University, Helsinki, Finland
valter.uotila@helsinki.fi
[2] Université Paris Cité, LIPADE, 75006 Paris, France
soror.sahri@u-paris.fr
[3] University of Lübeck, Lübeck, Germany
sven.groppe@uni-luebeck.de

Abstract. In today's data-driven world, ensuring data quality has become the key to success for organizations across industries and academia. Hence, this tutorial begins by exploring the foundational principles of data quality, emphasizing dimensions such as accuracy, completeness, consistency, timeliness, and validity. However, many strategies to improve data quality are costly in terms of processing time, computational resources, and/or the amount of training data required, where quantum computing promises advantages. Hence, this tutorial will briefly introduce quantum computing and focus on its applications to data quality, including a demonstration and open challenges that might open new research directions and business opportunities in improving data quality.

Keywords: Data Quality · Quantum Computing · Quantum Machine Learning · Quantum Optimization · Anomaly Detection

1 Introduction and Motivation

In the era of data-driven decision-making, the quality of data has become a key to the success of analytical models, machine learning systems, and business intelligence processes. Poor data quality, manifesting as inconsistencies, missing values, duplicates, or incorrect entries, can significantly undermine outcomes, leading to flawed insights and costly errors. Traditional methods for detecting and correcting these issues often rely on computationally intensive techniques, especially when applied to large-scale, high-dimensional, or unstructured datasets.

Quantum computing continues to advance at an accelerating pace, following decades of foundational research and technological development. Today, cloud-based services offer access to real quantum hardware and simulators, while error-mitigated and error-corrected quantum computers are currently under active development. Superpolynomial and polynomial speedups of many quantum algorithms have been proven [15], and recent studies like [19,20] claim solving real-world problems beyond the reach of classical devices. Quantum computers offer

P. K. Chrysanthis et al. (Eds.): ADBIS 2025, LNCS 16043, pp. 280–287, 2026.
https://doi.org/10.1007/978-3-032-05281-0_19

the potential to accelerate the core components of data quality management, such as data matching, anomaly detection and pattern recognition. Moreover, many data quality tasks can be framed as combinatorial optimization or probabilistic inference problems, both of which are natural candidates for quantum speedup. In this tutorial, we discuss the opportunities for addressing data quality problems on quantum computers.

Tutorial Overview: After introduction and motivation, we dive into the basics of quantum computing. We introduce the broad field of data quality and provide a literature review on quantum computing approaches for data management tasks. In the hands-on part, we focus on detecting anomalies with a specific quantum machine learning model. Finally, we discuss the opportunities and future work in quantum computing research for improving data quality.

Related Tutorials: Several tutorials have addressed the topic of applying quantum computing for databases at major conferences like SIGMOD [46] (with a focus on quantum machine learning) and ICDE [14] (providing a general overview of quantum computing for databases) and at the QCE conference on quantum computing [28] (discussing database optimizations utilizing quantum annealing). Additionally, several tutorials have addressed data quality challenges from various perspectives. Notably, [45] focused on quality challenges in deep learning contexts, [16] covered debugging data errors in ML pipelines, and [23] explored spatial data quality in IoT settings.

Contributions: This is the first tutorial on quantum computing for data quality. It supports database experts starting in quantum computing by providing a concise introduction and a comprehensive literature overview. Newbies and practitioners are shown how to interact with quantum computers through code examples on quantum anomaly detection, and researchers are encouraged to conduct further studies on improving data quality using quantum computing.

Target Audience: The tutorial serves students, academic researchers, industrial developers, and practitioners with a computer science background by exploring the fundamentals of quantum computing and data quality, offering a comprehensive literature review, and a practical demonstration.

2 Background on Data Quality and Quantum Computing

Background on Data Quality: Data quality refers to the extent to which data meet the requirements and expectations of its users, ensuring that it is suitable for its intended purposes, such as operations, decision-making and planning [30,44]. It is commonly assessed through key dimensions that include accuracy, completeness, consistency, and timeliness.

To address potential quality problems, various techniques including rule-based methods that enforce domain-specific constraints, statistical methods that identify deviations from expected distributions, and similarity-based approaches that detect duplicates or conflicts between records [4]. As data systems grow, data quality techniques emphasize scalability, adaptability, and responsiveness, often using machine learning for real-time anomaly detection and predictive assessment.

In AI and ML contexts, models learn data degradation patterns to detect shifts and quality issues like outliers or missing values before they affect downstream tasks. These techniques support continuous monitoring and dynamic correction, allowing data pipelines to automatically adapt to evolving quality challenges. Common data quality tasks also include data imputation, where missing values are estimated based on learned patterns in the data, and the use of constraints or learned rules to ensure consistency and prevent quality issues from propagating through models.

Quantum computing provides new capabilities for improving data quality techniques, particularly anomaly detection and data imputation, as we will present in this tutorial.

Background on Quantum Computing: Quantum computing is an emerging computational paradigm fundamentally different from classical computing [14,28,46]. In this tutorial, we introduce and approach quantum computing in relation to classical computing as part of a spectrum defined by the degree of classical computational support required. We note that while some quantum algorithms operate largely independently, others, such as hybrid quantum-classical algorithms, rely heavily on classical resources. More details on quantum computing are discussed in the demonstration part of this tutorial.

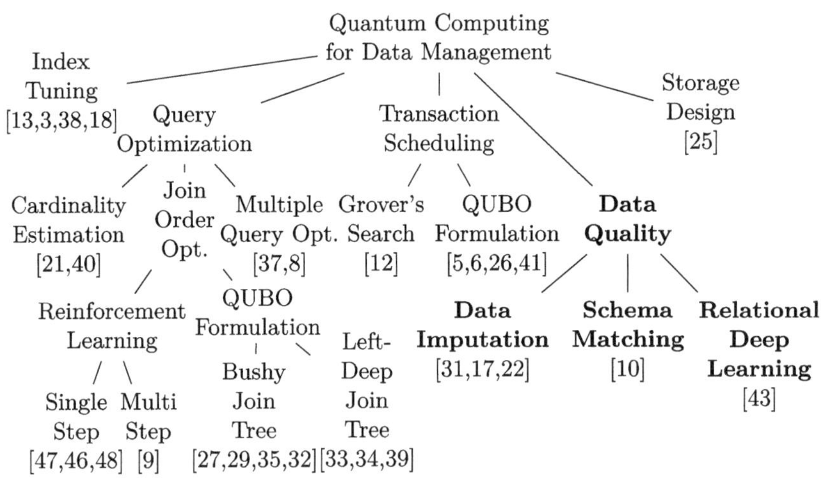

Fig. 1. Taxonomy of quantum computing approaches for data management tasks (inclusive references)

3 Potential of Quantum Computing for Data Quality

Figure 1 contains a taxonomy of quantum computing approaches for data management tasks (including references). These contributions primarily target data management, yet they also indicate that other areas, particularly data quality, could benefit from quantum computing. Approaches to relational deep learning may be useful for correcting, for example, foreign-key relationships in relational databases. Techniques proposed for tasks such as index tuning, join order optimization, multiple query optimization, and transaction scheduling demonstrate their potential for quantum acceleration. In the following, we present existing works that demonstrate the potential of quantum computing to improve key aspects of data quality, with a particular focus on data imputation and knowledge graph completion, where quantum computing techniques have been most actively explored to date.

Efficient data completion techniques leveraging quantum computing have been proposed to support imputation efforts. The contributions in [22,31] directly evaluate quantum computing solutions for missing data imputation. In [31], a quantum-inspired Hamiltonian Monte Carlo (HMC) approach is introduced, which applies quantum principles to enhance the classical HMC algorithm. This approach improves the sampling process, offering more accurate estimations and faster convergence compared to traditional Monte Carlo methods. Other approaches based on cardinality estimation, have demonstrated potential for addressing missing values and checking the plausibility of existing ones, though their applicability to broader domains remains an area for further exploration [21,40]. Quantum determinantal point processes (DPPs), proposed in [17], leverage quantum parallelism to efficiently explore missing data completions, capturing dependencies within datasets for more accurate imputations.

Beyond data imputation, methods for Knowledge Graph (KG) completion also benefit from the use of quantum computing. Approaches like QIQE-KGC [24] integrate quantum embeddings to capture semantic and logical relationships, improving link prediction. Similarly, QLogicE [7] combines quantum embedding techniques to enhance the accuracy of modeling KG facts. Variational quantum circuits and quantum negative sampling strategies [11] also contribute to more efficient KG embeddings. Additionally, QubitE applies quantum principles to tackle large-scale KG completion challenges, offering improved completion methods.

Anomaly detection, another central data quality task, can also benefit from quantum computing, as we will address in detail in the next section through a demonstration of a quantum machine learning model for time series data.

4 Demonstration: Anomaly Detection from Time Series

This demonstration is based on works [1,2], which develop a special type of quantum machine learning model for time series data and anomaly detection. The model is a variational, hybrid quantum-classical algorithm. The demonstrated model is trained on time series representing normal data and detects whether a series is anomalous.

To improve the work in [1,2], we utilize a video MNIST dataset [36], which comprises videos of black-and-white handwritten digits moving within a frame. We select a subset of videos considered normal. The model is trained using the normal dataset, where each video in this set is interpreted as a sequence of vectors of length $64 \times 64 = 4096$. Since every frame consists of 4096 pixels, we use so-called amplitude encoding and fit each vector in the time series into a 12 qubit quantum circuit ($\log_2(4096) = 12$). This is a different encoding compared to [1,2]. As noted in [42], we need to pay attention to the measurement operations in quantum machine learning. Hence, we measure a so-called global observable that better captures the system's global dynamics. After this, the model is trained as described in [1,2].

We test the model by inserting noise into the normal videos, while keeping the digits unchanged. Despite the noise, the model is still capable of detecting these videos as normal. Instead, if we insert videos with different digits in the frames compared to the normal set of videos, the model classifies them as anomalous.

5 Tutorial Presenters

Valter Uotila is a Ph.D. student at the University of Helsinki and a researcher at Aalto University, specializing in quantum computing applications for data management. He has previously presented tutorials at SIGMOD 2023 and IEEE Quantum Week 2024. For more details, see https://valteruo.github.io/.

Soror Sahri is an Associate Professor at Université Paris Cité. Her research interests focus on data management, including big data analytics, data quality, and query processing. For more details, visit, https://helios2.mi.parisdescartes.fr/~sseifedd/soror/

Sven Groppe is a Professor at the University of Lübeck and the project coordinator of the BMBF-funded project QC4DB âĂŞ Accelerating Relational Database Management Systems via Quantum Computing. For more details about his academic career, visit https://www.ifis.uni-luebeck.de/~groppe.

Acknowledgments. This work is funded by the German Federal Ministry of Education and Research within the funding program quantum technologies—from basic research to market—contract number 13N16090.

References

1. Baker, J., Radha, S.K.: Quantum detection of time series anomalies (2023). https://pennylane.ai/qml/demos/tutorial_univariate_qvr/. pennyLane Demo. Published: February 7, 2023. Last updated: Nov 6, 2024

2. Baker, J.S., et al.: Quantum variational rewinding for time series anomaly detection (2022). https://arxiv.org/abs/2210.16438
3. Barbosa, D., Gruenwald, L., DOrazio, L., Bernardino, J.: Qrlit: Quantum reinforcement learning for database index tuning. Future Internet 16(12), 439 (2024). https://doi.org/10.3390/fi16120439, http://dx.doi.org/10.3390/fi16120439
4. Batini, C., Cappiello, C., Francalanci, C., Maurino, A.: Methodologies for data quality assessment and improvement **41**(3) (2009)
5. Bittner, T., Groppe, S.: Avoiding blocking by scheduling transactions using quantum annealing. In: IDEAS, Seoul, Republic of Korea (virtual due to Covid-19 pandemic) (2020). https://doi.org/10.1145/3410566.3410593
6. Bittner, T., Groppe, S.: Hardware accelerating the optimization of transaction schedules via quantum annealing by avoiding blocking. OJCC **7**(1), 1–21 (2020). http://nbn-resolving.de/urn:nbn:de:101:1-2020112218332015343957
7. Chen, P., Wang, Y., Yu, X., Feng, R.: Qlogice: Quantum logic empowered embedding for knowledge graph completion. Knowledge-Based Systems **239**, 107963 (2021). https://doi.org/10.1016/j.knosys.2021.107963
8. Fankhauser, T., Solèr, M.E., Füchslin, R.M., Stockinger, K.: Multiple query optimization using a gate-based quantum computer. IEEE Access **11**, 114031–114043 (2023). https://doi.org/10.1109/access.2023.3324253
9. Franz, M., Winker, T., Groppe, S., Mauerer, W.: Hype or heuristic? Quantum reinforcement learning for join order optimisation. In: IEEE International Conference on Quantum Computing and Engineering (QCE) (2024). https://doi.org/10.1109/QCE60285.2024.00055
10. Fritsch, K., Scherzinger, S.: Solving hard variants of database schema matching on quantum computers. VLDB **16**(12), 3990–3993 (2023).https://doi.org/10.14778/3611540.3611603
11. Giri, P.R., Kurokawa, M., Saito, K.: Quantum negative sampling strategy for knowledge graph embedding with variational circuit. In: IEEE International Conference on Quantum Computing and Engineering (QCE). p. 280–281. IEEE (2023). https://doi.org/10.1109/qce57702.2023.10242
12. Groppe, S., Groppe, J.: Optimizing transaction schedules on universal quantum computers via code generation for grover's search algorithm. In: IDEAS, Montreal, QC, Canada (2021). https://doi.org/10.1145/3472163.3472164
13. Gruenwald, L., Winker, T., Çalkylmaz, U., Groppe, J., Groppe, S.: Index tuning with machine learning on quantum computers for large-scale database applications. In: QDSM@VLDB, Vancouver, Canada (2023). https://ceur-ws.org/Vol-3462/QDSM5.pdf
14. Hai, R., Hung, S.H., Feld, S.: Quantum data management: From theory to opportunities. In: ICDE, pp. 5376–5381. IEEE (2024). https://doi.org/10.1109/icde60146.2024.00410
15. Jordan, S.P.: Quantum algorithm zoo (2025). https://quantumalgorithmzoo.org/. created April 22, 2011; last updated Jun 16, 2025
16. Karlaš, B., Salimi, B., Schelter, S.: Navigating data errors in machine learning pipelines: Identify, debug, and learn. In: ICDE (2018)
17. Kazdaghli, S., Kerenidis, I., Kieckbusch, J., Teare, P.: Improved clinical data imputation via classical and quantum determinantal point processes. eLife **12** (2024). https://doi.org/10.7554/elife.89947
18. Kesarwani, M., Haritsa, J.R.: Index advisors on quantum platforms. VLDB **17**(11), 3615–3628 (2024). https://doi.org/10.14778/3681954.3682025

19. Kim, Y., Eddins, A., Anand, S., Wei, K.X., van den Berg, E., Rosenblatt, S., Nayfeh, H., Wu, Y., Zaletel, M., Temme, K., Kandala, A.: Evidence for the utility of quantum computing before fault tolerance. Nature **618**(79657965), 500–505 (2023). https://doi.org/10.1038/s41586-023-06096-3
20. King, A.D., Nocera, A., Rams, M.M., et al.: Beyond-classical computation in quantum simulation. Science **388**(6743), 199–204 (2025). https://doi.org/10.1126/science.ado6285, http://dx.doi.org/10.1126/science.ado6285
21. Kittelmann, F., Sulimov, P., Stockinger, K.: Qardest: Using quantum machine learning for cardinality estimation of join queries. In: Q-Data. p. 2–13. SIGMOD/PODS '24, ACM (2024). https://doi.org/10.1145/3665225.3665444
22. Kochan, D., Zhang, Z., Yang, X.: A quantum-inspired hamiltonian monte carlo method for missing data imputation. In: Dong, B., Li, Q., Wang, L., Xu, Z.Q.J. (eds.) Proceedings of Mathematical and Scientific Machine Learning. Proceedings of Machine Learning Research, vol. 190, pp. 17–32. PMLR (2022). https://proceedings.mlr.press/v190/kochan22a.html
23. Li, H., Tang, B., Lu, H., Cheema, M.A., Jensen, C.S.: Spatial data quality in the IoT era: Management and exploitation. In: SIGMOD, pp. 2474–2482. SIGMOD '22, Association for Computing Machinery
24. Li, L., et al.: Knowledge graph completion method based on quantum embedding and quaternion interaction enhancement. Information Sciences **648**, 119548 (2023). https://doi.org/10.1016/j.ins.2023.119548
25. Littau, T., Li, Z., Hai, R.: Quantum data structures for enhanced database performance. In: QDSM@VLDB, Guangzhou, China (2024). https://vldb.org/workshops/2024/proceedings/QDSM/QDSM.6.pdf
26. Nayak, N., Prisacaru, A., Çalkylmaz, U., Groppe, J., Groppe, S.: Quantum-enhanced transaction scheduling with reduced complexity via solving QUBO iteratively using a locking mechanism. In: 2nd International Workshop on Quantum Computing and Quantum-Inspired Technology for Data-Intensive Systems and Applications (Q-Data), Berlin, Germany (2025). https://doi.org/10.1145/3736393.3736701
27. Nayak, N., Rehfeld, J., Winker, T., Warnke, B., Çalkylmaz, U., Groppe, S.: Constructing optimal bushy join trees by solving QUBO problems on quantum hardware and simulators. In: BiDEDE, Seattle, WA, USA (2023). https://doi.org/10.1145/3579142.3594298
28. Nayak, N., et al.: Qce'24 tutorial: Quantum annealing – emerging exploration for database optimization. Tutorial at 2024 IEEE International Conference on Quantum Computing and Engineering (QCE) (2024). https://arxiv.org/abs/2411.04638
29. Nayak, N., Winker, T., Çalkylmaz, U., Groppe, S., Groppe, J.: Quantum join ordering by splitting the search space of qubo problems. Datenbank-Spektrum **24**(1), 21–32 (2024). https://doi.org/10.1007/s13222-024-00468-3
30. Oliveira, P., Rodrigues, F., Henriques, P.R.: A formal definition of data quality problems. In: ICIQ (2005)
31. Sanavio, C., Tibaldi, S., Tignone, E., Ercolessi, E.: Quantum circuit for imputation of missing data. IEEE Transactions on Quantum Engineering **5**, 1–12 (2024). https://doi.org/10.1109/tqe.2024.3447875
32. Saxena, P., Sabek, I., Spedalieri, F.: Constrained quadratic model for optimizing join orders. In: Q-Data, pp. 38–44. SIGMOD/PODS '24, ACM (2024). https://doi.org/10.1145/3665225.3665447

33. Schönberger, M., Scherzinger, S., Mauerer, W.: Ready To Leap (by Co-design)? Join Order Optimisation On Quantum Hardware, vol. 1. Association for Computing Machinery, New York, NY, USA (may (2023)
34. Schönberger, M., Trummer, I., Mauerer, W.: Quantum-inspired digital annealing for join ordering. In: VLDB, vol. 16 (2023)
35. Schönberger, M., Trummer, I., Mauerer, W.: Quantum optimisation of general join trees. In: VLDBW'23 (2023)
36. Srivastava, N., Mansimov, E., Salakhutdinov, R.: Unsupervised learning of video representations using LSTMs. In: Proceedings of the 32nd International Conference on International Conference on Machine Learning - Volume 37, pp. 843–852. ICML'15, JMLR.org (2015)
37. Trummer, I., Koch, C.: Multiple query optimization on the d-wave 2x adiabatic quantum computer. VLDB **9**(9), 648–659 (2016). https://doi.org/10.14778/2947618.2947621
38. Trummer, I., Venturelli, D.: Leveraging quantum computing for database index selection. In: Q-Data, pp. 14–26. SIGMOD/PODS '24, ACM (2024). https://doi.org/10.1145/3665225.3665445
39. Uotila, V.: Left-deep join order selection with higher-order unconstrained binary optimization on quantum computers (2025). https://arxiv.org/abs/2502.00362
40. Uotila, V.: Sql2circuits: Estimating cardinalities, execution times, and costs for sql queries with quantum natural language processing. IEEE International Conference on Quantum Computing and Engineering (QCE) (2025)
41. Uotila, V., Lu, J.: Quantum annealing method for dynamic virtual machine and task allocation in cloud infrastructures from sustainability perspective. In: 2023 IEEE 39th International Conference on Data Engineering Workshops (ICDEW), pp. 105–110 (2023). https://doi.org/10.1109/ICDEW58674.2023.00023
42. Uotila, V., Salmenperä, I., Becker, L., van de Griend, A.M., Shinde, A.R., Nurminen, J.K.: Perspectives on utilization of measurements in quantum algorithms. In: Proceedings of the 2025 IEEE International Conference on Quantum Software (QSW). IEEE, Helsinki, Finland (2025)
43. Vogrin, M., Vogrin, R., Groppe, S., Groppe, J.: Supervised learning on relational databases with quantum graph neural networks. In: QDSM@VLDB, Guangzhou, China (2024). https://vldb.org/workshops/2024/proceedings/QDSM/QDSM.5.pdf
44. Wang, R.Y., Strong, D.M.: Beyond accuracy: what data quality means to data consumers. J. Manag. Inf. Syst. **12**(4), 5–33 (1996)
45. Whang, S.E., Lee, J.G.: Data collection and quality challenges for deep learning. Proc. VLDB Endow. **13**(12), 3429–3432 (2020)
46. Winker, T., et al.: Quantum machine learning: Foundation, new techniques, and opportunities for database research. In: SIGMOD (2023). https://doi.org/10.1145/3555041.3589404
47. Winker, T., Çalkylmaz, U., Gruenwald, L., Groppe, S.: Quantum machine learning for join order optimization using variational quantum circuits. In: BiDEDE, Seattle, WA, USA (2023). https://doi.org/10.1145/3579142.3594299
48. Çalkylmaz, U., et al.: Opportunities for quantum acceleration of databases: Optimization of queries and transaction schedules. VLDB **16**(9), 2344–2353 (2023). https://doi.org/10.14778/3598581.3598603

Correction to: Independence Rules: Analysis of Nine Simple Cardinality Estimators and Their Impact on Plan Quality

Daniel Flachs and Guido Moerkotte

Correction to:
Chapter 2 in: P. K. Chrysanthis et al. (Eds.): *Advances in Databases and Information Systems*, LNCS 16043, https://doi.org/10.1007/978-3-032-05281-0_2

The original version of the chapter 2 has contained an error in Equation (3). This error has been corrected in both the PDF and HTML versions of the Article.

The updated version of this chapter can be found at
https://doi.org/10.1007/978-3-032-05281-0_2

P. K. Chrysanthis et al. (Eds.): ADBIS 2025, LNCS 16043, p. C1, 2026.
https://doi.org/10.1007/978-3-032-05281-0_20

Correction to: Bitemporal Property Graphs: Dealing with Both Valid and Transaction Time

Luca Anselma, Marco Ballerini, Paolo Giordano, Erica Raina, and Paolo Terenziani

Correction to: Chapter 15 in: P. K. Chrysanthis et al. (Eds.): *Advances in Databases and Information Systems*, LNCS 16043, https://doi.org/10.1007/978-3-032-05281-0_15

The original publication of Chapter 15 omitted the updated author's corrections. This has now been remedied by issuing the revised version.

The updated version of this chapter can be found at
https://doi.org/10.1007/978-3-032-05281-0_15

P. K. Chrysanthis et al. (Eds.): ADBIS 2025, LNCS 16043, p. C2, 2026.
https://doi.org/10.1007/978-3-032-05281-0_21

Author Index

A
Abelló, Alberto 147
Akter, Yeasmin Ara 147
Alarcan, Hugo 201
Aligon, Julien 79
Anselma, Luca 234

B
Ballerini, Marco 234
Barbon Junior, Sylvio 95
Bernhardt, Arthur 44
Bizer, Christian 111
Blasco, Hélène 201
Bøgsted, Martin 164
Bugiotti, Francesca 63

C
Chakravarthy, Sharma 217
Chanson, Alexandre 79
Christen, Peter 181

D
Dell'Aglio, Daniele 164
Dimitrieski, Vladimir 28
Dombrovskaya, Henrietta 264
Doumard, Emmanuel 79

E
Ehrlinger, Lisa 3

F
Flachs, Daniel 11

G
Garofalakis, Minos 164
Giordano, Paolo 234
Grigore, Iuliana Malina 95
Groppe, Sven 280

H
Harmouch, Hazar 3
Hose, Katja 147, 164
Hrnjak, Jelena 28

J
Josiowicz, Alejandra 253
Jovanovic, Petar 147

K
Kapenekakis, Antheas 164
Karapiperis, Dimitrios 131
Koch, Andreas 44
Korini, Keti 111

L
Labroche, Nicolas 79, 201
Lu, Jiaheng 272

M
Marcel, Patrick 201
Marques Tavares, Gabriel 95
Moerkotte, Guido 11
Mohammed, Sedir 3
Mohn, Matthias 181
Monsarrat, Paul 79

N
Nanayakkara, Charini 181
Naumann, Felix 3

P
Pasquadibisceglie, Vincenzo 95
Peralta, Veronika 201
Petrov, Ilia 44

R
Rahm, Erhard 181
Raina, Erica 234

P. K. Chrysanthis et al. (Eds.): ADBIS 2025, LNCS 16043, pp. 289–290, 2026.
https://doi.org/10.1007/978-3-032-05281-0

S
Sagi, Tomer 147
Sahri, Soror 280
Sakal Francišković, Teodor 28
Salimath, Shwetha 63
Santra, Abhishek 217
Schneider, Maja 181
Seidl, Thomas 95
Singh, Arshdeep 217
Srivastava, Divesh 3
Stock, Florian 44

T
Taipalus, Toni 272
Tamimi, Sajjad 44
Tejedor, Guillaume 201
Terenziani, Paolo 234
Tjortjis, Christos 131
Todorović, Nikola 28

U
Uotila, Valter 280

V
Vargas-Solar, Genoveva 253
Verykios, Vassilios 131

W
Wlodarczyk, Sylvain 63

MIX
Papier aus verantwortungsvollen Quellen
Paper from responsible sources
FSC® C105338

If you have any concerns about our products,
you can contact us on
ProductSafety@springernature.com

In case Publisher is established outside the EU,
the EU authorized representative is:
Springer Nature Customer Service Center GmbH
Europaplatz 3, 69115 Heidelberg, Germany

Printed by Libri Plureos GmbH
in Hamburg, Germany